T0313972

Frequency Stability

IEEE PRESS SERIES ON DIGITAL AND MOBILE COMMUNICATION

John B. Anderson, *Series Editor*
University of Lund

---

1. *Wireless Video Communications: Second to Third Generation and Beyond*
   Lajos Hanzo, Peter Cherriman, and Jurgen Streit
2. *Wireless Communications in the 2lst Century*
   Mansoor Sharif, Shigeaki Ogose, and Takeshi Hattori
3. *Introduction to WLLs: Application and Deployment for Fixed and Broadband Services*
   Raj Pandya
4. *Trellis and Turbo Coding*
   Christian Schlegel and Lance Perez
5. *Theory of Code Division Multiple Access Communication*
   Kamil Sh. Zigangirov
6. *Digital Transmission Engineering,* Second Edition
   John B. Anderson
7. *Wireless Broadband: Conflict and Convergence*
   Vern Fotheringham and Shamla Chetan
8. *Wireless LAN Radios: System Definition to Transistor Design*
   Arya Behzad
9. *Millimeter Wave Communication Systems*
   Kao-Cheng Huang and Zhaocheng Wang
10. *Channel Equalization for Wireless Communications: From Concepts to Detailed Mathematics*
    Gregory E. Bottomley
11. *Handbook of Position Location: Theory, Practice, and Advances*
    Edited by Seyed (Reza) Zekavat and R. Michael Buehrer
12. *Digital Filters: Principle and Applications with MATLAB*
    Fred J. Taylor
13. *Resource Allocation in Uplink OFDMA Wireless Systems: Optimal Solutions and Practical Implementations*
    Elias E. Yaacoub and Zaher Dawy
14. *Non-Gaussian Statistical Communication Theory*
    David Middleton
15. *Frequency Stabilization: Introduction and Applications*
    Venceslav F. Kroupa

Forthcoming Titles
Fundamentals of Convolutional Coding, Second Edition
Rolf Johannesson and Kamil Zigangirov

Frequency Stability

Introduction and Applications

Věnceslav F. Kroupa

IEEE Press

IEEE Press series on Digital and Mobile Communication

A John Wiley & Sons, Inc., Publication

IEEE Press
445 Hoes Lane
Piscataway, NJ 08855

Library of Congress Cataloging-in-Publication Data:

Kroupa, Venceslav F., 1923–
Frequency stability : introduction and applications / Venceslav F. Kroupa.
p. cm.
Includes bibliographical references.
ISBN 978-1-118-15912-5 (hardback)
1. Oscillators, Electric--Design and construction. 2. Frequency stability. I. Title.
TK7872.O7K76 2012
621.381'323—dc23 2011048135

10 9 8 7 6 5 4 3 2 1

To my wife Magda,
for her encouragement
in starting and finishing this work

Contents

Preface

Quid est ergo tempus?
Si nemo ex me quaerat, scio;
si quaerenti explicare velim,
nescio.

What then is time?
If no one asks me, I know,
if I want to explain it to someone
who asks, I do not know.

S. Augustine Confessions, Book XI

The flow of time is smooth and imperceptible—fluctuations originate in the measurement systems: Earth rotation is defined as a *day–night* cycle; Earth motion around the Sun is defined in *years*; astronomical observations of the motion in the universe last for centuries or for millennia, and so on. All these measurement systems introduce unpredictable imperfections and even errors designated as *noise*. The difficulty is also present in modern systems based on the atomic time definition.

The time fluctuations did not cause problems with ideas or their use until the twentieth century, with the introduction of modern technologies and the advent of the importance of the rapid delivery of messages, goods, even of people.

In today's methods of communications, our delivery channels are generally based on electromagnetic media that provide a sort of the common property that cannot be expanded. Only the channels perfected by technology and by reducing mutual interference or by extension of frequency ranges can be used (see Fig. P1). In this context, the time and frequency stability is of prime importance, and study of noise problems still proceeds. There are many papers, books, and even libraries about this topic of frequency and time stability.[1] Why a new one?

The intent of this book is twofold: to refresh students' memory of the field and provide additional information for engineers and practitioners in neighboring fields without recourse to complicated mathematics.

The first noise studies on frequency stability were based on earlier results of the probability of events and were connected with LC oscillators. However, their short-term time stability was soon insufficient even for the simplest radio traffic. Introduction of very stable crystal oscillators provided better stability (they even proved irregularities in the Earth's rotation), but their application on both transmitting and receiving positions, often with the assistance of frequency synthesizers, was soon insufficient for the increasing number of needed dependent communication channels.

To alleviate the situation, there were investigations of the frequency stability of local generators from both theoretical and practical points of approach on one hand, and the progress of technology toward the microwaves and miniaturization on the other. This situation is depicted in Fig. P1.

These problems are connected with the time and frequency stability, the extent of the used frequencies into microwave ranges, and the technology of application of miniaturization of integrated circuits on a large scale. In accordance with the intent of this book to refresh the memory of students in the field and provide additional information for engineers in neighboring fields, we divided the subject matter into six chapters.

Chapter 1 introduces the basic concepts of noise. It begins with the term *power spectral density* (PSD), that is, with the magnitude of fluctuations (phase, frequency, power, etc.) in the 1-Hz frequency span in the Fourier frequency ranges. The most important noises are:

1. **White noise,** with a constant PSD: $S(f)$ = constant. It is generated by black-body radiation (thermal noise) or on-spot fluctuations of the delivering media (as current shot noise).

[1]IEEE Standard Definitions of Physical Quantities for Fundamental Frequency and Time Metrology—IEEE Std 1139/1988. IEEE Standard Definitions of Physical Quantities for Fundamental Frequency and Time Metrology—Random Instabilities IEEE Std 1139/1999. IEEE Std 1139/2008.

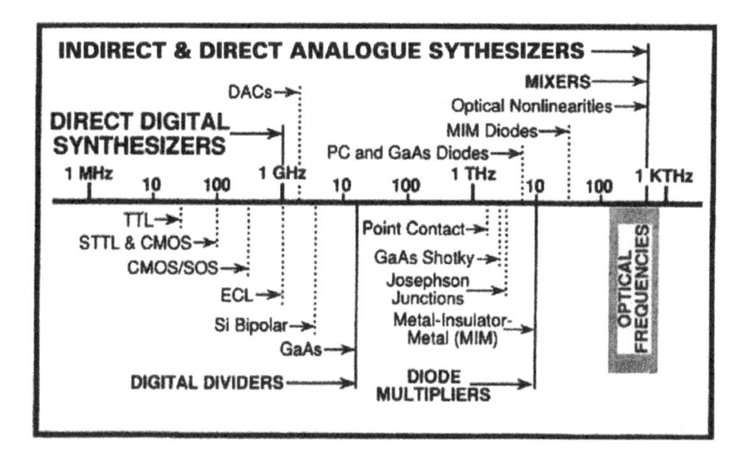

Fig. P1. The state of the art of frequency synthesis from the lowest frequency ranges to the optical frequencies; frequency division and multiplication.[2]

2. **Flicker noise** or $1/f$ noise with the PSD $S(f) \sim 1/f^{\alpha}$, where the superscript α is very close to one. It is generally present at low Fourier frequencies; its origin may be manifold in accordance with the performed studies.
3. **Random walk** with PSD $S(f) \sim f^{-2}$, represented by Brownian movement.

In the second part of Chapter 1, we investigate fluctuations from the probability point of view. The simplest is the rectangular distribution of events. A larger number of rectangular distributions results in the central limit or Gaussian distribution. Another approach provides the binomial distribution near to the Gaussian distribution or into the Poisson distribution. We mention the stochastic processes, the stationary processes invariant with respect to the time shift, and fractional integration, resulting in the possibility of explaining the flicker frequency phenomena and the random walk processes.

Chapter 2 investigates the noise generated in resonators, particularly, in quartz crystal resonators, with the assumption that $1/f$ noise is generated by material (dielectric) losses. It concludes that the product of the quality factor Q times the resonant frequency f_o is a constant. Its

[2]From V.F. Kroupa *Phase Lock Loops and Frequency Synthesis,* Wiley, 2003. Reproduced with permission.

validity was verified experimentally in the laboratory and by referring to the published noise data of crystal resonators in the entire frequency range from 5 MHz to nearly 1 GHz. Further, we discuss oscillating conditions and conclude, with the assistance of a sampling model, that the small phase error (e.g., noise) is compensated for by the integrated frequency shift of the resonant frequency. Vice versa, this experience is applied to the Leeson model.

Chapter 3 is dedicated to noise properties of very stable oscillators, quartz crystals, and new sapphire resonators (cryocooled), which are often used as a secondary frequency standard. We also include discussion of noise properties in oscillators stabilized by large-Q dielectric or optoelectronic resonators. Finally, the noise of integrated microwave oscillators in ranges designed for both LC- and RC-ring resonators is addressed. The advantage of the latter is simplicity and the problems are the large noise and power consumption.

Chapter 4 is dedicated to noises generated in individual elements and circuit blocks: resistors, inductances, capacitors, semiconductors, and amplifiers. A detailed discussion is dedicated to the different types of mixers, diode rings, and CMOS balanced and double-balanced systems. The spurious modulation signals with two-tone performance are investigated (third-order intercept points, IIP3), together with the expected noise performance.

Discussion of dividers is extended to the synchronized systems in the gigahertz ranges, to their noise, and to the regenerative division systems that provide the lowest additional noise.

Chapter 5 is devoted to the time measurements performed via the Allan variance. It discusses the reliability of the measurement and the connection of the slope of the measured characteristic with the type of investigated noise. Special attention is dedicated to the so-called modified Allan variance. The second part of this chapter deals with time jitter evaluations. Further, the probability of the time error, the bit error ratio (BER), is investigated. Eye diagrams and histograms are briefly discussed. Finally, we pay attention to the time jitter evaluation from the frequency and time domain measurements.

The last chapter, Chapter 6, deals in some detail with phase-locked loop (PLL) problems. It starts with a short introduction, proceeds to design, and stresses the importance of the design factors, such as natural frequency and damping. Further, we deal with the *or-*

der and the *type* of PLLs, and their transients and working ranges. We discuss the properties of digital loops and tristate-phase detectors in greater detail, and investigate the noises generated in individual parts of the PLL. Finally, we investigate the synchronized oscillators, on a PLL basis, for frequency division and multiplication applications.

Symbols

α	Phase constant, Bernamont superscript
α_F	Forward short-circuit current gain
α_R	Reverse short-circuit current gain
$\alpha(t)$	Normalized instantaneous amplitude a_o
β_{ko}	Normalized voltage gain
γ	Thermal noise constant in FET
γ_{var}	Doping profile constant in varactors
$\Gamma(\omega)$	Voltage transfer function
δ	Normalized time delay
$\delta(f_m)$	δ-Delta function
$\Delta f = f_{\text{high}} - f_{\text{low}}$	Frequency range (Hz)
Δv	Voltage step (V, μV)
$\Delta\omega_i$	Frequency step
$\Delta\omega_H = \lvert\omega_i - \omega_c\rvert$	Hold-in range
$\Delta\omega_p$	Pull-in range
$\Delta\omega_{PO}$	Pull-out range (frequency)
Δt	Time jitter
ε	Dielectric constant
ζ	Damping factor
η	Normalized time constant
θ	Field degeneration factor (MOS)
κ	Random variable, normalized time constant
φ_{PD}	Phase detector noise
φ_{osc}	Oscillator (VCO) noise
Φ	Phase, contact potential in varactors
$M(t)$	Instantaneous phase departure
$M_?(u)$	Characteristic function of the random variable ?
Φ_{pm}	Phase margin
$\lambda = pn$	Mean and variance of the Poisson distribution
χ^2	Chi squared statistics
ψ	Phase margin (PLL)
λ	Wavelength (m·km)

μ	Mobility
$\mu =<(\geq m_n)^n>$	Central moments
ζ	Random variable
ρ	Resistivity (Ωm)
Φ	Distribution, normalized frequency
$\Phi^2(n)$	Variance of event *n*
σ	Normalized frequency
$\sigma(\tau)$	Allan variance
σ^2	Variance
Σ	Summation sign
τ, τ_d	Time delay
$tg\ \delta$	Material losses (in dielectric)
ω_o, ω_m	Oscillation, modulation frequency
ω_n	Natural frequency (PLL)
ω_s	Sampling frequency
$<>^n$	*n*th-order moments
a_i	Designation of constants
$a_1, \ldots, a_4$	Oscillator noise constants
a_1 and a_2	Incident voltage waves in *S*-parameter techniques
$a_0, \ldots, a_{n-1}$	Weighting coefficients
a_E	Semiconductor flicker noise constant
a_R	Resonator flicker noise constant
a_o	White frequency noise constant
A	Amplifier gain (V or A)
AF	Empirical constant in flicker noise PSD
AM	Amplitude modulation
b_i ($i = -2, \ldots, 2$)	Constants for evaluation of the Allan variance
b_1 and b_2	Reflected voltage waves in *S*-parameter techniques
BAV	Resonators, adherent and nonadherent electrodes
BER	Bit error ratio
BJT	Bipolar junction transistors
B_W	Bandwidth
c	Speed of light
C	Capacity (F, μF, pF)
C_n	Cycle-to-cycle jitter
C_{ox}	Capacitance per unit area
CML	Current mode logic
CMOS	Complementary metal oxide
CW	Continuous wave
d.f.	Number of degrees of freedom
DAC	Digital-to-analogue convertor

DCD	Duty cycle distortion, jitter
DDJ	Data-dependent jitter
DDS	Direct digital synthesizers
DR	Dielectric resonators
DRO	Dielectric resonator oscillator
D_y	Aging (as per day)
$1/f^2$, $1/f^1$	Flicker noise or $1/f$ noise with the PSD
$1/f^2$	Random walk or Brownian movement (often as Wiener–Levy process) with $S(f) - f^{-2}$
e_n, e_n	Root-mean-square (rms) noise voltage (V)
EMI	Electromagnetic interference
$E(x)$	Mean or expected value
ECL	Emitter coupled logic
E_{diss}	Energy loss during one time period
f_B	Frequency bandwidth, often upper-bound frequency
f_c	Intersection frequency between flicker and white noise PSD
f_m	Modulation frequency
f_n	Natural frequency (PLL)
f_o	Resonant frequency
f_T	Cut-off frequency of transistors
f_{LO}	Local oscillator frequency
f_H	Upper-bound frequency
f_L	Lower-bound frequency
$f(t)$	General time function, frequency time function
f_c	Intersection frequency
f_{PLL}	Band-pass frequency (PLL)
f_n	Natural frequency (PLL)
f_r	Reference frequency (in PLL)
$f(x)$, $F(x)$	Probability density function or, simply, probability density
F	Noise factor, noise figure
FET	Field-effect transistors
FOM	Figure of merit
FPN	Flicker phase noise
$F(s)$	Fourier transform
$F_L(s)$	Loop-filter function (in Fourier transform)
$F_{\text{res}}(s)$	Resonator filter function (in Fourier transform)
FFN	Flicker frequency noise
FPN	Flicker phase noise
FFN	Flicker frequency noise
FFT	Fast Fourier transform

g	Conductance, transconductance
g_{m3}	Transconductance of current mirors
$G = 1/R$,	Conductivity [Ω]
$G(s)$, $G(\sigma)$	Open-loop gain (in PLL)
G_p	Amplifier power gain
G_v	Amplifier voltage gain
G_M	Mixer-conversion gain
GPS	Global positioning system
GSM	Global system for mobile communications
HBT	Heterojunction bipolar transistors
HMT	Modulation-doped (MODFET)
$H(s)$	Transfer function, filter function
h	Planck constant ($h = 6.625 \times 10^{-35}$ Js)
h_{-1}	Flicker noise fractional frequency noise constant
h_o	White noise fractional frequency noise constant
h_i	Fractional frequency noise constant ($i = -2, -1, \ldots, 2$)
ISI	Intersymbol interference
i, I	Current (A)
I_B	Current of the current mirror
I_{BE}	Saturation current
I_{BC}	Collector saturation current
I_{ES}	Forward short-circuit saturation current
I_{CS}	Reverse short-circuit saturation current
I_D	Drain current
I_{ds}	Saturation current
I_p	Peak phase detector current
IC	Integrated circuit
I_{pD}, I_{pU}	Charging and discharging peak currents
IIP3	Third-order intercept point
JFET	Field-effect transistors with junction gates
J_D	Deterministic parts representing the time jitter
$J_n(m_f)$	Bessel functions of the first kind and *n*th order
J_R	Unbounded random time jitter
J^D_{App} and J^D_{Dpp}	Advanced and delayed peak–peak deterministic jitter
J(UI)	Unit interval
$J(t)$	Jitter in units of time
k	Boltzmann constant
K	Overall gain (in PLL)
K_A	Amplifier gain
K_d	Phase detector gain (coefficient)
K_{di}	Phase detector gain (current)

K_o Oscillator gain (Hz/V)

K_T Temperature coefficient

K_v Velocity error constant

KF Empirical constant in flicker noise PSD

L Inductance (H, mH, μH), length (m, km), channel length; conversion loss

L_{eff} Effective channel length

$\mathcal{L}\,(ft)$ Laplace transform of the time function $f(t)$

$\mathcal{L}\,(f)$ Power spectral density $[= S(f)\,(\frac{1}{2})]$

m_n nth-order moments

Mod Modified (Allan variance)

MESFET Schottky barrier gates

MIM Metal–insulatior–metal

MODFET Modulation-doped field-effect transistors

MOSFET Metal oxide semiconductor

$M(s)$ Mixer gain (in Fourier transform)

n Effective refractive index, index of noise variables; number of variables

N Division factor

N_{min} to N_{max} Division factors from min to max

OEO Opto-electronic oscillators

$p(t)$ Periodic rectangular function, unit rectangular pulse

$p(n)$ Probability distribution

$p(x)$ Probability density function or, simply, probability density

PJ Periodic jitter

$P(a < x, < b)$ Probability for x between a and b

P_{DC} DC (input) power

P_n, Noise power (W, J/s)

$P_n(s)$ Polynomial

P_o, P_r Effective power in the system (W, mW)

$P_o/P_{\text{diss}} = Q$ Presents the device (resonator) quality factor

$P(\omega)$ Fourier transform of the rectangular unit pulse $p(t)$

PDF Probability density function

PLL Phase-locked loop

PJ Periodic jitter

PM Phase modulation

Q Resonator device quality factor

Qf_{o} Product as a material constant

q Electron or hole charge (1.6×10^{-19} C)

rhs Right-hand side

rms Root mean square

R_x	Resistance of the conductor in (Ω)
$R_x(\tau)$	Autocorrelation of the process $x(t)$
RC	Time constant (s)
RF	Radio frequency
RW	Random walk
RWF	Random walk frequency
R_L, R_S	Load, source resistance
$R_{s,\text{leak}}$	Leaking resistor
R_p, R_s	Parallel, series resistance
r.v.	Random variable
S	Tangent slope close to the zero level
SCL	Source-coupled logic
SPICE	Simulation computer programs
$S_x(f)$	Power spectral density (PSD) of $x(t)$ (generally in 1 – Hz bandwidth)
$S_y(f)$	Power spectral density of the fractional frequency noise
$S_\varphi(f)$	Power spectral density of the phase noise
$S(f)$	Constant white noise power spectral density (PSD) (in 1 – Hz bandwidth)
$S_{i,n}$	PSD of the noise current (A^2 in 1-Hz bandwidth)
$S_{m.n}$	s-factors in S-parameter techniques
$s_r(t_k)$	Time modulation function
$S_1(x)$	Function of the cosine integral
Sp_1	PSD of spurious signals
SAW	Surface acoustic wave
STALO	Stabilized microwave local oscillator
T	Absolute temperature (K)
T_{obs}	Observation time
T_1, T_2	Time constants (in PLL)
T_o	Time period
T_{01x}	Resonance mode
TTL	Transitor–transistor logic
UI	Unit interval
V_D, V_{DD}	Drain votage (V)
V_G	Gate voltage
V_{GS}	Gate source voltage
V_T	Threshold voltage
VCO	Voltage control oscillator
WFN	White frequency noise
WPN	White phase noise
$y(t)$	Fractional frequency fluctuations

Y Admittance ($1/\Omega$)

W Channel width

$<z^2(t)>$ Power (Parceval's theorem)

Z_o Load impedance

$Z(s)$ Impedance (of the loop filter)

$\otimes$ Convolution symbol

CHAPTER 1

Noise and Frequency Stability

Frequency stability or instability is a very important parameter in both modern terrestrial and space communications, in high-performance computers, in GPS (global positional system), and many other digital systems. In this connection, even very small frequency or phase changes of steering frequency generators (exciting oscillators, clock generators, frequency synthesizers, amplifiers, etc.) are of fundamental importance. Since all physical processes are subject to some sort of uncertainties due to fluctuations of individual internal or external parameters, generally designated as noise, the investigation of the overall noise properties is of the highest importance for the analysis of frequency stability.

In practice, we encounter three fundamental types of noises that differ by the power in the time or frequency unit $S(f)$ (generally in the 1 Hz bandwidth), the latter being called the power spectral density (PSD—see Fig. 1.1).

There are three major types of noises:

1. **White noise** with a constant PSD: $S(f) \sim$ const.
2. **Flicker noise** or $1/f$ noise with the PSD $S(f) \sim 1/f^{\alpha}$, where the power α is very close to one.
3. **Random walk** or **Brownian motion** (often as the Wiener–Levy process) with $S(f) \sim f^{-2}$.

Frequency Stability. By Venceslav F. Kroupa

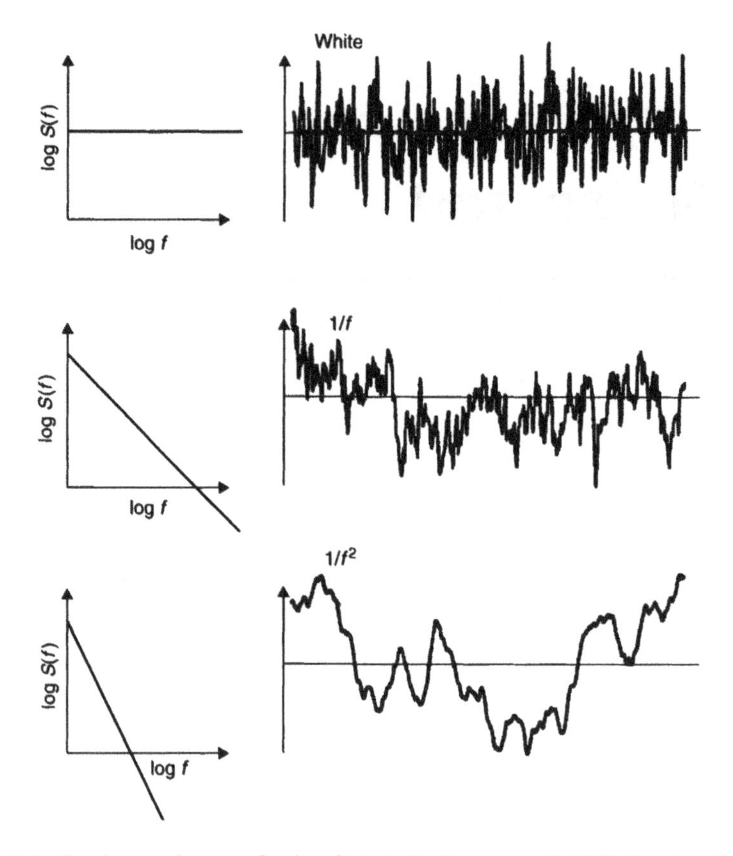

Fig. 1.1 Fundamental types of noises [1.1, 1.2]. (Copyright © IEEE. Reprinted with permission.)

The last two noise processes with their integrals [generating PSD $S(f)$ proportional to $\sim f^3$, $\sim f^4$, etc.] are often called *colored noises.*

1.1 WHITE NOISE

Typical representation of white noise consists of *black body radiation* or the thermal noise of resistors, or shot noise, in electronic devices.

1.1.1 Thermal Noise

In 1928, Johnson [1.3] and Nyquist [1.4] published a theory explaining the existence of thermal noise in conductors. It is caused by short current pulses generated by collisions of a large number of electrons. The result is such that a noiseless conductor is connected in series with a generator with a root mean square (rms) noise voltage, e_n (see Fig. 1.2)

$$e_n^2 = 4kTR\Delta f \tag{1.1}$$

where k is the Boltzmann constant, T the absolute temperature (see Table 1.1), R is the resistance of the conductor (Ω), and Δf is the frequency bandwidth (in Hz) used for the appreciation of the noise action.

After dividing (1.1) by the frequency bandwidth Δf, we arrive at the PSD in 1-Hz bandwidth, that is,

$$S_{e,n}(f) = 4kTR \qquad (\mathrm{V}^2/\mathrm{Hz}) \tag{1.2}$$

Similarly, with the assistance of the Thevenin theorem we get the noise current, i_n, flowing into the resistance R or the conductivity $G = 1/R$, that is,

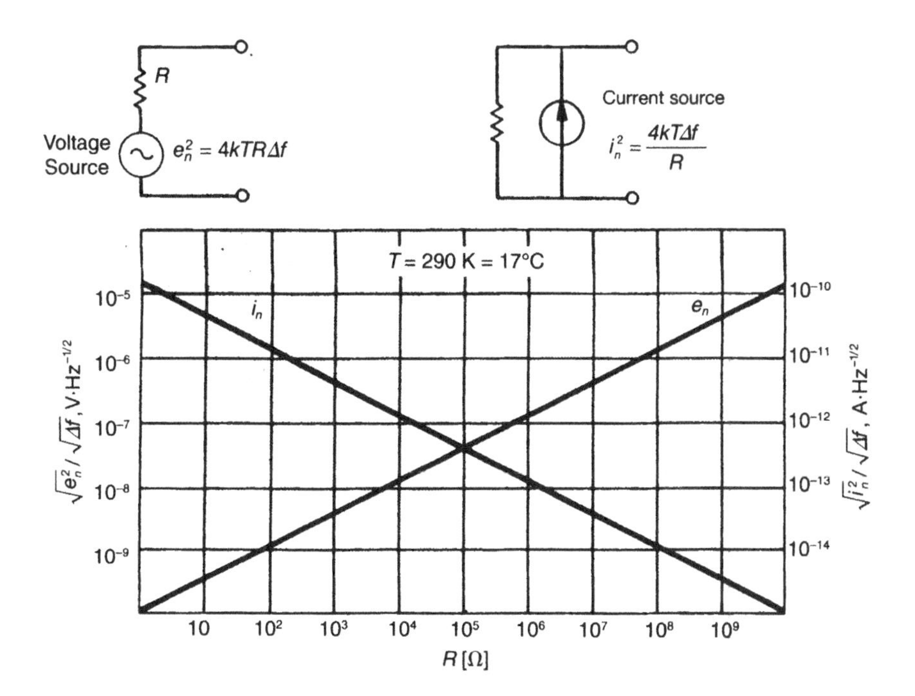

Fig. 1.2 Thermal noises of conductors [1.2]. (Copyright © IEEE. Reprinted with permission.)

Table 1.1 Several physical constants

Physical constants	Symbols	Numerical Values
Planck constant	h	6.625×10^{-35} (Js)
Boltzmann constant	k	1.380×10^{-23} (J/K)
Electron charge	q	1.6×10^{-19} (C)
Speed of light	c	299,792,458 (m/s)
Noise voltage	$4kTR$; $T = 296$ K, $R = 1\ \Omega$	$10^{-19.8}$ (V^2, rms)
Noise voltage	$4kTR$; $T = 296$ K, $R = 50\ \Omega$	$10^{-18.1}$ (V^2, rms)

$$S_{i,n}(f) = 4kTG \qquad (\mathrm{A^2/Hz}) \tag{1.3}$$

(See Fig. 1.2.) In instances with a general impedance or admittance, we introduce only the real parts into the above equations. Further, since both PSDs (1.2) and (1.3) are constant in a very large frequency bandwidth (with no filter at the output), we call this type of noise *white* (in accordance with optical physics). By considering the noise power in a frequency range $\Delta f = f_{\text{high}} - f_{\text{low}}$, we get

$$P_n = \int_{f_l}^{f_h} \frac{S_{e,n}(f)}{R}\, df = 4kT(f_{\text{high}} - f_{\text{low}}) \approx 4kTf_h \qquad (\mathrm{Ws}) \tag{1.4}$$

However, by increasing the upper bound f_{high} above all limits the noise power P_n would also increase above all limits. But this is not possible and the correct solution is provided by quantum mechanics, which changes noise PSD for extremely high frequencies into relation (1.5), where h is the Planck constant, $h = 6.625 \times 10^{-35}$ (Js).

EXAMPLE 1.1

Compute the thermal noise generated in the 1 Hz bandwidth in the $R = 50$-Ω resistor placed at room temperature

$$P_{n,f} = \frac{4hf}{e^{hf/kT} - 1} \qquad (\mathrm{Ws}) \tag{1.5}$$

$k = 1.38 \times 10^{-23}$ (J/K)

$T = 296$ (K)

$R = 50$ (Ω)

$\langle e_n^2 \rangle = 4 \times 1.38 \times 10^{-23} \times 296 \times 50 = 8.17 \times 10^{-19} = 10^{-18.1}$ (V^2)

or $e_n \approx 1.10^{-9}$ mV

1.1.2 Shot Noise

In all cases where the output current is composed of random arrivals of a large number of particles, we again witness fluctuations of the white noise type [1.5].

By considering an idealized transition in Fig. 1.3, where electrons flow randomly from A to B and holes flow from B to A, in a negligible transit time, each particle arrival is connected with transport of a current pulse. Consequently, in a time unit τ (s) the number of n charges generates the current

$$i = \frac{q}{\tau} n \tag{1.6}$$

where q is the electron or hole charge, $q = 1.6 \times 10^{-19}$ (C). It was shown earlier [1.5] that the probability of the transition of the charge carriers was subjected to the Poisson distribution (see also Section 1.4.2.3):

$$p(n) = \frac{<n>^n}{n!} e^{-<n>} \tag{1.7}$$

where $<n>$ is the mean value of the number of carriers in the time unit. In such cases, the variance is equal to

$$\sigma^2(n) = <n> \tag{1.8}$$

By reverting to (1.6), we get for the mean current

$$I = \frac{q}{\tau} <n> \tag{1.9}$$

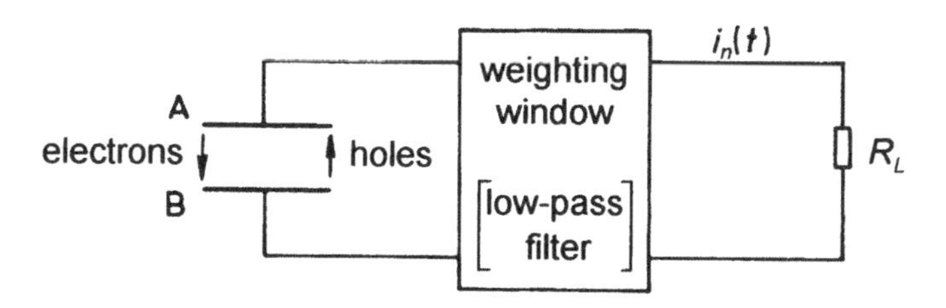

Fig. 1.3 Circuit model for the shot noise [1.2]. (Copyright © IEEE. Reprinted with permission.)

and for its variance value,

$$\sigma^2(I) = \frac{q^2}{\tau^2} < n > = \frac{q}{\tau} I \qquad 1.10$$

To arrive at the PSD, we use a bit heuristic approach with the assistance of the autocorrelation [cf. (1.94)]

$$S_{i,n} = 2\int_0^{\infty} \sigma^2 \cos(\omega t)dt \approx 2\int_0^{\tau} \sigma^2(i)\cos(\omega t)dt = \qquad 1.11$$
$$2\sigma^2(i)\frac{\sin \omega t}{\omega} /_0^{\tau} \approx 2\frac{q}{\tau} I \frac{\omega\tau}{\omega} = 2qI$$

EXAMPLE 1.2

Find the PSD $S_{i,n}(f)$ of the shot noise for the transistor current

$I = 1$ mA $\qquad S_{i,n} = 2 \times 1.6 \times 10^{-19} \times 10^{-3} = 3.2 \times 10^{-22}$

For PSD of other currents, see Fig. 1.4.

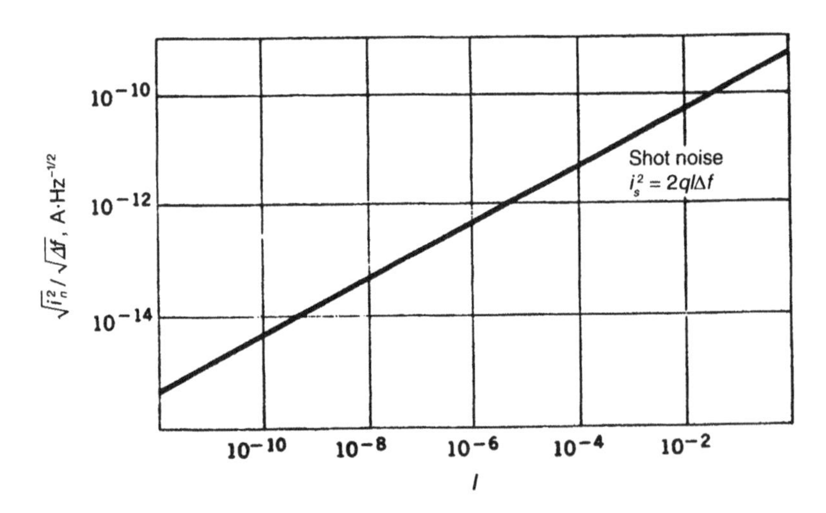

Fig. 1.4 Noise current through a semiconductor junction [1.2]. (Copyright © IEEE. Reprinted with permission.)

1.2 COLORED NOISES

Until now we have discussed PSD of noises generated by slow time-independent fluctuations, that is, with constant PSD over a large Fourier frequency range. However, with oscillators and other frequency generators, we encounter phase fluctuations with frequency-dependent PSDs proportional to $1/f$, $1/f^2$, or even to $1/f^3$, $1/f^4$, at very low Fourier frequencies that are often called *colored noises.*

In the mid-1920s, Johnson [1.3] found that at very low frequencies the shot noise in vacuum tubes did not follow white noise at low frequencies and he introduced for the additive noise the name flicker noise. This name is still used. Subsequent observations proved the $1/f$ law for a much larger set of physical phenomena on one hand and its validity at very low frequencies on the other hand. Some years later, Bernamont [1.6] suggested a law for its PSD:

$$S_n(f) \approx \frac{1}{f^{\alpha}} \qquad 1.12$$

where the power of α was in the vicinity of *one.* In electronic devices, the higher order noises are often generated by integration in the corresponding Fourier transform division by s (cf. Table 1.2). The only exception presents $1/f$ noise fluctuations encountered both in crystal resonators and oscillators, and in many other physical systems [1.7] (dispersions of cars on highways [1.8], frequency change around 50 or 60 Hz in power line systems [1.9], or even flooding in the Nile river valley [1.1]; the latter reference is based on the time dependence of generating fluctuations). Note that all are based on the time.

The problem of colored noises was investigated by many authors in the past from different points of approach and often with different results; particularly, with the ever-present $1/f$ noise. For example, Keshner [1.10] investigated noises with different slopes, $1/f^{\alpha}$, and arrived at a number of variables needed for generation of the desired colored noise (cf. Fig. 1.5). His finding for $1/f$ noise is one degree of freedom per decade.

1.2.1 Mathematical Models of $1/f^{\alpha}$ Processes

Characterization of the frequency stability of all types of generators, inclusive of phase-locked loops (PLLs), is important for applications;

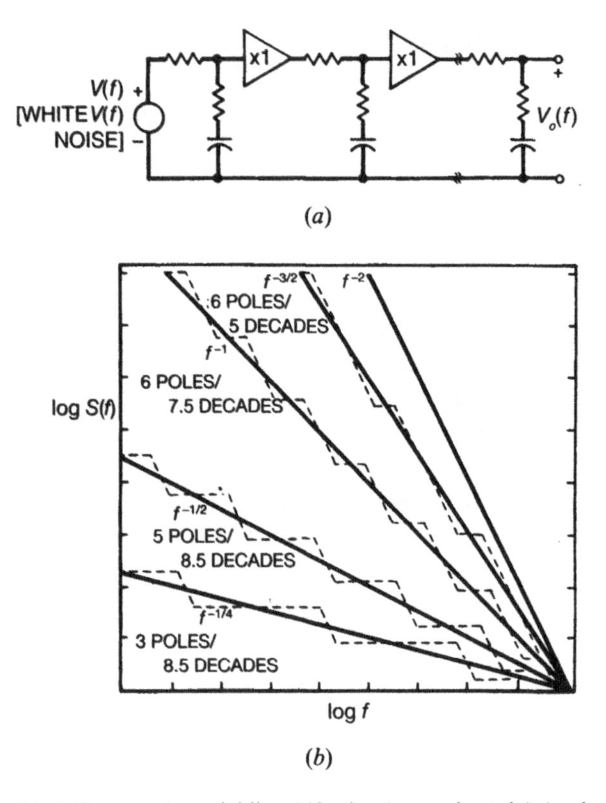

Fig. 1.5 (*a*) A linear system yielding $1/f$ noise (approximately) (each section has a one-state variable, which is the capacitor voltage). (*b*) A curve fit of the power spectral densities of an approximating linear system to obtain $1/f^{\alpha}$ ($\alpha = 0.25, 0.5, 1, 1.5, 2$) [Adapted from 1.10.]

in the first place for their designers, and vice versa for users. In the mid-1960s, theoretical principles of the phase noise theory in frequency generators were established [1.11] and later a number of practical papers were published (e.g. [1.12, 1.13]). Here, we will briefly recall the corresponding theory.

Solution of the noise problems is performed with the assistance of statistics by investigating correlations and by the transformation of the time domain processes into the (complex) frequency domain via the Laplace transform (cf. Appendix at the end of this chapter).

$$\pounds(f(t)) = \int_0^{\infty} f(t)e^{-st}dt \qquad 1.13$$

In instances where the lower bound of the above integral is $-\infty$ the Laplace transform changes into the Fourier transform, the corresponding pairs for important time functions encountered in practice are summarized in Table 1.2.

Reverting to the investigated time function, $f(t)$ in (1.13), the process can be represented as a power series:

$$f(t) = a_0 + a_1 t + a_2 t^2 + \cdots; \qquad \lim_{n\to\infty}(a_n t^n) = \text{const} \tag{1.14}$$

By retaining only the first two terms, we arrive at the exponential approximation that represents a large set of actual situations of the time domain fluctuations,

$$f(t) \approx a_0 - a_1 + n(t) \approx a_o e^{(-a_1/a_o)t} = a_o e^{-at} \tag{1.15}$$

with the respective Fourier transform (cf. Table 1.2),

$$F(s) = \frac{a_o}{s+a} \tag{1.16}$$

After multiplication with the complex conjugate of $F(s)$, we arrive at the so-called Lorenzian PSD (cf. Section 1.5.1, Brownian Motion):

$$S(f) = \frac{a_o^2}{f^2 + a^2} \tag{1.17}$$

Table 1.2 Fourier and Laplace transform pairs for important time functions, encountered in practice

Type	$f(t)$	$F(s)$	Process
Unit step	$u(t)$	$1/s\ [F(s)/s]$	Integration
Ramp	t	$1/s^2$	Aging
Differentiation	$df(t)/dt$	$sF(s)$	
Time delay	$f(t-\tau)$	$F(s)e^{-s\tau}$	
	$1/\sqrt{\pi t}$	$1/\sqrt{s}$	
	$2/\sqrt{t/\pi}$	$s^{-3/2}$	
	t^{k-1}	$\Gamma(k)/s^k\ (k>0)$	$\Gamma(k) = (k-1)!$
	e^{-at}	$1/(a+s)$	Exponential decay
	te^{-at}	$1/(a+s)^2$	Exponential decay with aging
	$\sin(at)$	$a/(a^2+s^2)$	

1.2.2 1/f Noise (Flicker Noise)

To generate the PSD of the $1/f$ slope, so often observed in practice, we encounter a large number of approaches. McWorther [1.14] suggested the mathematical model (for the flicker noise generated in semiconductors) as a multistep process composed of single events (cf. Fig. 1.6a):

$$\sum_{\tau_i} \frac{\tau_i}{1+(2\pi f \tau_i)^2} \tag{1.18}$$

By assuming that in the time domain we have a set of events of the type in equation (1.15), the PSD will retain the shape as in equation (1.17) as long as the time constants, a, do not change appreciably from one. The final amplitude of the PSD is then still a_o^2 at low Fourier frequencies. To arrive at the flicker noise behavior, we start with inspection of the PSD, $S(f)$, in (1.17) and find that in the neighborhood of the corner frequency, $2\pi f \approx 1/\tau$, its slope is approximately proportional to $1/f$. Evidently, by proportionally increasing the time constant and decreasing the amplitude in the corresponding series,

$$S(f) = \sum_{\tau_i} \frac{a_o^2 / \tau_i}{(2\pi f + 1/\tau_i)^2} \tag{1.19}$$

The summation reveals a slope of $1/f$ (see the example in Fig. 1.6*b*), where we have chosen $\tau_i/\tau_{i+1} \approx 10$ and arrived at a nearly perfect slope of $1/f$. This finding is in a good agreement with a discussion by Keshner [1.10]. However, note a rather forceful, not random, condition on the amplitudes and time constants in the set of the Lorentzian noise characteristics (1.19) needed for the generation of the flicker noise system. The difficulty is that this is true for voltage or current fluctuations (e.g., [1.15, 1.16]), however, in instances of other physical quantities (transfer of power, flow of cars on a highway, etc.) the rms of (1.19) must be used. Effectively, we face a fractional integration discussed in connection with the $1/f$ fluctuations by Halford [1.17] or suggested by Radeka [1.18].

In another approach, let us again consider a flow (e.g., of power) with losses during defined time periods (cf. Fig. 1.7). In that case, we introduce a sampling process with the dissipated energy, P_{diss}, during

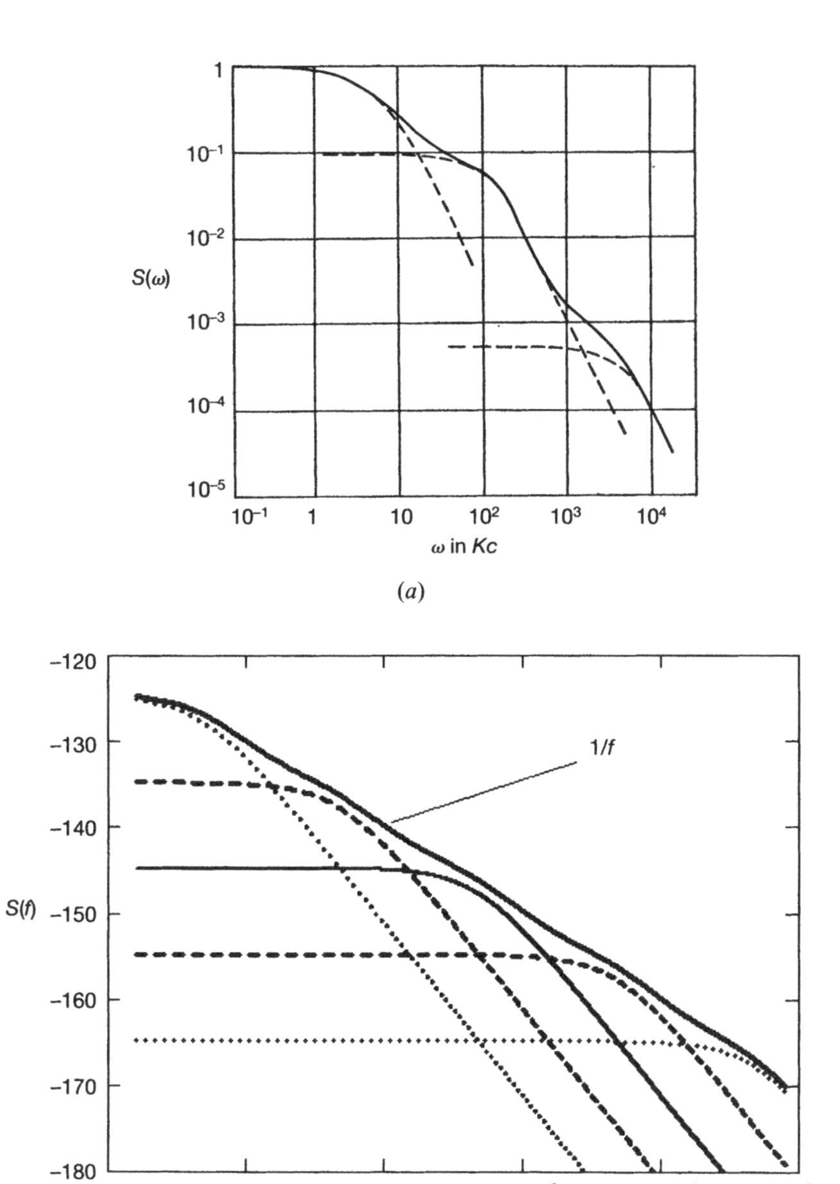

Fig. 1.6 (a) Flicker phase noise generated by a set of several $1/f^2$ noises. Their summation (the solid line) presents the ideal slope $1/f$ [1.14]. (*b*) The simulated slope $1/f$, with five $1/f^2$ noise characteristics providing the background set.

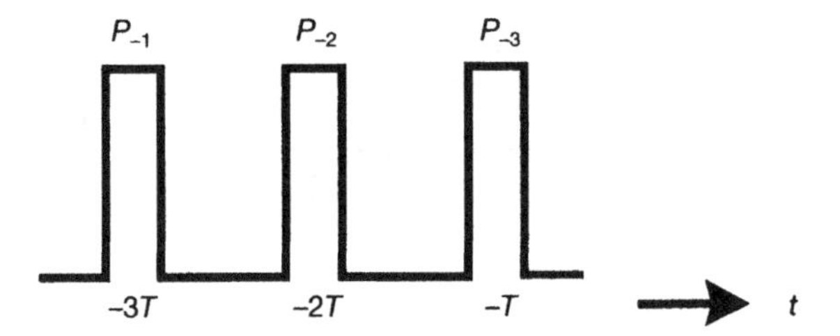

Fig. 1.7 Generation of the phase noise in the sampled form reduced to a set of pulses.

one sampling period T_o. In the next period, we encounter nearly the same energy losses, and so on. Generalization reveals a sampling process whose noise model in the z-transform is (cf. [1.19, 1.20])

$$P_{\text{noise}}(z) = \frac{1}{P_o}(P_{\text{diss},0} + P_{\text{diss},1}z^{-1} + P_{\text{diss},2}z^{-2} + P_{\text{diss},3}z^{-3} + \cdots) \approx \frac{P_{\text{diss}}}{P_o} \cdot \frac{1}{1 - z^{-1}} \qquad 1.20$$

where P_o is the energy of the flux. To get the corresponding Fourier transform, we have to replace z^{-1} with e^{-sT_o} and multiply by the transfer function $H(s)$

$$H(s) = \frac{1 - e^{-sT_o}}{s} \qquad 1.21$$

with the result

$$P_{\text{noise}}(s) = \frac{P_{\text{diss}}}{P_o} \cdot \frac{1}{s} \qquad 1.22$$

However, to get the PSD of the noise power we must apply on the complex product of $P_{\text{noise}} \times P^*_{\text{noise}}$ the rms operation and thus we arrive at the $1/f$ slope (i.e., at the slope of PSD 10 dB/dec in 1-Hz bandwidth):

$$S_{\text{noise}}(f) = \frac{P_{\text{diss}}}{P_o} \cdot \frac{1}{f} \qquad 1.23$$

Where $P_{diss,i}$ are losses in individual periods and P_o the overall power in the steady-state flow.

In this connection, we recall the paper by Kasdin and Walter [1.19] who suggested the sampling process for generation of the flicker noise. By assuming the memory system shown schematically in Fig. 1.8 and the corresponding z-transform, they assumed both $X(z)$ and $Y(z)$ to be energy during one sampling period. After very complicated computations, they arrived at the desired $1/f$ slope and found the approach acceptable from the stochastic point of view but had to apply the fractional integration (i.e., to arrive at the PSD, application of the rms operation on the respective transfer function).

Finally, we have to mention Hooge's formula presented 1969 [1.21], since it was intensively studied, for relation of the power spectral density of current or resistance fluctuations:

$$\frac{S_i(f)}{I^2} = \frac{S_{\Delta R}(f)}{R^2} = \text{const}\,\frac{1}{f} \qquad 1.24$$

(see Section 1.5.2.2.)

1.2.3 $1/f^2$, $1/f^3$, and $1/f^4$ Noises

The PSD of the first type of noises, also designated as the so-called random walk, is generated by the randomly distributed pulses of the type (1.15). With the assistance of (1.17) we get for the system of n pulses the PSD:

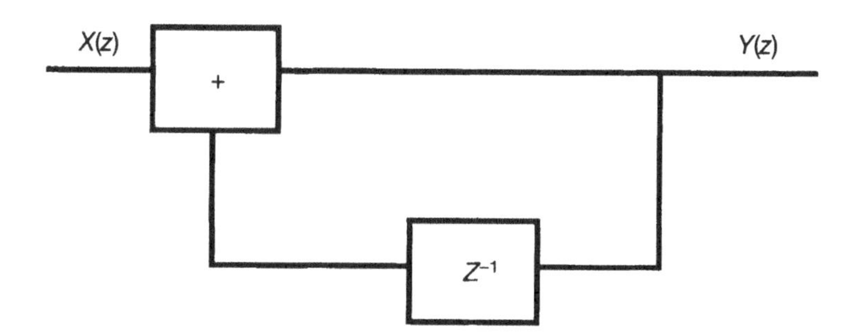

Fig. 1.8 Basic block diagram of the z-transform approach for generation of $1/f$ noise.

$$S(f)=\sum_n \frac{a_o^2}{f^2+a_n^2} \qquad 1.25$$

Note that the above PSD changes into a pure $1/f^2$ spectrum for high Fourier frequencies ($f > a_n$). To this class of noise generators, the stochastic Wiener–Levy process (Sec. 1.5.2) may be enclosed. Another origin of the $1/f^2$ noise is integration of white noise (e.g., in oscillators) and due to the Laplace transform the process as realized by dividing by s (cf. Table 1.2).

EXAMPLE 1.3
In oscillators, the thermal white noise originating in the maintaining electronics generates frequency fluctuations with PSD, as in (1.2):

$$S_\varphi(f)\approx\frac{4kTR}{V_o^2}=\frac{2kT}{P_o} \qquad 1.26$$

However, the oscillating condition requires that the phase around the loop is equal to $2\pi n$ ($n = 1, 2, \ldots$). But this condition is connected with integration (i.e., division by s in the Fourier transform), which changes the white phase noise inside the resonance range into the random walk with the PSD (cf. Chapter 2, Sec. 2.1.3):

$$S_\varphi(f)=\frac{2kT}{P_o f^2} \qquad 1.27$$

Similarly, the higher order noises ($1/f$ and $1/f^2$) in the oscillator maintaining electronics generate the phase noise with PSD inversely proportional to f^{-3} or f^{-4} due to the integration process.

1.3 SMALL AND BAND LIMITED PERTURBATIONS OF SINUSOIDAL SIGNALS

Till now, we have considered single-frequency generators, that is, oscillators with rather small and continuous amplitude and phase perturbations. However, in frequency synthesizers, particularly *direct digital synthesizers* (DDS), we encounter many spurious signals. In the following sections, we will investigate some of their properties [1.2, 1.22].

1.3.1 Superposition of One Large and a Set of Small Signals

In actual frequency synthesizers, PLL systems not excluded, we always encounter many generally small spurious signals accompanying the carrier. The composite signal may be written as

$$v(t) = \sum_{n=1}^{N} V_n \cos(\omega_n t + \phi_n) \qquad 1.28$$

After introducing

$$\omega_n t + \phi_n = \omega_1 + (\omega_n - \omega_1)t + \phi_n = \omega_1 t + \Phi_n(t) \qquad 1.29$$

and after putting

$$\alpha_n = \frac{V_n}{V_1} \qquad 1.30$$

we get

$$\begin{aligned} v(t) &= V_1 \sum_{n=1}^{N} \alpha_n \cos[\omega_1 t + \Phi_n(t)] = \\ &V_1 \cos(\omega_1 t) \sum_{n=1}^{N} \alpha_n \cos[\Phi_n(t)] - V_1 \sin(\omega_1 t) \sum_{n=1}^{N} \alpha_n \sin[\Phi_n(t)] \end{aligned} \qquad 1.31$$

or

$$v(t) = V_1 \alpha(t) \cos[\omega_1 t + \Phi(t)] \qquad 1.32$$

where $\alpha(t)$ is the normalized instantaneous amplitude

$$\alpha^2(t) = \left[\sum_{n=1}^{N} \alpha_n \cos \Phi_n(t) \right]^2 + \left[\sum_{n=1}^{N} \alpha_n \sin \Phi_n(t) \right]^2 \qquad 1.33$$

and $\Phi(t)$ is the instantaneous phase departure

$$\Phi(t) = \arctan \frac{\sum_{n=1}^{N} \alpha_n \sin \Phi_n(t)}{\sum_{n=1}^{N} \alpha_n \cos \Phi_n(t)} \qquad 1.34$$

Without any loss of generality, it is possible to choose the time scale in such a way that

$$\phi_1 = \Phi_1(t) = 0 \qquad 1.35$$

If only small perturbations are assumed one may put

$$\alpha_1 \gg \alpha_n;\ (n = 2,3,\ldots,N)\ \ (\text{note } \alpha_1 = 1) \qquad 1.36$$

With this situation, the spurious amplitude is

$$\alpha(t) \approx 1 + \sum_{n=1}^{N} \alpha_n \cos \Phi_n(t) \qquad 1.37$$

and the spurious phase is

$$\Phi(t) = \arctan \sum_{n=1}^{N} \alpha_n \sin \Phi_n(t) \approx \sum_{n=1}^{N} \alpha_n \sin \Phi_n(t) \qquad 1.38$$

EXAMPLE 1.4

For the superposition of one strong signal $V_1 \cos(\omega_1 t)$ and one weak signal $V_2 \cos(\omega_2 t + \varphi_2)$ (i.e., $V_1 \gg V_2$) we get

$$v(t) \approx V_1 \left[1 + \frac{V_2}{V_1} \cos(\Omega t + \phi_2)\right] \cos\left[\omega_1 t + \frac{V_2}{V_1} \sin(\Omega t + \phi_2)\right] \qquad 1.39$$

Evidently, in the first approximation we face a simultaneous amplitude and phase modulation of the stronger signal at the rate of difference frequency, $\Omega = |\omega_2 - \omega_1|$, with the modulation indexes V_2/V_1. In the case of its larger value, higher order terms must be added.

1.3.2 Narrow Bandwidth Noise

In instances where the noise power is concentrated in a relatively narrow band around the frequency ω_1, the noise voltage can be expressed as

$$e(t) = e_c(t)\cos\omega_1 t - e_s(t)\sin\omega_1 t \qquad 1.40$$

where the slowly varying time functions $e_c(t)$ and $e_s(t)$ are statistically independent. Note that (1.40) resembles (1.31). Consequently, the product of the mean values is zero if $<e(t)>$ is zero, that is,

$$< e_c(t)e_s(t) > = < e_c(t) >< e_s(t) > = 0 \qquad 1.41$$

and also

$$< e_c(t) > = < e_s(t) > = 0 \text{ and } < e^2(t) > = < e_c^2(t) > = < e_s^2(t) > \qquad 1.42$$

1.4 STATISTICAL APPROACH

In the previous sections, frequency stability was discussed from the point of view of common noises. However, the problem is much more complicated since actual situations may contain both continuous and sampled systems, both random and discrete fluctuations, and even other processes solved with the advantage of statistical approaches whose basic properties are discussed briefly in the following sections.

1.4.1 Probability

When inspecting physical, biological, economical, and many other processes, we find either a deterministic model or start from experimental observations and guess the details. One of the tools we use is the appreciation of the outcome with the assistance of probability theory, which is the ratio of the positive outcomes of an experiment n_p to all trials n_m:

$$P = \frac{n_p}{n_m} \qquad 1.43$$

The corresponding theory was well established in the past (e.g., [1.22]). Here, we recall the three axioms, namely, that the result is always a positive number between zero and one, that the probability of mutually independent events is equal to the sum of individual probabilities, and that the probability of the whole set of events is equal to 1 (cf. Fig. 1.9)

$$P(e_i < x) \geq 0;\ P(x = S) = 1$$
$$P(A_x + B_x) = P(A_x) + P(B_x) - P(A_x)P(B_x) \qquad 1.44$$

In addition, conditional probability of mutually independent events is equal to their product

$$P(A_x \mid B_x) = P(A_x) \cdot P(B_x) \qquad 1.45$$

Note that all operations are performed on sets subjected to the Boole summations and multiplications.

1.4.2 Random Variables, Distribution Function, Density of Probability

Let us assume an experiment E with events e_i identified by real or complex numbers, $\xi(e_i)$, which will be designated as *random variables*. Next, we define the probability of a set of events meeting condition

$$\xi(e_i) \leq x \quad P(\xi \leq x) = F_\xi(x) \qquad 1.46$$

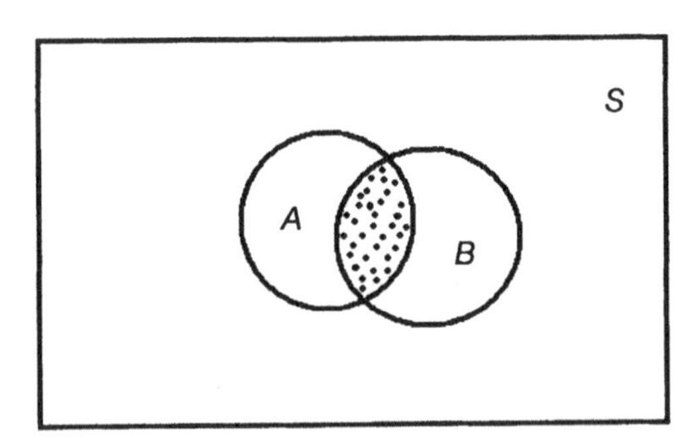

Fig. 1.9 Idealization of the probability space $S = 1$ of all the events e_n. A represents the set of the e_{A_i} events and B the set of the e_{B_i} events.

where $F_\xi(x)$ defines the *distribution of the probability,* which is a monotonic nondeclining function from 0 to 1. A typical behavior is presented in Fig. 1.10 for both continuous and discrete variables.

By reverting to the continuous variable, we define its derivative and designate it as the *probability density function* or simply *probability density f(x)* or *p(x),*

$$f(x) = \frac{\delta F(x)}{\delta x} \qquad 1.47$$

if this derivative exists. Further, the probability for x between a and b is

$$P(a \le x \le b) = F(b) - F(a) \qquad 1.48$$

The mean or the expected value (the moment of the first order) is

$$E(x) = \mu(x) = \int_{-\infty}^{\infty} x f(x) dx \quad \text{or} \quad E(x) = \frac{1}{n}\sum_{i=1}^{n} x_i \qquad 1.49$$

Generally, the mean values of the nth order of random variables are designated as the nth-order moments:

$$m_{n\xi} = <\xi^n> = \int_{-\infty}^{\infty} x^n f(x) dx \qquad 1.50$$

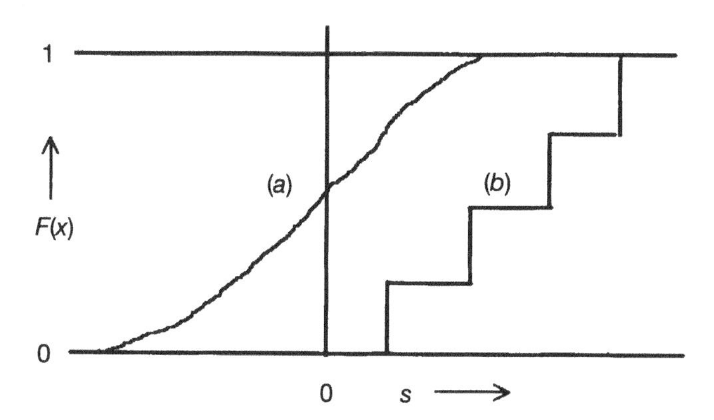

Fig. 1.10 Probability distribution function: (*a*) a continuous variable and (*b*) a discrete variable.

Central moments are important:

$$\mu_n = \langle (\xi - m_{n,\xi})^n \rangle = \int_{-\infty}^{\infty} (x - m_{n,\xi})^n f(x)dx \quad 1.51$$

The central moment of the second order defines the *variance*, σ^2, or distribution and the corresponding rms, σ, the so-called dispersion:

$$\mu_2 = \sigma^2 = m_{2\zeta} - m_{1\zeta}^2 \quad 1.52$$

1.4.2.1 The Uniform Distribution

The simplest *probability density function f(x)* is assumed to be constant between a and b on the x axis and zero otherwise (cf. Fig. 1.11):

$$f(x) = \frac{1}{b-a} \quad \text{and} \quad 0 \text{ elswere} \quad 1.53$$

The corresponding mean and variance are

$$\mu = \frac{b-a}{2} \qquad \sigma^2 = \frac{1}{12}(b-a)^2 \quad 1.54$$

1.4.2.2 Binomial Distribution

We face a discrete distribution and the task is to compute the probability of the event ξ in n trials. Let p be the probability of the positive

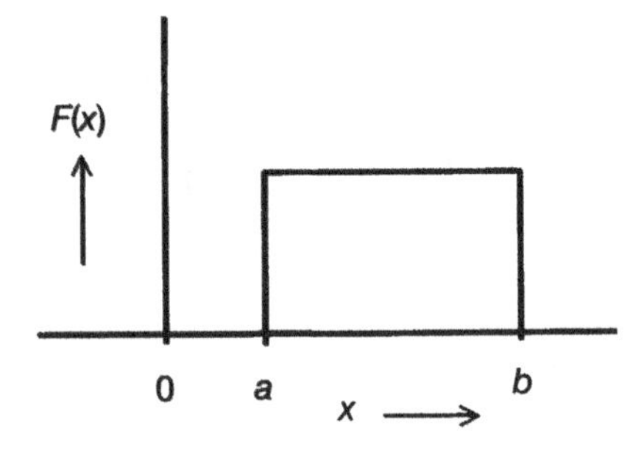

Fig. 1.11 The probability density function $f(x)$ of the uniform distribution.

outcome and $q = 1 - p$ the probability of the opposite. The probability of the occurrence of the event ξ occurring only once in a sequence of trials is

$$P_n(\xi)' = p \cdot q^{n-1} \tag{1.55}$$

In instances where the sequence is unimportant, the probability is

$$P_n(\xi) = n \cdot p \cdot q^{n-1} \tag{1.56}$$

Finally, if the event ξ should repeat *k times* the probability is

$$P_{n,k}(\xi) = \binom{n}{k} p^k q^{n-k} \tag{1.57}$$

The first and second moments are computed in Section 1.4.3.1:

$$\mu = pn \qquad \sigma^2 = pn \tag{1.58}$$

1.4.2.3 Poisson Distribution

For a very large number of trials, $n \to \infty$, and, $p \to 0$, the binomial distribution passes into the Poisson distribution [1.22 p. 72] from (1.57):

$$\begin{aligned} p_{n,k}(\xi) = \binom{n}{k} p^k q^{n-k} \approx \frac{n!}{k!(n-k)!} p^k q^n \approx \frac{(np)^k}{k!}(1 - pn) \approx \\ \frac{(np)^k}{k!} e^{-pn} = \frac{\lambda^k}{k!} e^{-\lambda} \end{aligned} \tag{1.59}$$

The mean and variance are the same as above, namely, equal to $\lambda = pn$ (cf. shot noise).

1.4.2.4 Gaussian Distribution

Another limiting process of the binomial distribution for large n results in the *Gaussian* or *normal distribution.* Computation of the probability density is a cumbersome one. The asymptotic solution is based on introduction of a new variable, $k = np + x$, and application of the Stirling approximation for factorials, and after approximating powers of binomials close to one,

$$n! = n^n e^{-n}\sqrt{2\pi n}\left(1+\frac{1}{12n}\right) \qquad (1+\alpha)^\beta \approx e^{\beta lg(1+\alpha)} \approx e^{\beta\alpha(1-\alpha/2)} \tag{1.60}$$

After introduction of these approximations into (1.59), we finally arrive at the *density function:*

$$\begin{aligned} p_{n,k} &\approx \frac{(pn)^{(pn+x)}}{(pn+x)^{(pn+x)} e^{-(pn+x)}\sqrt{2\pi(pn+x)}} e^{-pn} \\ &\approx \frac{1}{\left(1+\frac{x}{pn}\right)^{(pn+x)} e^{-x}\sqrt{2\pi(pn+x)}} \\ &\approx \frac{1}{e^{[(pn+x)\frac{x}{pn}\left(1-\frac{1}{2pn}\right)-x]} * \sqrt{2\pi(pn+x)}} \approx \frac{e^{\frac{x^2}{2pn}}}{\sqrt{2\pi pn}} = \frac{e^{-\frac{(x-x_o)^2}{2\sigma^2}}}{\sigma\sqrt{2\pi}} = f(x) \end{aligned} \tag{1.61}$$

and after integration we arrive at the Gaussian *distribution function F(x):*

$$F(x) = \frac{1}{\sigma\sqrt{2\pi}}\int_{-\infty}^{x} e^{(x-x_o)^2/(2\sigma^2)}dx \quad x_o = \mu \tag{1.62}$$

where σ^2 is the variance, σ is the dispersion, and μ is the mean value of the process. Note that this distribution is also designated as the *normal distribution* and is, by far, the most important. It is often postulated in concrete situations when solving actual probability problems. Numerical values of $F(x)$ are published in tables or on-spot computed. The normalized distribution function $F(x)$ for $\sigma = 1$ and $\mu = 0$ is depicted in Fig. 1.12. Note that there are other distributions; however, we feel that those mentioned here are sufficient for information needed in this book. The sample distributions χ^2 is discussed in Chapter 5, Sec. 5.2.4.

EXAMPLE 1.5

Determination of the density function of $f_y(y) = g(x)$, where x is randomly distributed in the interval $(-\pi, \pi)$. Papoulis [1.22] pos-

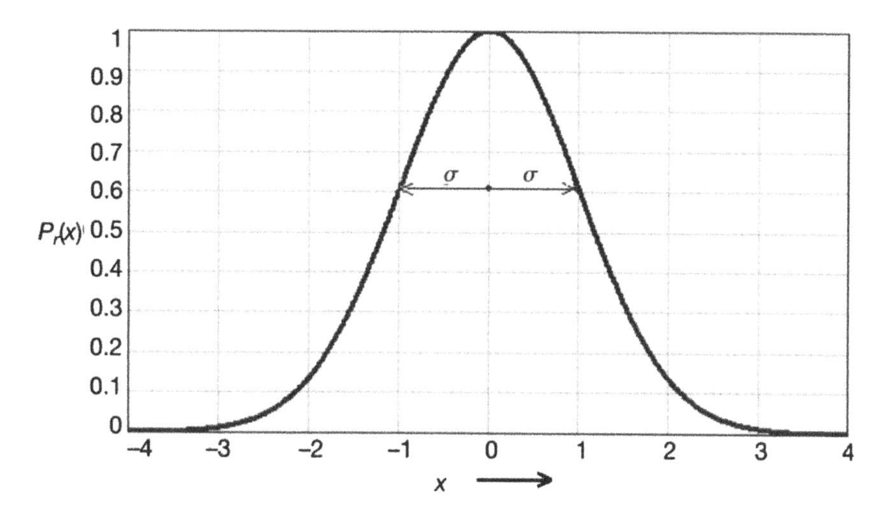

Fig. 1.12 The normal distribution for $\sigma = 1$ and $\mu = 0$.

tulated the fundamental theorem: If $x_1, x_2, \ldots, x_n, \ldots$ are all real roots of $f_y(y) = g(x)$, then the density function is

$$f_y(y) = \frac{f_x(x_1)}{g'(x_1)} + \cdots + \frac{f_x(x_n)}{g'(x_n)} + \cdots + \qquad g'(x) = \frac{dg(x)}{dx} \tag{1.63}$$

For an important case (cf. Fig. 5.16*b*),

$$y = \sin(x + \theta) \qquad f_y(y) = \frac{2}{2\pi\sqrt{1 - y^2}} \tag{1.64}$$

1.4.3 Characteristic Functions

The Fourier transform of the probability density $f_\xi(x)$ is the so-called characteristic function of the random variable ξ:

$$\Phi_\xi(u) = < e^{ju\xi} > = \int_{-\infty}^{\infty} f_\xi(x) e^{jux} dx \tag{1.65}$$

We have seen that the introduction of the moments simplified some conclusions and certain computations encountered in the probability applications. Similarly, adoption of the characteristic function and its logarithm simplifies some statements about moments. Expansion of the e-funtion in (1.65) in series reveals

$$\Phi_\xi(u) = \int_{-\infty}^{\infty} f_\xi(x)\left[1 + jux + \cdots + \frac{(jux)^k}{k!} + \cdots + \right] dx = \\ 1 + ju<\xi> - \frac{u^2<\xi^2>}{2} - \cdots + \frac{(ju)^k}{k!}<\xi^k> + \cdots \qquad 1.66$$

Note that the characteristic function of the random variable ξ is easily evaluated from the knowledge of moments, the first and second order often suffice.

1.4.3.1 Characteristic Function of the Binomial Distribution

For a large n, the integration in (1.65) can be approximated with the summation of the binomial series:

$$\Phi_\xi(u) \approx \sum_{k=0}^{n} \binom{n}{k} p^k q^{n-k} e^{juk} = (pe^{ju} + q)^n \qquad 1.67$$

The first derivation reveals the mean value (the first moment):

$$m_{1,\xi} = -jn(pe^{ju} + q)^{n-1}(pe^{ju}j)/_{u=0} = np = <\xi> \qquad 1.68$$

Similarly, the second moment and the variance are:

$$m_{2,\xi} = <\xi^2> = np(np+1) \qquad \sigma^2 = <\xi^2> - <\xi>^2 = np \qquad 1.69$$

1.4.3.2 Characteristic Function of the Gaussian Distribution

After introducing the probability density function of the Gaussian distribution $f(x)$ (1.62) into the characteristic function definition (1.65), we have

$$\Phi(u) = \frac{1}{\sigma\sqrt{2\pi}} \int_{-\infty}^{\infty} e^{-(x-x_o)^2/(2\sigma^2)} e^{jxu} dx \approx$$
$$\frac{1}{\sigma\sqrt{2\pi}} \int_{-\infty}^{\infty} e^{-(x-x_o)^2/(2\sigma^2)} \cos(jxu) dx \qquad 1.70$$

For the evaluation, we used the property that $f(x)$ is an even function, that is, $f(-x) = f(x)$, and that it is concentrated around $x = 0$. With the assistance of the Dwight formula 861.20 [1.23] we arrive at

$$\Phi(u) = \frac{1}{2} e^{-u^2\sigma^2/2 + jux_o} \qquad 1.71$$

1.4.3.3 Characteristic Function of the Sum of Distributions

The Fourier transform of the probability density simplifies some solutions, for example, of sums of random variables:

$$\eta = a \cdot \xi + b$$
$$\Phi_\eta(u) = < e^{ju\eta} > = < e^{ju(a\xi + b)} > = e^{jub} \Phi_\xi(au) \qquad 1.72$$

or generally for

$$\zeta = \xi_1 + \xi_2 + \cdots + \xi_n$$
$$\Phi_\zeta(u) = \Phi_{\xi,1}(u) \Phi_{\xi,2}(u) \ldots \Phi_{\xi,n}(u) \qquad 1.73$$

Note, that limitation of the characteristic function to the two first moments reveals

$$\Phi_\zeta(u) \approx \left(1 + \sum_i^n ju\xi_i\right) - \frac{u^2}{2} \sum_i^n < \xi_i^2 > + \text{higher order terms}) \qquad 1.74$$

from which it follows that the mean value is the sum of the individual mean values and the variance is the sum of individual variances. Now, let us calculate the distribution for n-Gaussian probability distributions:

$$F(x) = \frac{1}{2\pi} \int_{-\infty} e^{-(u^2/2)(\sigma_1^2+\sigma_2^2+...+\sigma_n^2)+ju(x_{o1}+x_{o2}+...+x_{on})} du =$$

$$\frac{1}{\sigma\sqrt{2\ \pi}} e^{-(x-x_o)^2/(2\sigma)^2} \quad \left[x_o = \sum_i x_{oi} \quad \sigma^2 = \sum_i \sigma_{oi}^2 \right] \qquad 1.75$$

After comparing (1.74) with (1.75), we conclude that the sum of partial distributions, irrespective of the type, finally results in the Gaussian or normal probability distribution. The process is often designated as the central limit theorem.

1.4.4 Stochastic Processes

In instances where the evaluated system is accompanied with a lot of disturbing signals subjected, in addition, to time fluctuations, it is labeled as a stochastic process encountered in different fields of engineering. The situation is generally so complicated that it is difficult to solve in the closed form; the proper situation is with frequency stability of oscillators, frequency synthesizers, communications channels, and so on. Now, let us assume an experiment formed by a set of e_i events with assigned time functions $x_i(t)$; in such a case, we face a class forming a stochastic process $\mathbf{x}(t)$, where time t may have any value, either continuous or discrete. In instances where the time is fixed, $t = t_i$, then $x_i(t)$ is a random variable of the event e_i.

1.4.4.1 Distribution Functions and Probability Density

Similarly, as with the time-independent system, the probability distribution function is

$$F((x_1 \ldots x_n; t_1 \ldots t_n)) = P[x(t_1) \leq x_1 \ldots x(t_n) \leq x_n] \qquad 1.76$$

with the probability density

$$f(x_1 \ldots x_n; t_1 \ldots t_n) = \frac{\delta^n F(x_1 \ldots x_n; t_1 \ldots t_n t)}{\delta x_1 \ldots \delta x_n} \qquad 1.77$$

It is evident that the above equations are not suitable for solving stochastic processes. The solution provides either an analytical description, where the parameter is the random variable as, for example,

$$v(t) = A\sin(\omega_o t + \xi) \qquad 1.78$$

or we must content ourselves with partial information about the investigated stochastic process, such as the knowledge of the moments. Generally, the mean or the expected value which, however, remains a function of time is

$$m_{1x}(t) = < x(t) > = \int_{-\infty}^{\infty} x f(x,t) dx \qquad 1.79$$

Similarly, the autocorrelation is the moment of the second order of the random variables $x(t_1)$ and $x(t_2)$, that is,

$$R(t_1,t_2) = < x(t_1)x(t_2) \geq \int_{-\infty}^{\infty} x_1 x_2 f(x_1,x_2;t_1,t_2) dx_1 dx_2 \qquad 1.80$$

Furthermore, the autocovariance is

$$C(t_1,t_2) = < [x(t_1) - \eta(t_1)][x(t_2) - \eta(t_2)] > \qquad 1.81$$

and the variance of the random variable (r.v.) $x(t)$ is given by

$$\sigma_{x(t)}^2 = C(t,t) = R(t,t) - \eta^2(t,t) \qquad 1.82$$

Autocorrelation or, eventually, autocovariance characterizes the statistical relationship of both random variables $x(t_1)$ and $x(t_2)$ for any times t_1 and t_2.

1.4.4.2 Stationary Stochastic Processes

Very important are stochastic processes that are invariant with respect to shift on the time axis and are designated as stationary in the strict sense. In this case, the probability density for any time shift δ is

$$f(x_1 \ldots x_n; t_1 + \delta \ldots t_n + \delta) = f(x_1 \ldots x_n; t_1 \ldots t_n) \qquad 1.83$$

When choosing $\delta = -t_1$ we find out that the *probability density* is a *constant,* and consequently the mean or expected value is also a constant

$$< x(t) > = E[x(t)] = m_{x,1} = \text{const} \qquad 1.84$$

However, the second-order moment is a function of the time difference:

$$< [x(t_1)x(t_2)] > = < [x(t)x(t+\tau)] > = R_{xx}(\tau) \qquad 1.85$$

In instances where $x(t)$ is real, the autocorrelation is also real and in addition is an even function:

$$R_{xx}(\tau) = R_{xx}(-\tau) \qquad 1.86$$

and the autocovariance is

$$C_{xx}(\tau) = R_{xx}(\tau) - m_{x,1}^2 \qquad 1.87$$

Furthermore, it follows that autocorrelation of a sum

$$z(t) = x(t) + y(t) \qquad 1.88$$

may be expressed as a sum of autocorrelations:

$$R_{zz} = R_{xx}(\tau) + R_{yy}(\tau) + R_{xy}(\tau) + R_{yx}(\tau) \qquad 1.89$$

and for any time shift τ, the probability density is constant. However, the autocorrelation of a product

$$w(t) = x(t)y(t) \qquad 1.90$$

generally cannot be expressed as a function of the second-order moments. Only in instances where both time processes are independent is the autocorrelation equal to the product of partial autocorrelations

$$R_{ww}(\tau) = R_{xx}(\tau) \cdot R_{yy}(\tau) < R_{ww}(0) \qquad 1.91$$

Finally, we conclude with the fact that usually we do not know all the information needed for stationary processes (cf. 1.83). In such cases, we must be content with the statement that *t*he process is stationary in the wide sense or weakly stationary.

1.4.4.3 Random Walk

The random walk is the sampling process taking equal steps either in the positive or negative sense (direction). By taking advantage of the central limit theorem, we may assume that the corresponding probability density of individual steps is

$$p(x) = xe^{-x^2/\sigma_i^2} \tag{1.92}$$

The corresponding variance after *n* steps is (Dwight [1.23], Eq. 860.12)

$$\sigma^2 = \sum_{i=1}^{n} \int_{-\infty}^{\infty} x\, e^{-x^2/\sigma_i^2} dx = n\sigma_i^2 \tag{1.93}$$

1.4.5 Ergodicity

Ergodicity deals with problems of determining the statistics of the process ***x(t)***: The process is ergodic in the most general form if all its statistics can be determined from a single function $x(t, \xi)$ of the process, or the process is ergodic if the time averages equal ensemble averages. The various criteria for ergodicity are discussed in detail by Papoulis in [1.22].

1.5 POWER SPECTRA OF STOCHASTIC PROCESSES

Power spectra or spectral density of the process ***x(t)*** is the Fourier transformation of its autocorrelation:

$$S(\omega) = \int_{-\infty}^{\infty} R(\tau) e^{-j\omega\tau} d\tau \tag{1.94}$$

with the inversion formula

$$R(\tau) = \frac{1}{2\pi} \int_{-\infty}^{\infty} S(\omega) e^{j\omega\tau} d\omega \tag{1.95}$$

and with the results for real processes

$$S(-\omega) = S(\omega)$$

$$S(\omega) = \int_{-\infty}^{\infty} R(\tau)\cos(\omega\tau)d\tau \qquad R(\tau) = \frac{1}{2\pi}\int_{-\infty}^{\infty} S(\omega)\cos(\omega\tau)d\omega \tag{1.96}$$

Table 1.3 Shows the correspondence between a process ***x*(*t*)**, its autocorrelation $R(\tau)$, and the power spectrum $S(\omega)$.

1.5.1 Brownian Motion

The random movement of particles immersed in liquids is referred to as Brownian motion. The first observations (1827) were provided with mechanical particles, however, later studies proved a more general process. Let us start with the velocity of a free particle in a viscous medium. With the assistance of the laws of motion, we arrive at the following differential equation:

$$m\frac{dv(t)}{dt} + bv(t) = B(t) \equiv m \cdot n(t) \tag{1.97}$$

where m is its mass, b is the friction force proportional to the velocity $v(t)$, and $B(t)$ represents the collision force [1.22]. In cases where the observation time is long, one may assume that $v(t)$, $B(t)$, and $n(t)$ are

Table 1.3 Correspondence between a process $x(t)$, its autocorrelation $(R)t$, and power spectrum $S(\omega)$ [1.22]

$x(t)$	$R(\tau)$	$S(\omega)$
$ax(t)$	$\lvert a\rvert^2 R(\tau)$	$\lvert a\rvert^2 S(\omega)$
$\frac{dx(t)}{dt}$	$-\frac{d^2R(\tau)}{d\tau^2}$	$\omega^2 S(\omega)$
$\frac{d^n x(t)}{dt^n}$	$(-1)^n\frac{d^{2n}R(\tau)}{d\tau^{2n}}$	$\omega^{2n} S(\omega)$
$x(t)e^{\pm j\omega_0 t}$	$R(\tau)e^{\pm j\omega_0\tau}$	$S(\omega \mp \omega_0)$
	$R(\tau)\cos\omega_0\tau$	$\frac{1}{2}[S(\omega+\omega_0) + S(\omega-\omega_0)]$

stochastic processes but $n(t)$ is normal white noise with a zero mean and spectrum $S_n(f) = a$. Now let us assume a long observation time, that is,

$$t \gg \frac{m}{b} = \frac{1}{\beta} \qquad 1.98$$

one may consider $v(t)$ as a stationary process and (1.97) as stochastic. With the rules for the derivation of the spectra of stochastic processes (cf. Table 1.3) we get

$$\omega^2 S_v(\omega) + \beta^2 S_v(\omega) = S_n(\omega) \qquad 1.99$$

Consequently, we arrive at the Lorezian spectrum

$$S_v(f) = \frac{a}{f^2 + (\beta/2\pi)^2} \approx \frac{a}{f^2} \qquad (\omega \gg \beta) \qquad 1.100$$

1.5.2 Fractional Integration (Wiener–Levy Process)

Let us consider the situation where the output events are random, with nearly equal changes in one or the opposite direction, with the nearly constant variances $(\Delta e_i)^2$ in each period or time span. By taking into account the central limit theorem, the variance of the expected change, after n steps, will be

$$< (\Delta e)^2 > \approx nT_o < (\Delta e_{\text{one step}})^2 > \qquad 1.101$$

The situation is explained in the following example:

EXAMPLE 1.6

One hundred years ago, K. Pearson and Lord Rayleigh [1.24] presented the following random-walk problem: A man (presumably very drunk) takes steps of equal length m from a starting point O, one after the other in successively random directions. Where will he likely be after n steps? If n is large, the probability that he is at a distance r and $r + dr$ from the starting point is

$$P(r)dr = \frac{2r}{nm^2} e^{-r^2/(nm^2)} dr \qquad 1.102$$

His average distance is equal to the distribution σ, that is,

$$r_{av} = \int_0^\infty rP(r)dr = \frac{\sqrt{\pi}}{2}\sqrt{nm} \qquad 1.103$$

[cf. Eq. (1.93)] with the assistance of Dwight formula 860.12. Since each step takes some time, n is a measure of time, and so the distance will increase with the square root of time (see Fig. 1.13).

1.5.2.1 Power Spectra with Fractional Integration Proportional to $\sqrt{t}$

Reverting to relation (1.102) and considering the time dependence of the final σ, we may also suppose Δe_i to be a function of time without any appreciable error. In the first approximation, we propose for its time dependence

$$\Delta e(t) \approx \sigma(\Delta e)\sqrt{t} \qquad 1.104$$

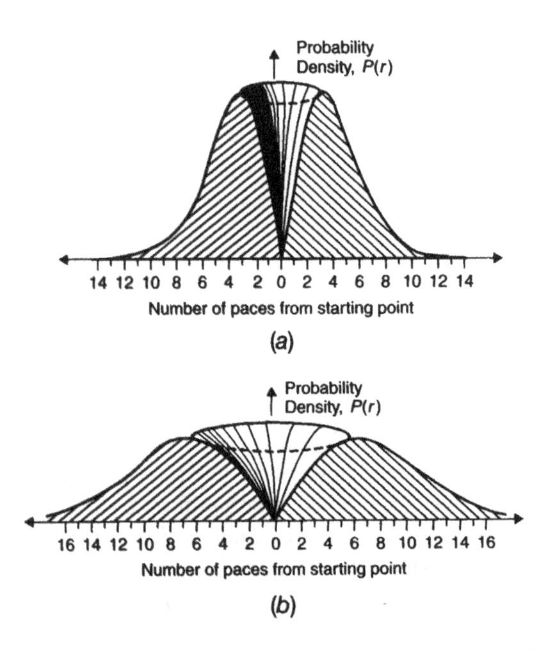

Fig. 1.13 Probability surface of a random walk (position of the drunken man): (*a*) after 18 random paces, and (*b*) after 72 random paces) [1.24].

With the assistance of the Table 1.2, we get for the Fourier transform

$$F(s)_{\Delta e} = s^{-3/2}\sigma(\Delta_e) \tag{1.105}$$

and for the corresponding PSD

$$S_{\Delta e}(f) = \frac{1}{f^3}\sigma^2(\Delta e) \tag{1.106}$$

1.5.2.2 Power Spectra with Fractional Integration Proportional to $1/\sqrt{t}$

In Section 1.2, we have showed that the origin of the flicker fluctuations in physical systems is based on the loss of energy [1.20]. Here we recall, once more, the problem in a much more general way, where the resonator (oscillator) system is supplied from an ideal voltage source V with current I. However, during its passage through the system some energy is lost; let it be E_{diss} during one time segment. With the assistance of the corresponding power, the loss is equal to

$$E_{\text{diss}} = P_{\text{dis}} T_o \tag{1.107}$$

Since the dissipated energy in each period (or time span) is rather constant, the effective power decreases and after n periods it may be equal to

$$P_{\text{diss},n} \approx < i^2_{\text{noise}} R > \approx \frac{nP_{\text{diss},n}T_o - (n-1)P_{\text{diss},n-1}T_o}{nT_o} = \frac{1}{t}P_{\text{diss}}T_o \tag{1.108}$$

which is inversely proportional to the elapsed time. Hence, the noise current is also a function of time and without any appreciable error, in the first approximation, we propose for its time dependence

$$i_{\text{noise}}(t) \approx \sqrt{\frac{1}{t} \cdot \frac{P_{\text{diss}}T_o}{R}} \tag{1.109}$$

whose Fourier transform is (Table 1.2)

$$I_{\text{noise}}(s) \approx \sqrt{\frac{\pi}{s} \cdot \frac{P_{\text{diss}}T_o}{R_{\text{diss}}}} \tag{1.110}$$

and the corresponding PSD with respect to the current I^2 (i.e., the phase noise) is

$$S_\phi(f) = \frac{\pi}{f} \cdot \frac{P_{\text{diss}} T_o}{I^2 R} \approx \frac{\pi}{f} \cdot \frac{P_{\text{diss}} T_o}{P_o} \approx \frac{1}{f} \cdot \frac{\pi}{f_o Q} = a_R \frac{1}{f} \qquad 1.111$$

where P_o is the effective power in the system and $P_o/P_{\text{diss}} = Q$ is the device quality factor. The last expression in (1.111) is valid for the quartz crystal resonators or oscillators (cf. Chapter 2). Since the process is much more general, here we recall an earlier example shown in Fig. 1.14, presenting the PSD $S_R(f)/R^2$ for an India ink resistor in accordance with (1.109). The validity extends for more than 10 decades with $\alpha \approx 1.21$ [1.1].

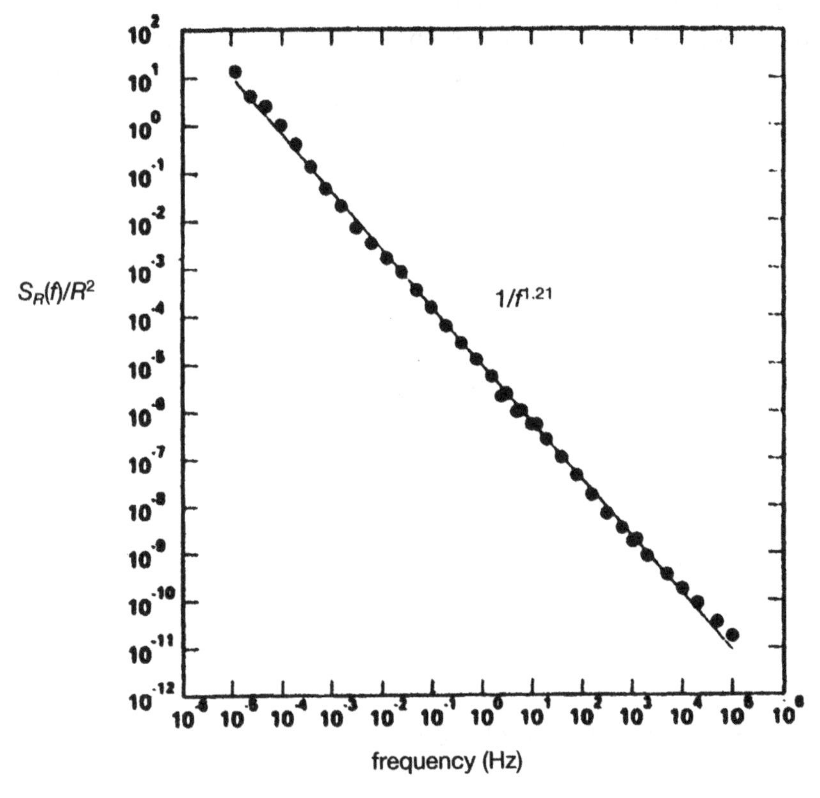

Fig. 1.14 Relative resistance fluctuation spectrum, $S_R(f)/R^2$, for India ink resistor.

REFERENCES

1.1. R.F. Voss, 1/*f* (flicker) *Noise: A Brief Review, Annual Frequency Control Symposium* (1979), pp. 40–46.
1.2. V.F. Kroupa, *Phase Lock Loops and Frequency Synthesis,* New York: Wiley, 2003.
1.3. J.B. Johnson, Thermal Agitation of Electricity in Conductors, *Phys. Rev.*, **32** (July 1928), pp. 97–109.
1.4. H. Nyquist, Thermal Agitation of Electric Charge in Conductors, *Phys. Rev.*, **32** (July 1928), pp. 110–113.
1.5. B.M. Olivier, Thermal and Quantum Noise, *Proc. IEEE,* **53,** (May 1965) pp. 436–454. [Reprinted in Gupta, *Electrical Noise: Fundamentals and Sources,* IEEE Press, 1977, p. 129.]
1.6. J. Bernamont, *Ann. Phys.* [Paris], **7**, (1937), p. 71.
1.7. D. Wolf, ed. 1/*f* noise, Noise in Physical Systems, in *Proceedings of the Fifth Conference on Noise,* Bad Nauheim, Germany, 1978, Springer-Verlag, p. 122.
1.8. T. Musha, The 1/*f* Frequency Fluctuation of the Traffic Current on an Expressway, *Jpn. J. Appl. Phys.* **15**, (July 1976), pp. 1271–1275.
1.9. V.F. Kroupa and J. Cernak, Power Line Stability, C.R. University of Ostrava, Workbook VII, 1998.
1.10. M.S. Keshner, 1/*f* Noise, *Proc. IEEE,* **70**, No. 3, (March 1982), pp. 212–218.
1.11. *Proc. IEEE*, Special Issue on Frequency Stability (February 1966), **54**.
1.12. J.Rutman, Characterization of Phase and Frequency Instabilities in Precision Sources: Fifteen Years of Progress, *Proc. IEEE*, **66**, (Sept. 78), pp. 1048–75.
1.13. V.F. Kroupa, *Frequency Stability: Fundamentals and Measurements,* IEEE Press, (1983).
1.14. A. L. McWorther, 1/*f* Noise and Related Surface Effects in Germanium, Lincoln Laboratory, Massachusetts Institute of Technology, Lexington, MA, Report No. 80 (May, 1955), unpublished.
1.15. K.M. van Vliet, Noise in Semiconductors and Photoconductors, *Proc. IEEE,* **46**, June 1958, pp. 1004–18.
1.16. A. Van der Ziel, Noise in solid-state devices and lasers, *Proc. IEEE*, **58** (Aug. 1970), pp. 1178–1206. [Reprinted in Gupta, *Electrical Noise: Fundamentals and Sources*, IEEE Press, 1977, p. 129.]
1.17. D. Halford, A General Mechanical Model for $|f|^{\alpha}$ Spectral Density Random Noise with Special Reference to Flicker Noise $1/|f|$. *Proc. IEEE,* **56**, (Mar. 1968), pp. 251–258.
1.18. V. Radeka, 1/*f* Noise in Physical Measurements, *IEEE Trans. Nucl. Sci.* (Oct. 1969), **NS-16**, pp. 17–35. [Reprinted in Gupta, *Electrical Noise: Fundamentals and Sources*, IEEE Press, 1977, p. 129].
1.19. N.J. Kasdin and T. Walter, Discrete Simulation of Power Law Noise, in *1992 IEEE Annual Frequency Control Symposium* (1992), pp. 274–283.
1.20. V.F. Kroupa, Theory of 1/*f* Noise—A New Approach, *Phys. Lett.,* **A 336** (2005), pp. 126–132.
1.21. F.N. Hooge, *Phys. Lett.*, **29 A** (1969), pp. 139–140.

1.22. A. Papoulis, *Probability, Random Variables, and Stochastic Processes,* New York: McGraw-Hill (1965).
1.23. H.B. Dwight, *Tables of Integrals and Other Mathematical Data,* 4th ed. New York, The MacMillan Company.
1.24. D.K.C. MacDonald. The Brownian Movement and Spontaneous Fluctuations of Electricity, *Res. Appl. Ind.,* **1** (Feb. 1948), pp. 194–203. [*Nature* (London), **72** (1905)], Lord Rayleigh p. 318, K. Pearson p. 342, probable place of the drunken man).

APPENDIX

Throughout the entire book we refer to the instantaneous phase or frequency,

$$\omega(t) \approx \frac{d}{dt}\left(\omega_o t + \varphi(t)\right) = \left(\omega_o + \dot{\varphi}(t)\right) \qquad \text{A.1}$$

and to their power spectral densities (PSD) $S_\varphi(f)$ or the normalized fractional frequency PSD $S_\varphi(f) = (f/f_o)^2 S_\varphi(f)$ (i.e., to the one-sided PSPs). Actually, however, we deal with the double-sided PSD:

$$S(\omega) \approx \frac{P_s^2}{2}[S_\varphi(\omega + \omega_o) + S_\varphi(\omega - \omega_o)] \qquad \text{A.2}$$

where P_s is the carrier power. The approximation is valid except for $\pm\,\omega_o$ and the surrounding narrow bands containing all the high modulation index, low frequency side bands. In the case of the validity of (A.2), we define the two-sided PSD $\mathcal{L}(\omega)$ as

$$\mathcal{L}(\omega) \approx S_\varphi(\omega + \omega_o) \approx S_\varphi(\omega - \omega_o) \qquad \text{A.3}$$

CHAPTER 2

Noise in Resonators and Oscillators

Noise in oscillators, particularly in crystal oscillators, has been studied for some 50 years. The progress is closely connected with perfection of the manufacturing processes on the one hand and reduction of the noise in maintaining electronics on the other hand. In addition, precision crystal oscillators are indispensable in standard time and frequency systems and laboratories.

2.1 NOISE GENERATED IN RESONATORS

Investigation of the noise in crystal resonators started in the mid-1970s [2.1, 2.2]. Many papers were to follow [e.g., 2.3, 2.4]. With the help of Fig. 2.1, we can express the noise current of the investigated system in the Laplace notation as

$$i_n(s) = \frac{e_n(s)}{sL + \frac{1}{sC} + R} = \frac{e_n(s)}{\frac{1}{sC} \cdot \frac{-\omega^2 + \omega_o^2}{\omega_o^2} + R} \approx \frac{e_n}{R} \qquad 2.1$$

In accordance with discussions in Chapter 1, the noise can be generated by temperature fluctuations, power losses in the system, and by fluctuations (random walk) of the resonance:

$$<\phi^2> = \frac{<e_n^2>}{V_{\text{osc}}^2} + \frac{1}{2} < \frac{P_{\text{dis}}}{P_o} > + < \Delta f^2 \tau^2 > \qquad 2.2$$

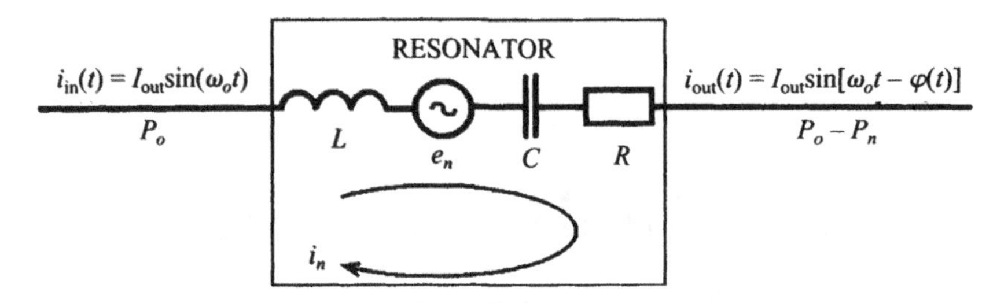

Fig. 2.1 A schematic arrangement for investigation of the noise generated in a resonator [2.5].

where the noise voltage e_n is generated in the effective resistence of the investigated circuit (1.1), the losses are defined by (1.22 or 1.108), and the random walk of the resonant frequency by (1.97 or 1.104). Note that the middle term in (2.2) is divided by 2 since one-half of the noise is phase noise and one-half is the amplitude noise.

2.1.1 White Frequency Noise (WFN) Generated in Resonators

Inspection of (2.1) reveals that outside of the passband the noise is generated mainly in the circuit resistance, and for the phase noise power spectral density (PSD) we get

$$<\phi^2> \approx \frac{<i_n^2>}{I^2} \approx \frac{4kTRB_w}{(RI)^2} \approx \frac{2kT\,B_w}{P_o} = S_\phi(f)B_v \qquad 2.3$$

By assuming the effective power in the resonator to be 1 dBm, we get for the PSD at room temperature (in 1 Hz bandwidth) approximately

$$S_\phi(f) \approx \frac{10^{-20.1}}{10^{-3}} \approx 10^{-17} \qquad 2.4$$

2.1.2 Flicker Frequency Noise Generated in Resonators

The resonator, as an isolated system, would reveal only the thermal noise. However, a continuous flow of the current generates the flicker frequency components or flicker phase noise fluctuations [2.5]. To get more insight into the problem, we will investigate the resonator alone; the situation was explored in Chapter 1 [cf. relation (1.111)] with the conclusion that

$$S_\phi(f) \approx \frac{\pi}{f} \cdot \frac{P_{\text{diss}} T_o}{2P_o} \approx \frac{1}{f} \cdot \frac{\pi}{2f_o Q} = a_R \frac{1}{f} \qquad 2.5$$

where P_o is the effective power in the system and $P_o/P_{\text{diss}} = Q$ represents the resonator quality factor. Figure 2.1 illustrates the situation for the case where we have at its input the current $i(t)$,

$$i(t) = I_o \sin(\omega_o t) \qquad 2.6$$

When the sinusoidal carrier is supplying energy during one period, T_o, the corresponding power, P_o, is equal to the power P_o:

$$P_o = \frac{1}{2} I_o^2 \cdot R \qquad 2.7$$

At the output of the resonator, we meet a reduced current and a bit smaller output power due to the losses in the body of the resonator:

$$i_{\text{out}}(t) = I_{\text{out}} \cos(\omega_o t) - \Sigma i_n(t) \qquad 2.8$$

where $\Sigma i_n(t)$ is a set of noise side bands close to the carrier. Evidently, at the output we meet a reduced carrier current and the output power is lowered due to the losses, P_n, with the simultaneous amplitude and phase modulation [therefore division by 2 in (2.5) refers to the phase noise only, since half of the noise is phase noise and half is the amplitude noise]. Note that the thermal wastes are orders smaller [2.3]; however, the product Qf_o is a material constant. For the quartz [2.6, 2.7] (see also, Fig. 2.2 and Section 1.5.2.2),

$$i_{\text{out}}(t) = I_{\text{out}} \cos(\omega_o t) - \Sigma i_n(t) \qquad 2.9$$

and after introducing (2.9) into (2.5) we get for the $1/f$ PSD of the quartz resonators

$$S_{\phi,\text{res}}(f) = a_R \frac{1}{f} = \frac{\pi (Qf_o)^{-1}}{2f} \approx \frac{10^{-13}}{f} \qquad 2.10$$

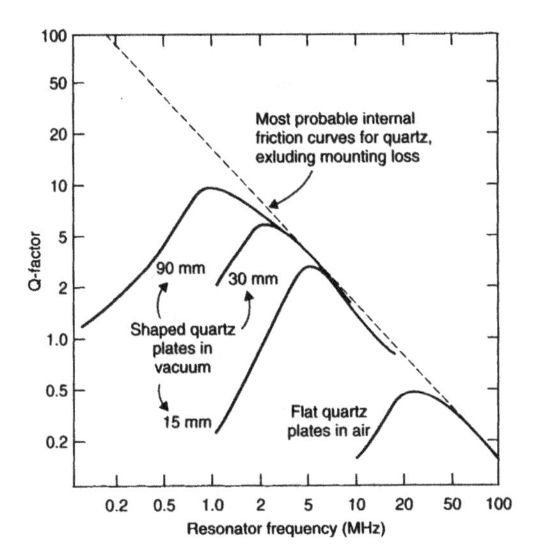

Fig. 2.2 The frequency dependence of the quality factor Q of the crystal resonators as found by Warner [2.7].

Next, we evaluate the important *fractional frequency fluctuations* by using the fact that in the vicinity of the resonant frequency, f_o, the relation between phase and frequency fluctuations is nearly linear and equal to the time delay τ (cf. Fig. 2.3):

$$\frac{\Delta\phi}{\Delta\omega} \approx \frac{2Q}{\omega_o} = \tau \qquad 2.11$$

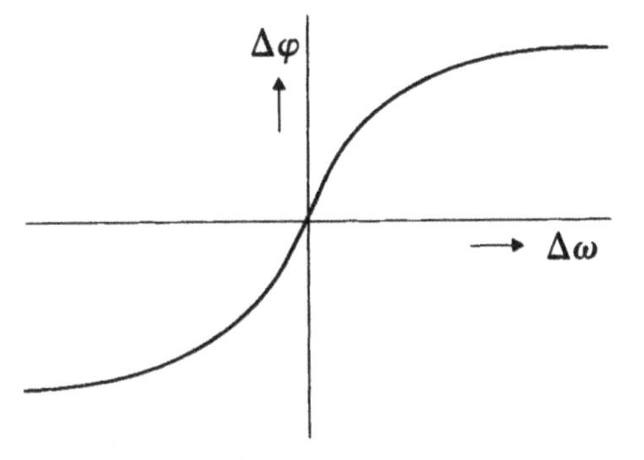

Fig. 2.3 Relation between phase and frequency in the resonant circuit.

and the important *fractional frequency fluctuations* follow:

$$S_y(f) = S_\phi(f)\frac{1}{4Q^2} = \frac{h_{-1}}{f} = \frac{(Q \cdot f_o)^{-1}}{8Q^2 f} \qquad 2.12$$

By introducing (2.9) into the above relation, we arrive at

$$S_y(f) \approx \frac{10^{-14}}{Q^2 f} \qquad 2.13$$

Finally, we compute the noise constant h_{-1} for the quartz crystal resonators as a function of the resonant frequency f_o:

$$S_y(f=1) = h_{-1} = 10^{-14} \times 10^{-26} f_o^2 \approx 10^{-40} f_o^2 \qquad 2.14$$

Note that the flicker phase noise (FPN) PSD $S_\varphi(f)$ of the quartz resonators is independent both of the carrier frequency, f_o, and of the resonator quality factor Q (cf. 2.10). Further, note that (2.13) is in good agreement with an earlier value for the PSD of the fractional frequency, S_y, found by Parker [2.3] for surface acoustic wave (SAW) oscillators.

2.1.3 Random Walk of the Resonant Frequency

By assuming that the resonant frequency f_o is also subjected to the random changes $\Delta\omega_i$, either in each period or larger time spans caused by voltage or temperature steps in the oscillating circuit, we face another contribution to the effective noise (cf. also Section 1.5), which we begin to investigate. Let us start with a voltage separation, ΔV. In accordance with Kurokawa [2.8.] and (6.161), the corresponding frequency change is

$$\Delta\omega_o \approx \frac{a_o}{2LA_o}\sin(\phi_o) \le \frac{a_o}{2LA_o} \approx \omega_o \frac{i_{syn}}{2QI_{osc}} \approx \frac{\omega_o}{2Q}\frac{\Delta V}{V_o} \qquad 2.15$$

and the corresponding Fourier transform of individual steps is (Table 1.2)

$$\Delta\omega(s) = \frac{1}{s}\frac{\omega_o}{2Q}\frac{\Delta V}{V_o} \qquad 2.16$$

By supposing that the noise voltage is generated by temperature fluctuations in the resonator band pass, the variance of the fractional frequency will be

$$\left\langle \left(\frac{\Delta\omega}{\omega_o} \right)^2 \right\rangle \approx \frac{1}{f^2} \left(\frac{1}{2Q} \right)^2 \frac{4kTRB_w}{V_o^2} \qquad 2.17$$

After introducing P_o, that is, the radio frequency (RF) power in the resonator, we get for the PSD of the fractional frequency fluctuations

$$S_y(f) \approx \frac{1}{f^2} \left(\frac{1}{2Q} \right)^2 \frac{2kT}{P_o} = \frac{h_{-2}}{f^2} \qquad 2.18$$

EXAMPLE 2.1

Let us investigate the expected random walk noise of the quartz crystal resonator $f_o = 10^7$ MHz with $Q = 10^6$. When introducing the thermal noise $2kT = 10^{-20.1}$ from Table 1.1 and 1 mW for P_{out}, we arrive at fractional PSD:

$$S_y(f) \approx \frac{10^{-29}}{f^2} \qquad 2.19$$

which is close to the actually measured values (cf. Table 5.5).

2.1.4 Spurious Frequency Modulation Generated in Resonators

By considering that resonators of stable oscillators are placed in ovens, we may expect, in addition, rather small temperature fluctuations originating in the regulation system causing frequency variations around the resonant frequency of the resonator. Starting from this position, we can estimate the temperature variations to be nearly periodic with approximately the same amplitudes $\Delta\omega$ and frequency f_m. By considering a sinusoidal frequency modulation, the corresponding PSD is

$$S_y(f) = \frac{1}{2} \left(\frac{\Delta\omega}{\omega_o} \right)^2 \times \delta(f_m) \qquad 2.20$$

Frequency variations due to temperature changes are generally described by a polynomial. However, if the fluctuations are small we can retain the linear term only, that is,

$$K_T \approx \frac{df}{dT} \qquad 2.21$$

By introducing the expected temperature variations ΔT and the fractional frequency temperature coefficient K_T, we get a PSD of the fractional frequency noise in a general form:

$$S_y(f) = \frac{(K_T \Delta T)^2}{f_o^2} \qquad 2.22$$

2.1.5 The Aging and Drift of the Frequency of Resonators

The long-term time dependence of a frequency generator is often called *frequency aging*. In accordance with [2.9, 2.10] its main causes are mass transfer due to the contamination between the resonator and the envelope, stress relief of the fastening mechanics, changes in the sustaining circuitry and oven control, and so on. Note that according to earlier recommendations, *aging* is the systematic change in frequency with time due to internal changes in the oscillator, particularly when factors external to the oscillator (environment, power supply, etc.) are kept constant, whereas *drift* is defined as the systematic change in frequency with the time of an oscillator (i.e., it encloses all frequency changes). Even if the sources of the frequency instability in frequency resonators are so different, they are generally controlled by an exponential law:

$$f_o(t) \approx f_o e^{t \cdot D} \approx f_o(1 + D \cdot t) \qquad 2.23$$

However, in tens, hundreds, or thousands of seconds the change is nearly linear with time and is generally expressed by fractional frequency in 1 day. For a 5- or 10-MHz crystal oscillator, practical values are

$$D = 10^{-11} \text{ per day} \quad \text{or} \quad D_y \approx 10^{-16} \text{ per second} \qquad 2.24$$

Cryocooled sapphire oscillators reveal aging as low as $D_y = 10^{-17}$/day (see Chapter 3).

2.2 PHASE NOISE OF RESONATORS: EXPERIMENTAL RESULTS

In the presented theory of the resonator, particularly the crystal resonator, phase noise will be compared with actual measurements performed in the last 30 years [2.1–2.5 and 2.11–2.17]. However, certain caution with the published results is necessary because additional losses or sources of noise may distort conclusions. This is the case with the earlier measurements performed with a rather low-frequency, nonevacuated resonator in which the energy is wasted in much higher proportions than expected for the material limit only (see the noise data for different oscillators in Fig. 2.4). The other difficulty in comparing earlier and actual noise measurements might be due to the technology and mechanical construction used: quality of the surface

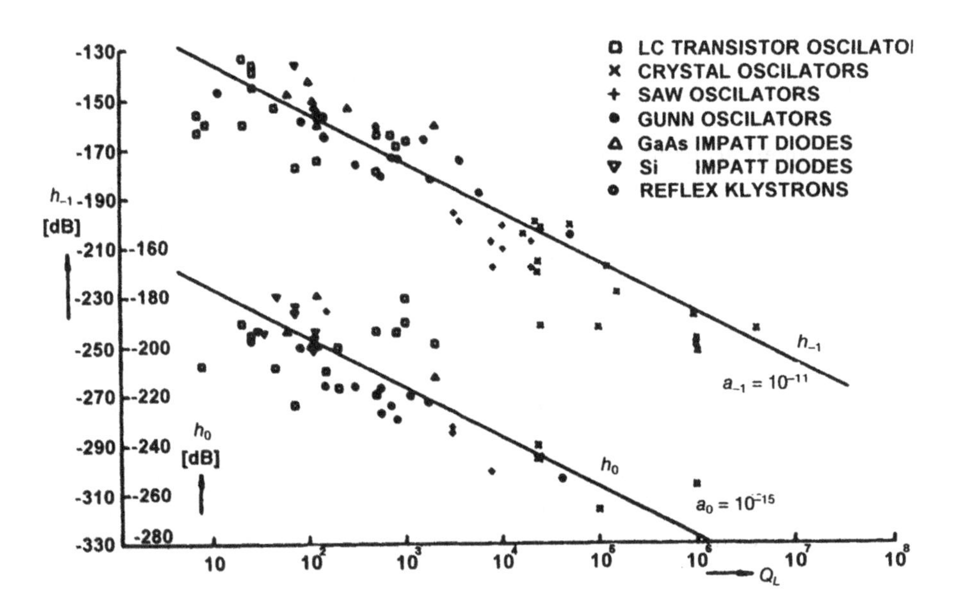

Fig. 2.4 The earlier measurements of the noise constant h_{-1} and h_o for different oscillators as functions of load Q_L [2.18].

of resonators, nonadherent and adherent electrodes (BAV), and so on [2.19, 2.20].

2.2.1 Flicker Phase Noise in Crystal Resonators: Experimental Results

The flicker phase noise of quartz crystal resonators has been investigated since the early 1960s [cf. 2.7]. From the wealth of performed measurements, we have summarized a few results published by different authors from different laboratories over the last 50 years. We assembled these important results in Table 2.1. First, we investigated the validity of (2.9) both for resonators and oscillators. Results gained with oscillators are enclosed in the data and designated by an asterisk [2.16, 2.17]. The data reveal that the flicker frequency constant a_{R1} is nearly independent of the carrier frequency f_o over the entire investigated frequency range of 5–1000 MHz, as expected. From column 7, we find the mean value and the dispersion of constant a_{R1} to be

$$a_{R1} = -130 \pm 1.2 \qquad \text{(dB)} \tag{2.25}$$

Table 2.1 Summery of crystal resonator properties in the frequency range from 5 MHz to 1 GHz[a]

f_o (MHz)	$Q*10^3$	$Q*f_o*10^{13}$	$S_\varphi(1)$ (dB) (oscillator)[a]	$S_y(1)$ ($\times 10^{-26}$)	h_{-1} (dB)	a_{R1} (dB)	Q_L/Q	Reference
5	2600	1.3	−111*	31.6	−245	−131	0.16	[2.1]
5	2600	1.3	−128	0.8	−261	−132	0.6	[2.5]
5	2700	1.35	−135	0.25	−266	−134	0.8	[2.12]
10	1200	1.2	−123*	0.63	−262	−131	0.47	[2.13]
10	1320	1.32	−120	1	−261	−131	0.48	[2.14]
10	1320	1.32	−123*	0.63	−262	−131	0.53	[2.15]
40	250	1	−80*	100	−240	−130	0.54	[2.16]
80	125	1	−74*	631	−232	−130	0.5	[2.16]
100	119	1.19	−76*	200	−237	−132	0.71	[2.16]
160	75	1.2	−72*	252	−236	−134	0.9	[2.16]
401	24	1.04	−55*	1800	−227	−133	1.07	[2.17]
450	22	0.98	−52*	2850	−227	−133	0.84	[2.17]
800	10.7	0.856	−45*	2800	−226	−132	1.7	[2.17]
919	9.3	1.7	−42*	6800	−222	−132	1.4	[2.17]

[a]Asterisk indicate computed values from f_o and h_{-1} [see (2.14)].

Furthermore, we learn that the noise constant h_{-1} is inversely proportional to Q^{-2}, which is in agreement with the earlier data in Fig. 2.4 and the later data in Fig. 2.5.

More important, to get additional information about problems of the flicker noise in quartz crystal resonators, several authors investigated nonoscillating systems. Here, we reproduce only two measurements in Figs. 2.6 and 2.7. Note that for larger Fourier frequencies they reveal a steeper slope of $1/f^2$. The difficulty is easily explained by using the simplification of (2.1). After reverting to the noise equation

$$i_n = \frac{e_n}{R} \cdot \frac{1}{1 + j\Delta\omega\tau} \tag{2.26}$$

we get for the phase noise power in one period T_o

$$P_n = \frac{e_n^2 T_o}{R[1 + (\Delta\omega\tau)^2]} = \frac{P_{\text{diss}} T_o}{1 + (\Delta\omega\tau)^2} = \frac{(Qf_o)^{-1}}{1 + (\Delta\omega\tau)^2} \tag{2.27}$$

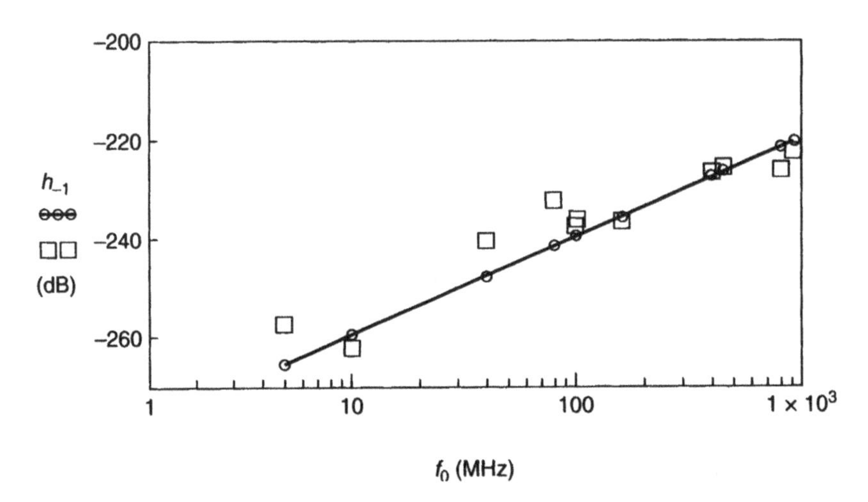

Fig. 2.5 Summary of crystal resonator properties in the frequency range from 5 MHz to 1 GHz. Points on the full line are theoretical values of h_{-1} in accordance with (2.14), whereas rectangles represent values taken from Table 2.1.

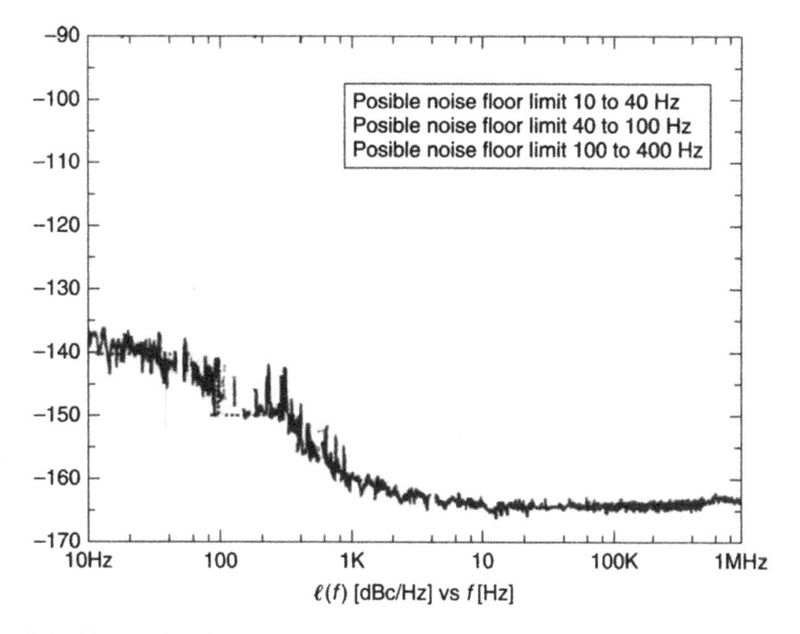

Fig. 2.6 Phase noise characteristic of a 80-MHz resonator reproduced from [2.11]. (Copyright © IEEE. Reproduced with permission.)

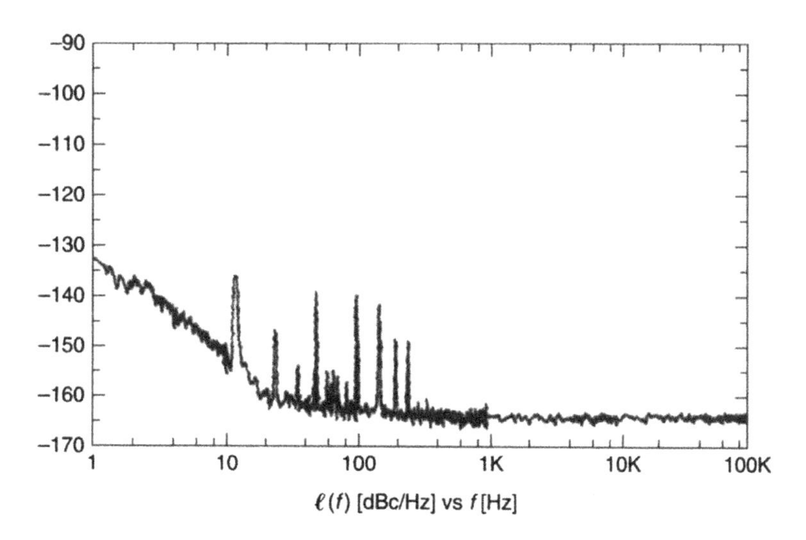

Fig. 2.7 Phase noise characteristic of a 10-MHz resonator reproduced from [2.13]. (Copyright © IEEE. Reproduced with permission.)

and further at the PSD of the output phase noise of the resonator

$$S_{\phi,res}(f)=\frac{(Qf_o)^{-1}}{2f}\cdot\frac{1}{1+(2\pi f\tau)^2}=a_R\frac{1}{f}\cdot\frac{1}{1+(2\pi f\tau)^2} \qquad 2.28$$

Evidently, in the resonator bandpass we might expect the PSD of the slope $1/f$ (i.e., 10 dB/decade) and outside $1/f^3$ to be as predicted by Gagnepain [2.2, cf. Fig. 2.8]. However, our own experience with noise measurements [2.4] and many earlier measurements [e.g., 2.11] and later ones [2.15] reveal slopes of the PSD's $1/f$ and $1/f^2$ only. The difficulty is easily solved by introduction of additional white (thermal) losses:

$$a_o=\frac{2kT}{P_o} \qquad 2.29$$

In that case, the output phase noise characteristic of the resonators can be simulated as follows:

$$S_{\phi,\text{res}}(f)=\left(\frac{a_R}{f}+a_o\right)\cdot\frac{1}{1+(2\pi f\tau)^2} \qquad 2.30$$

which agrees with observations (see Example 2.2).

EXAMPLE 2.2

Let us examine the phase noise characteristic from [2.13], which is reproduced in Fig. 2.7:

$f_o = 10$ MHz; $Q = 1.2 \times 10^6$; $Qf_o = 1.2 \times 10^{13}$

$S_v(1) = -131$ dB or the $1/f$ noise (from the published characteristic [2.13])

$S_v(100) = -158$ dB for the white noise (from the published characteristic [2.13])

$\tau = 0.038$ (from [2.11])

After introduction into (2.30), we arrive at

$$S_\phi(f)=\left(\frac{10^{-13.1}}{f}+10^{-15.8}\right)\frac{1}{1+(0.25f)^2}+10^{-16} \qquad 2.31$$

The above resonator noise characteristic is plotted in Fig. 2.9 with several points read from the original characteristic in Fig. 2.7

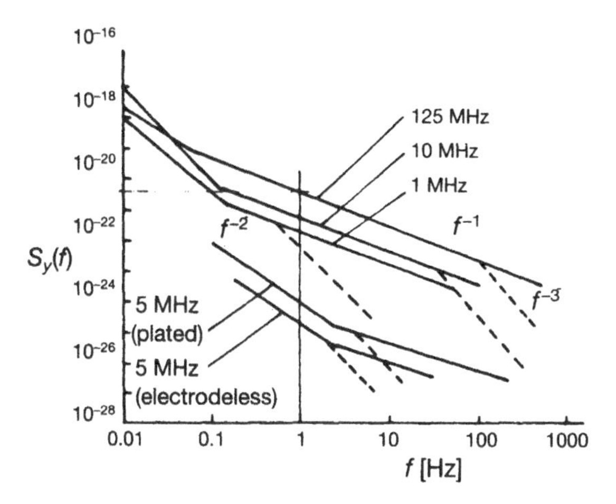

Fig. 2.8 The expected power spectra of fractional frequency fluctuations of quartz crystal resonators at different frequencies [2.2]. (Copyright © IEEE. Reproduced with permission.)

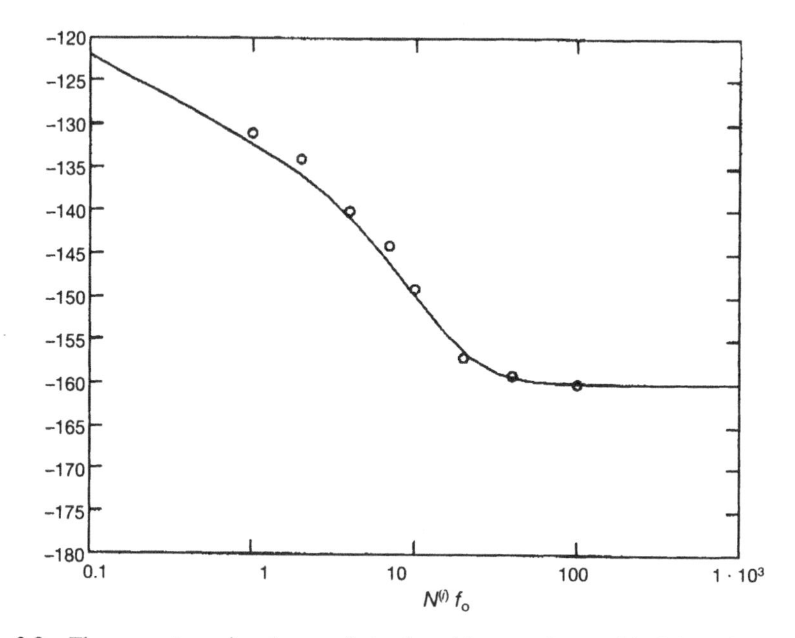

Fig. 2.9 The resonator noise characteristic plotted in accordance with (2.31); circles have been read from the original characteristic in Fig. 2.7.

2.2.2 Random Walk of the Frequency Noise Generated in Resonators

From the measurements performed with resonators, Gagnepain [2.2] predicted the noise slope $1/f^2$ at very low Fourier frequencies (cf. Fig. 2.7). However, the problem is that neither of the investigated phase noise characteristics of stable resonators reveals this slope. On the other hand, analysis of the fractional frequency noise characteristics of oscillators, particularly of crystal oscillators (e.g., [2.3] and Section 3.1) indicates the presence of this type of frequency noise. Its origin may result from the random walk noise, RW, due to environmental influences or the resonant frequency.

Here, we encounter several difficulties. One of them is the dependence of the noise constant h_{-2} on the magnitude of the effective power P_o in the resonators [cf. relation (2.18)]. Another is the need of the noise measurements at low Fourier frequencies in the frequency domain (below ~ 1 Hz). The expected behavior of the phase noise characteristics at these low frequencies were provided by Gagnepain [2.2] (Fig. 2.7) but, many years ago, Parker [2.17] published frequency fluctuations of SAW resonators operating in the 400- and 900-MHz range (see Fig. 2.10). In Table 2.2, we tried to summarize data from several earlier publications in which the random walk was noticeable.

2.3 NOISE IN OSCILLATORS

There exist a large number of papers dealing with the fundamental oscillator theory. However, discussion of noise problems is limited mostly to only white noise [e.g., 2.8, 2.20], with different approaches to the solution of the problem. Here, we repeat only the leading ideas, recall our own investigations [2.4], and thereafter compare the theoretical results with experimental findings.

In principle, the oscillator is formed by a feedback system satisfying the Barkhausen condition of the overall gain to be equal to *one* and the all-round phase to be equal to $2\pi N$ (N being an integer).

2.3.1 Analogue Arrangement

An analogue arrangement is depicted in Fig. 2.11; the oscillation is started either by ever-present noise or by switching-in. The amplifier

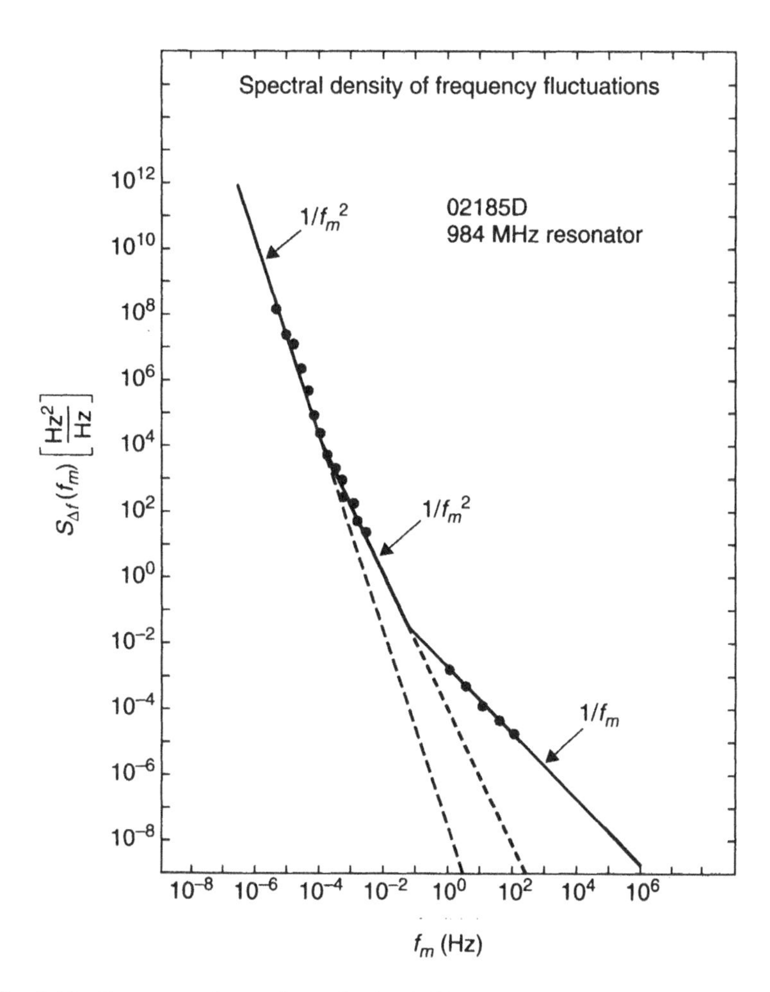

Fig. 2.10 Frequency fluctuations of a SAW 984-MHz resonator exhibiting random walk fluctuations with slopes $1/f^2$ and $1 \approx f^3$ [2.21]. (Copyright © IEEE. Reproduced with permission.)

supplies the necessary oscillation power and its nonlinearity is responsible for keeping the overall gain equal to 1, whereas the remaining excess power is pushed into harmonics. The condition of zero phase shifts ($2\pi N$) around the loop is accomplished via the small compensating shift of the resonant frequency. The basic phase noise equation in Fourier transform notation recalls that of the PLL (phase-locked loop). The starting linearized relation is

Table 2.2 Summery of crystal resonators (oscillators) exhibiting random walk noise fluctuations

f_o (MHz)	h_{-2} (dB)	h_{-1} (dB)	h_o (dB)	Reference
5	−267	−266	−275	[2.20]
5	−291	−267	−264	[2.21]
10		−252	−250	[2.22]
100		−238	−275	[2.16]
400	−230	−212		[2.17]
400	−240	−180		[2.17]
984	−220	−210		[2.23]

$$V_{out}(s) = [V_{in}(s) - V_{out}(s)B(s)]R(s)A(s) \quad 2.32$$

The transfer function follows:

$$\frac{V_{out}(s)}{V_{in}(s)} = H(s) = \frac{R(s)A(s)}{1 - A(s)B(s)R(s)} \quad 2.33$$

The oscillation condition is met if $|AB| = 1$ and $\varphi_A + \varphi_B + (d\varphi_R/dt)\tau = 0$.

2.3.2 Sampling Arrangement

A schematic sampling arrangement is shown in Fig. 2.12. In the first time segment, the output noise is equal only to the noise of the maintaining electronics,

$$\phi_{out} = \phi_e \quad 2.34$$

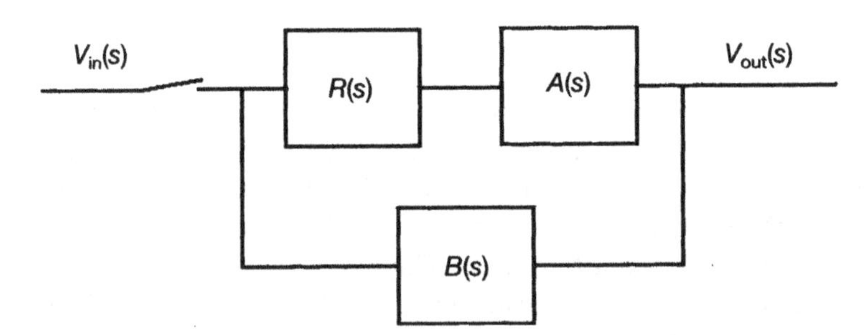

Fig. 2.11 Block diagram of the oscillator: principle of analogue feedback.

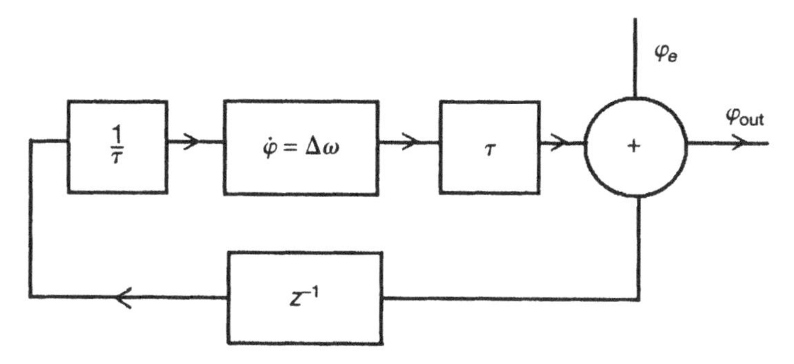

Fig. 2.12 Block diagram of the oscillator as the sampled noise generator.

After one delay period, the output noise is changed into a frequency correction term by division with τ (cf. 2.11) and decreases or increases the instantaneous frequency deviation generated in the resonator itself:

$$\phi_{\text{out},1} = \phi_{e,1} + \left(\frac{\phi_{\text{out},0}\, z^{-1}}{\tau} + \dot{\phi}_1 \right) \tau = \phi_{e,1} + \phi_{e,0}\, z^{-1} + \dot{\phi}_1\, \tau \qquad 2.35$$

The next sampling period results in

$$\begin{aligned} \phi_{\text{out},2} &= \phi_{e,2} + \left(\frac{\phi_{out,1}\, z^{-1}}{\tau} + \dot{\phi}_2 \right) \tau = \\ &\phi_{e,2} + \phi_{e,1}\, z^{-1} + \phi_{e,0} z^{-2} + (\dot{\phi}_2 z^{-1} + \dot{\phi}_1)\tau \end{aligned} \qquad 2.36$$

and so on. Since individual phase and frequency contributions are nearly of the same magnitude, we get, after summation of the series, for the oscillator output phase noise

$$\phi_{\text{out}} = \phi_e + \frac{\phi_e + \dot{\phi}\tau}{1 - z^{-1}} \qquad 2.37$$

Finally, by changing the z-transform into the Fourier transform [see, e.g., 2.18], we arrive at

$$\begin{aligned} \phi_{\text{out}}(s) &= \phi_e(s) + \frac{1}{s\tau}[\phi_e(s) + \dot{\varphi}(s)\tau] = \\ &\phi_e(s) + \frac{1}{s\tau}[\phi_e(s) + \phi_{\text{res}}(s)] \end{aligned} \qquad 2.38$$

Note that (2.38) recalls the well-known relation for combination of the phase noises generated in the resonator and maintaining circuit (cf. Section 3.1). The corresponding PSD of the oscillator output phase noise is

$$S_{\phi,\mathrm{out}}(f)=\left(\frac{\omega_o}{2Q\omega}\right)^2[S_{\phi,\mathrm{res}}(f)+S_{\phi,e}(f)]+S_{\phi,e}(f) \qquad 2.39$$

Note that we have limited our investigation to the first-order terms and omitted all nonlinearity.

2.3.3 Evaluation of the Oscillator Output Phase Noise

The phase noise introduced by the maintaining electronics is composed of the flicker and random walk noise contributions at low Fourier frequencies and the inevitable white noise at higher frequencies:

$$S_{\phi,e}(f)=\frac{a_E}{f}+\frac{2kTF}{P_o}=\frac{a_{1E}}{f}+a_o \qquad 2.40$$

where P_o is the oscillator output power and F is the noise factor. To the electronics phase noise, we must add the resonator noise (cf. 2.30),

$$S_{\phi,R}(f)=\frac{a_R}{f}+\frac{2kT}{P_r} \qquad 2.41$$

and after their combination together with the noise of the maintaining output electronics, we finally arrive at the oscillator output phase noise

$$S_{\phi,\mathrm{osc}}(f)=\left(\frac{\omega_o}{2Q\omega}\right)^2\left[\frac{a_{R2}}{f^2}+\frac{a_{R1}+a_E}{f}+\frac{2kT}{P_r}(1+F)\right]+ \frac{a_E}{f}+\frac{2kT}{P_r}(1+F)+S_{\phi,\mathrm{add}}(f) \qquad 2.42$$

which can be expressed as a polynomial in f:

$$S_{\phi,\mathrm{osc}}(f)=\frac{a_4}{f^4}+\frac{a_3}{f^3}+\frac{a_2}{f^2}+\frac{a_1}{f^1}+a_o \qquad 2.43$$

and the coefficients of the noise characteristic asymptotes are (see Fig. 2.13)

$$a_4 = \left(\frac{f_o}{2Q}\right)^2 a_{R2} \qquad a_3 = \left(\frac{f_o}{2Q}\right)^2 (a_{R1} + a_E)$$

$$a_2 = \left(\frac{f_o}{2Q}\right)^2 \frac{2kT}{P_r}(1+F) \qquad a_1 = a_E \qquad a_o = \frac{2kT}{P_r}(1+F)] \qquad 2.44$$

Note that constants, $a_4 \ldots a_o$, are easily found by asymptotic approximation applied on the measured oscillator-phase noise characteristic.

EXAMPLE 2.3

We performed a phase noise measurement of the HP crystal oscillator at 10 MHz, type 10811. Values for several Fourier frequencies are summarized in Table 2.3 (2nd col.). With the assistance of the computer, we plotted the corresponding phase noise characteristic (see Fig. 2.13). Next, we evaluated the slopes of three asymptotes with the following result:

$$S_\phi(f) = \frac{10^{-10.3}}{f^3} + \frac{10^{-11.3}}{f^2} + \frac{10^{-13.5}}{f} + 10^{-16} \qquad 2.45$$

Table 2.3 Phase noise PSD of the HP crystal oscillator[a]

Frequency (Hz)	PSD (dBc/Hz)	PSD (dBc/Hz)	PSD (dBc/Hz)
1	–102	–98	–120
2			–132
3			–137
5			–147
10	–132	–133	–154
20			–159
30			–161
50			–163
100	–160	–148	–168
200			–170
1,000	–160	–153	–175
10,000	–160	–158	–179
80,000	–160	–160	–181

[a]At 10 MHz, type 10811, of a technical crystal oscillator 5 MHz, and noise constants of a 5-MHz crystal oscillator with SC cut resonator with the FPN characteristic exhibiting slope $1/f^4$ at low Fourier frequencies [2.20].

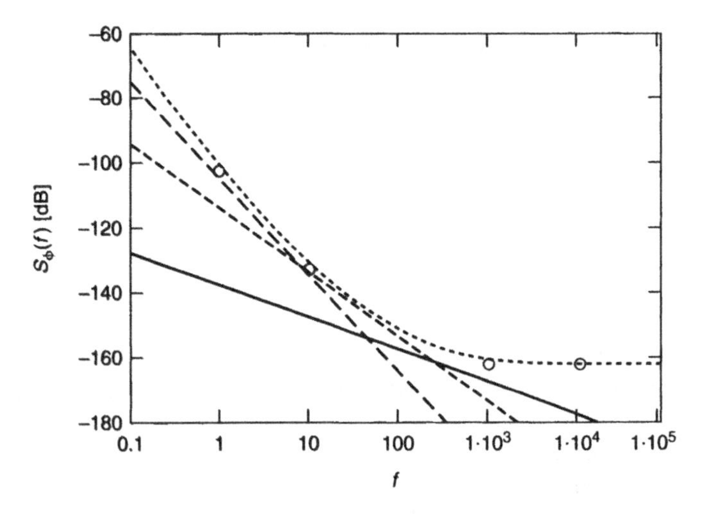

Fig. 2.13 Asymptotic approximation of the oscillator phase noise characterics.

For information, we reproduce in Fig. 2.14 the phase nose characteristics of crystal oscillators for important carrier frequencies over the entire frequency range from 5 to 500 MHz.

2.3.4 Evaluation of the Oscillator Output Fractional Frequency Noise

In the above sections, we introduced the concept of fractional frequency as the noise measure by multiplying the phase noise by the factor $(f/f_o)^2$:

$$S_y(f) = \left(\frac{f}{f_o}\right)^2 S_{\phi,\mathrm{osc}} = \frac{h_{-2}}{f^2} + \frac{h_{-1}}{f} + h_o + h_1 f + h_2 f^2 \qquad 2.46$$

Inspection of the PSD of the fractional frequency noise reveals the invariance with respect to the carrier frequency f_o, that is, in instances of frequency multiplications or divisions no changes are expected. In addition, in the same diagram we can compare noise properties of oscillators with different resonant frequencies. With the assistance of (2.44) and (2.42), we can express h_i constants as functions of the frequency noise constants a_i as

$$h_i = \frac{a_{2-i}}{f_o^2} \qquad 2.47$$

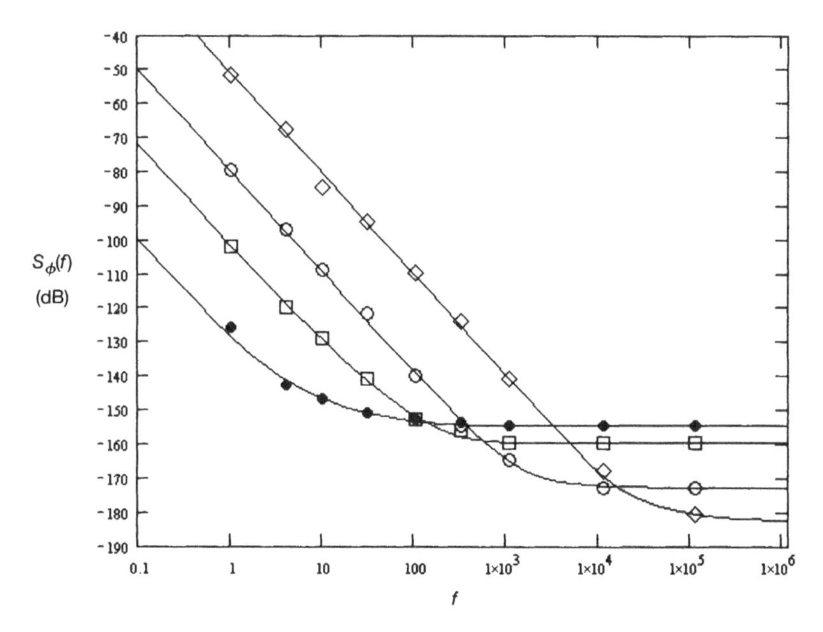

Fig. 2.14 Phase noise characteristic adapted from technical data of the present-day crystal oscillators. (•) Oscillaoquartz 5-MHz type 8607, (□) HP 10 MHz, type 10811, (○) Frequency Electronics 100 MHz, and (◇) SAV 500 MHz. Adapted from [2.18].

Values of the h_{-1} coefficient for both resonators and oscillators form column 6 in Table 2.1 allow us to provide at least a partial check on the resonator Q; see column 8 in Table 2.1.

$$Q = \left(\frac{1}{8h_{-1}f_o} \right)^{1/3} \qquad 2.48$$

This is particularly useful when investigating crystal oscillators, since the constant h_{-1} is read from the analyzed phase noise characteristic with the assistance of the a_3 coefficients [(2.44)]. From this follows, vice versa, other important information about the expected output flicker phase noise characteristic as a function of the fundamental frequency, f_o, of the crystal oscillators, namely,

$$S_{\phi,\text{out}}(f) \approx \frac{10^{-40} f_o^4}{f^3} \qquad 2.49$$

(cf. column 4 in Table 2.1). However, in instances in which multiplication is applied, we get a lower flicker phase noise, since the output frequency f_o is replaced with $f_o = N \times f_{inp}$:

$$S_{\phi,osc}(f) \approx \frac{10^{-40} f_{inp}^{4} \times N^{2}}{f^{3}} \tag{2.50}$$

EXAMPLE 2.4
For a 1-GHz crystal oscillator, the flicker phase noise at 1 Hz would be from Table 2.1 and from (2.49)

$$S_{\varphi}(1) = -400 + 360 = -40 \text{ (dB/Hz)}$$

However, multiplication of the 10-MHz oscillator in the 1-GHz range will result in a much larger noise. With the use of (2.50), we arrive at

$$S_{\varphi}(1) = -400 + 280 + 40 = -80 \text{ (dB/Hz)}$$

2.3.5 Asymptotic Evaluation of the Fractional Frequency Noise Characteristics

The slope of the asymptotes in the phase noise characteristics of stable oscillators are very steep at low Fourier frequencies, making an estimation of noise constants, a_{-3}, and so on, rather vague. The difficulty may be alleviated with the assistance of the PSD of the fractional frequency noise [relation (2.40)]: The corresponding approximating asymptotes provide more reliable results. We will explain the procedure in Example 2.5.

EXAMPLE 2.5
In this case, we will evaluate the noise constants of a 5-MHz crystal oscillator with an SC-cut resonator with the PN-characteristic exhibiting a slope $1/f^4$ at low Fourier frequencies on the one hand and low white noise on the other hand, as depicted in Figure 2.15. From this plot, we have read phase noise values for several Fourier frequencies and summarized them in Table 2.2 (4th col.). Then, we plotted the $S_y(f)$ characteristic, multiplied by f_o^2. The result is illustrated in Fig. 2.16. From intersections with the vertical line $f = 1$, we read the values for all a_i, and the evaluated PN-characteristic is

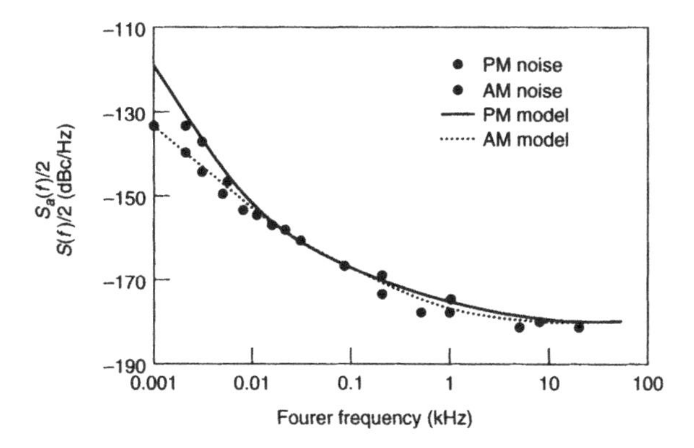

Fig. 2.15 Phase noise characteristic of a 5-MHz crystal oscillator with SC-cut resonator [2.20]. (Copyright © IEEE. Reproduced with permission.)

$$S_\phi(f) = \frac{10^{-12}}{f^4} + \frac{10^{-13}}{f^3} + \frac{10^{-13.7}}{f^2} + \frac{10^{-14.7}}{f} + 10^{-18.1} \qquad 2.51$$

and compared with the authors' published fitting model using minimal variations:

$$S_\phi(f) = \frac{10^{-12.1}}{f^4} + \frac{10^{-12.8}}{f^3} + \frac{10^{-14.7}}{f} + 10^{-18.01} \qquad 2.52$$

There is good agreement for both results. A detailed inspection of Fig. 2.16 reveals that the flicker-phase noise contribution is minimal. This confirms the asymptotic approximation to the original phase-noise characteristic shown in Fig. 2.15. Only three asymptotes are of importance.

Further, we mention important information about crystal resonators and oscillators, namely, the value of the expected Allan variance in its lowest plateau. From its definition (5.27), we obtain the following equation:

$$\sigma(\tau) = \sqrt{2 \times \ln(2) \times h_{-1}} \qquad 2.53$$

Application of (2.14) reveals the best value:

$$\sigma(\tau) \approx 10^{-20} : f_o \qquad 2.54$$

a value suggested by Parker [2.23].

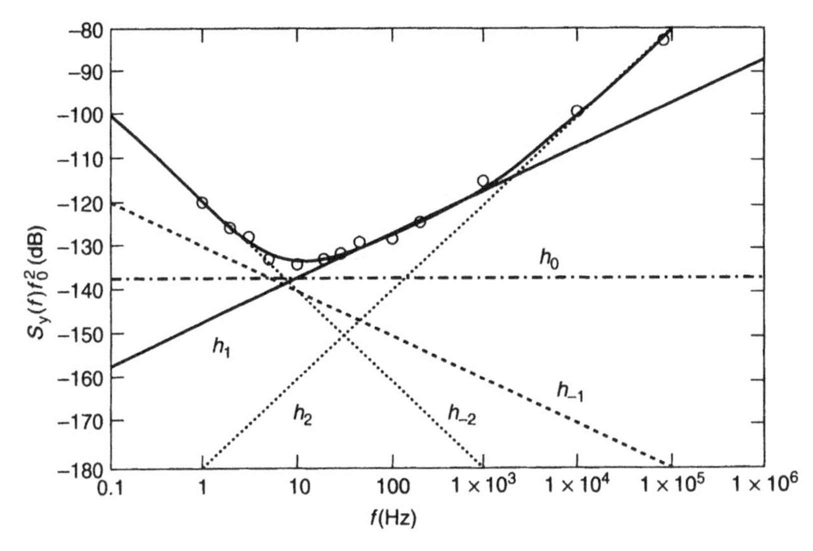

Fig. 2.16 Plot of the fractional frequency noise characteristic $S_y(f)$ of a 5 MHz SC oscillator (solid line; all other lines are marked with their respective constants).

2.3.6 Higher Order Noise Terms

In some instances, interpretation of the oscillator phase noise PSD characteristics reveals higher order terms of the type $1/f^4$ or even *higher,* expressed as a polynomial in f:

$$S_{\phi,\mathrm{osc}}(f)=\frac{a_5}{f^5}+\frac{a_4}{f^4}+\frac{a_3}{f^3}+\frac{a_2}{f^2}+\frac{a_1}{f^1}+a_o \qquad 2.55$$

however, their origin is difficult to trace since they may be attributed both to the additional noises generated in the resonators or in amplifiers and integrated, often with the additional filtering in phase noise measuring systems. One example is the term $1/f^5$ in Fig. 2.17. One explanation might involve additional integration of the $1/f^3$ term, misinterpretation of the phase-noise characteristics at very low Fourier frequencies; another possibility is its connection with aging.

2.4 LEESON MODEL

On the occasion of the special issue of the IEEE proceedings devoted to the problems of *frequency stability* [2.25] Leeson [2.26], published a

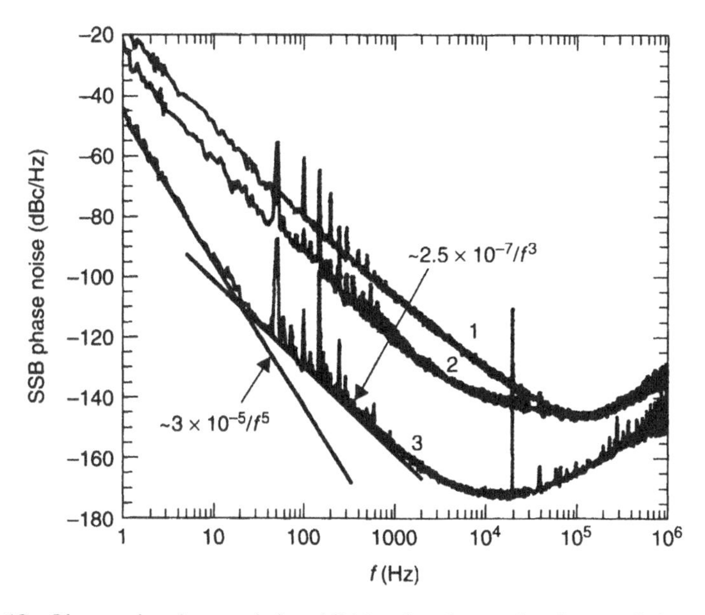

Fig. 2.17 Phase noise characteristic exhibiting the phase noise characteristic with the slope $1/f^5$ close to the 1-Hz Fourier frequency [2.24]. (Copyright © IEEE. Reproduced with permission.)

letter discussing the oscillator phase-noise model: *A heuristic derivation without formal proof.* His reasoning starts from our (2.37) and (2.38), that is, from the mean value of the electronics phase noise that is compensated by frequency fluctuations. However, they are only important in the passband of the resonator and are changed into additional phase noise as explained in Section 2.2.2. After summation of both components, he arrived at his famous relation:

$$S_\phi(\omega_m) = S_{\Delta\theta}\left[1+\left(\frac{\omega_o}{2Q\omega_m}\right)^2\right] \qquad 2.56$$

After comparing the above results with (2.41), we conclude that the Leeson relation deals with only the electronics noise. But some 50 years ago, the noise generated in electronic circuits was much larger than the frequency noise originating in the resonators themselves. So far, the application of the Leeson formula is correct. However, in both modern crystal and SAW oscillators, the resonator noise often predominates and application of (2.56) is not justified. One must revert to a more precise definition by (2.38). Unfortunately, we often witness efforts to amend Leeson's results with more sophisticated oscillator networks, which in many cases is not justified.

REFERENCES

2.1. F.L. Walls and A.E. Wainwright, Measurement of the Short-Term Stability of Quartz Crystal Resonators and the implications for Crystal Oscillator Design and Applications, *IEEE Tr.*, **IM-24,** (Mar. 1975), pp. 15–20.

2.2 J.J. Gagnepain, Fundamental Noise Studies of Quartz Crystal Resonators, in *Proceedings of the 37th Frequency Control Symposium* (1976), pp. 84–91.

2.3 T.E. Parker, Characteristics and Sources of Phase Noise in Stable Oscillators, in *Proceedings of the 37th Frequency Control Symposium* (1987), pp. 99–100

2.4. V. F. Kroupa, Flicker Frequency Noise in BAW and SAW Quartz Resonators, IEEE Tr., *UFFC-35* (1988), pp. 406–420.

2.5. V.F. Kroupa, Theory of 1/*f* Noise—A New Approach, *Phys. Lett. A,* **336** (2005), pp. 126–32.

2.6. L.E. Halliburton and D.R. Koehler, Properties of Piozoelectric Materials, in *Precision Frequency Control,* Academic Press, Orlando, 1985 (eds. E.A. Gerber and A. Ballato), 1985, pp. 1–45.

2.7. A.W. Warner, Design and Performance of Ultra Precise 2.5-mc Quartz Crystal units, *Bell System Tech. J.,* **33** (Aug. 1960), pp. 1193–1217.

2.8. K. Kurokawa, Noise in Synchronized Oscillators, *IEEE Tr.,* **MTT-16** (Apr. 1968), pp. 234–240.

2.9. J.R. Vig and T.R. Meeker, The Aging of Bulk Acoustic Wave Resonators, Filters and Oscillators, in *Proceedings of the 445th Frequency Control Symposium* (1991), pp. 77–101.

2.10. J.R. Vig and F.L. Walls, Fundamental Limits on the Frequency Instabilities of Quartz Crystal Resonators, in *Proceedings of 1994 International Frequency Control Symposium* (1994), pp. 506–523.

2.11. M.M. Driscoll, Low Noise Crystal Oscillators using 50 Ohm Modular Amplifier Sustaining Stages, in *Proceedings of the 40th Frequency Control Symposium* (1986), pp. 329–35.

2.12. V. Candelier, P. Canzian, J. Lamboley, G. Marotel, and P. Poulain, State of the Art in Ultra Stabile Oscillators for Accurate/Precise on Board and Ground Applications, in *2001 IEEE International Frequency Control Symposium* (2001), pp. 767–73.

2.13. E. Rubiola, J. Groslambert, M. Brunet, and V. Giordano, Flicker Noise Measurement of HF Quartz Resonators. *IEEE Tr.,* **UFFC-47** (2000), pp. 361–368.

2.14. F. Stahl, M. Mourey, S. Galliou, F. Marionnet, and R.J. Besson, Characterization of Quartz Crystal Resonators on Phase Modulation Noise Without an Oscillator, in *2000 IEEE International Frequency Control Symposium* (2000), pp. 393–396.

2.15. S. Galliou, F. Sthal, N. Gufflet, and M. Mourey, Predicting Phase Noise in Crystal Oscillators, in *2003 IEEE International Frequency Control Symposium* (2003), pp. 499–501.
2.16. M.M. Driscoll and W.P. Hanson, Measured vs. Volume Model-Predicted Flicker Frequency Instability in VHF Quartz Crystal Resonators, in *1993 IEEE International Frequency Control Symposium* (1993), pp. 186–92.
2.17. T.E. Parker, D. Andres, J. A. Greer, and G.K. Montrress, $1/f$ Noise in Etched Groove Surface Acoustic Wave (SAW) Resonators, *IEEE Tr.,* **UFFC-41** (1994), pp. 853–862.
2.18. V.F. Kroupa *Phase Lock Loops and Frequency Synthesis,* New York: Wiley, 2003.
2.19. F. Stahl, J.J. Boy M, Mourey, and F. Marionnet, Phase Noise Study of Quartz Crystal Resonators Versus the Radius of Curvature, in *2001 IEEE International Frequency Control Symposium* (2001), pp. 639–642.
2.20. M.L. Nelson, C.W. Nelson, and F.L. Walls, Relationship of AM to PM Noise in Selected RF Oscillators, *IEEE Tr.,* **UFFC-41** (1994), pp. 680–406–420.
2.21. V. Candelier, P.Canzian, J. Lamboley, M. Brunet, and G. Santarelli, Space Qualified Ultra Stabile Oscillators, in *2003 IEEE International Frequency Control Symposium* (2003), pp. 575–582.
2.22. S. Galliou, M. Mourrey, F. Marionnet, R.J. Besson and P. Guillemot, An Oscillator for Space, in *Proceedings of the 2003 IEEE International Frequency Control Symposium* (2003), pp. 430–434.
2.23. T. E. Parker, $1/f$ Fluctuations in Quartz Acoustic Resonators, *Appl. Phys. Lett.,* **46** (3), 1985, pp. 246–248.
2.24. E.N. Ivanov and M.E. Tabor, Low Phase-Noise Sapphire Crystal Microwave Oscillators: Current Statutes, *IEEE Tr.UFFC,* **56** (Feb. 2009), pp. 263–269.
2.25. Proceedings of the IEEE, Special Issue on Frequency Stability, **54** (Feb. 1966).
2.26. D.B. Leeson, A Simple Model of Feedback Oscillator Noise Spectrum, *Proceedings of the IEEE* (Feb. 1966), pp. 329–330.

CHAPTER 3

Noise Properties of Practical Oscillators

The need to designate information channels began at the time electromagnetic waves began to be used for communications. First it started with the length of the respective waves, and then with carrier frequencies, which were generated both in transmitter exciters and receiver local oscillators. Their reliability and precision soon became indispensable for dependable transfer of information. In the beginning, and often to the present time, the role was dominated by LC oscillators with the resonant circuit built from an inductor and a capacitor. Since the early 1930s, we encountered more precise quartz oscillators whose stability was able to demonstrate irregularities in the Earth's rotation. Their use as frequency and time standards was supplanted in the 1950s by atomic devices. Nevertheless, the area between the primary standards and practical applications still calls for stable crystal oscillators. At the same time, we witnessed efforts to reduce instability, frequency, and phase fluctuations of the generated carrier frequencies for long-distance communication, particularly in space activities (global positioning systems, GPS).

3.1 PRECISION OSCILLATORS

Investigations performed in the second half of the twentieth century provided evidence that the larger the quality factor, Q, of the resonator,

Frequency Stability. By Venceslav F. Kroupa

the higher the stability of the respective oscillator. In Chapter 2, we established the following relation for the noise in oscillators:

$$S_{\phi,\mathrm{osc}}(f)=\left(\frac{\omega_o}{2Q\omega}\right)^2\left[\frac{a_{R2}}{f^2}+\frac{a_{R1}+a_E}{f}+\frac{2kT}{P_r}(1+F)\right] \tag{3.1}$$

$$+\frac{a_E}{f}+\frac{2kT}{P_r}(1+F)+S_{\phi,\mathrm{add}}(f)$$

Note that this relation is valid even in instances in which the flicker noise generated in maintaining electronics exceeds the corresponding components generated in the resonator. The search for larger Q values requires dielectric materials with smaller losses, that is, with smaller $tg(\delta)$, on the one hand and application of cryogenic temperatures on the other hand.

3.1.1 Quartz Crystal Oscillators

Precision quartz crystal oscillators are manufactured predominantly with the 5- or 10-MHz carrier frequencies. This choice is preferred for several reasons. First, in Chapter 2 we saw that the crystal resonator flicker noise constant a_{R1} is nearly the same over the entire frequency range (cf. 2.9) and equal to the product $1/Qf_o$, irrespective of the resonant frequency f_o in the entire frequency range from 1 to 1000 MHz [3.1]:

$$a_{R1}\approx\frac{1}{Qf_o}\approx 10^{-13} \tag{3.2}$$

Further, the phase noise is proportional to the square of the resonant frequency. The third motive is that RF and low GHz frequencies, the flicker noise generated in the maintaining electronics a_E (transistors and amplifiers) is $< a_{R1}$, that is,

$$a_E\approx 10^{-14} \tag{3.3}$$

and values even more than one order lower were reported [3.2]. This is not the case with higher carrier frequencies. In the range from 500 to 1000 MHz, the flicker noise constant a_E starts exceeding a_{R1}. By applying what was discussed in Chapter 2, we have, for the fractional frequency PSDs,

$$S_{\frac{\Delta\omega}{\omega}}(f) = S_{y,\text{osc}}(f) = \left(\frac{f}{f_o}\right)^2 S_\phi(f) =$$
$$\left(\frac{1}{2Q}\right)^2 \left[\frac{a_{R2}}{f^2} + \frac{a_{R1} + a_E}{f} + \frac{2kT}{P_r}(1+F)\right] +$$
$$\left[\frac{a_E}{f} + \frac{2kT}{P_r}(1+F) + S_{\phi,\text{add}}(f)\right]\left(\frac{\omega}{\omega_o}\right)^2 \tag{3.4}$$

from which, with the assistance of (3.2), the most important noise coefficients are found:

$$h_{-2} \approx \frac{1}{(2Q)^2} \cdot \frac{2kT}{P_o} \approx 10^{-43.7\pm1} f_o^2$$
$$h_{-1} = \frac{a_{R1}}{(2Q)^2} \approx \frac{10^{-13.1}}{4Q^2} = \frac{10^{-13.7} f_o^2}{(Qf_o)^2} \approx 10^{-40\pm1} \cdot f_o^2 \tag{3.5}$$
$$h_o \approx \frac{2kT}{P_{\text{res}} 4Q^2} = \frac{10^{-20.7} f_o^2}{P_{\text{res}} (Qf_o)^2} \approx \frac{10^{-43.7\pm1} f_o^2}{P_{\text{res}}} \quad (P_o = P_{\text{res}} = 10^{-3})$$

and

$$h_1 = \frac{a_E}{f} \cdot \frac{1}{f_o^2} \quad \text{and} \quad h_2 = \frac{2kT}{P_{\text{res}}}(1+F) \cdot \frac{1}{f_o^2} \tag{3.6}$$

Since the first three noise coefficients are proportional to the square of the carrier frequency, they provide the second reason for building stable oscillators in the low megahertz (MHz) ranges. The increasing mechanical dimensions keep 5 MHz as the lower bound. Reverting to the fundamental measure of stability in the time domain [the Allan variance (5.27)], we get for the flicker frequency noise the plateau

$$\sigma(\tau) \approx \sqrt{2\ln(2) h_{-1}} \approx \sqrt{\frac{1}{Qf_o} \cdot \frac{1}{(2Q)^2}} \approx 10^{-20} f_o \gtrsim 10^{-13} \tag{3.7}$$

In addition, resonators contribute another source of noise due to medium fluctuations in the environment, which in most instances are temperature, humidity, and so on, that are responsible for the random walk coefficient, a_{R2}. However, after placing resonators in vacuum enclosures and ovens, the corresponding noise can generally be neglected. This conclusion is confirmed by Walls and Vig [3.2], particularly if the oven temperature is placed into the inflection point of the resonator frequency characteristic (T vs. f_o). Here, the first difference $\Delta f/\Delta T$ is zero (SC cuts) and the oven is carefully designed with a very high thermal quality factor, $Q_{T,\text{oven}}$. Note that the above indicated guidelines for the stability of crystal oscillators require a careful selection of both of the resonators [3.3, 3.4] and a cautious design of the whole oscillating network. In summarizing the results, we may expect the following relation for the output phase noise power spectral density (PSD) both of BAV as well as SAW crystal oscillators in the entire frequency range from 5 to 500 MHz (see Fig. 3.1).

$$S_\phi(f) \approx \left(\frac{f_o}{2Q}\right)^2 \left[\frac{10^{-17.1}}{f^4} + \frac{10^{-13}}{f^3} + \frac{10^{-17.6}}{f^2}\right] + S_{\phi,\text{add}} \approx$$
$$f_o^4 \cdot \left[\frac{10^{-44.5\pm0.5}}{f^4} + \frac{10^{-39.5\pm0.5}}{f^3} + \frac{10^{-44\pm1}}{f^2}\right] + \frac{10^{-14.5\pm0.5}}{f} + 10^{-17\pm1} \tag{3.8}$$

Note that we have also enclosed the term exhibited at the very low Fourier frequencies, which is the random walk, that is, the term is inversely proportional to f^4 (cf. Section 2.1.3 and the discussion in Chapter 5, Section 5.2.2), the part of the Allan variance proportional to $\tau^{0.5}$ (cf. Fig. 3.2).

3.1.2 Precision Microwave Oscillators

Nowadays, there are a number of scientific and technological applications requiring generators with very high frequency stability even in the microwave ranges. Mere multiplication of lower radio frequency (RF) carrier frequencies does not solve the problem because of the phase-noise multiplication in the entire Fourier frequency range (see Section 2.3.4 and cf. Fig. 3.3 lines a and b). Application of special microwave oscillators with high Q resonators also does not alleviate the problem,

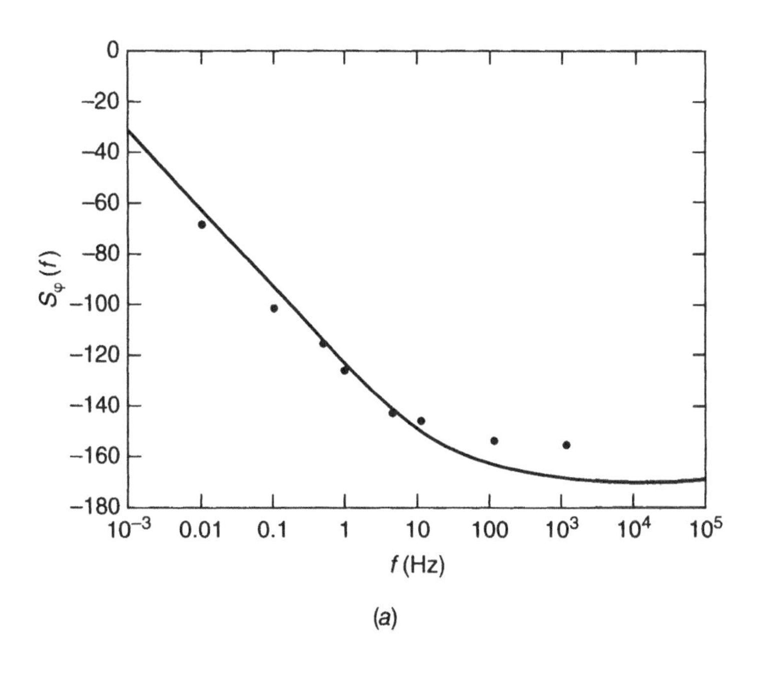

(*a*)

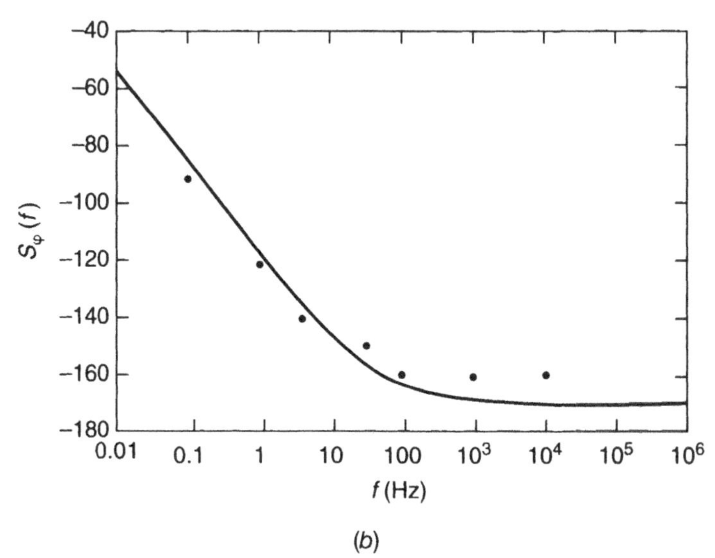

(*b*)

Fig. 3.1 The expected hase noise characteristics of precise quartz crystal oscillators: (*a*) 5 MHz and (*b*) 10 MHz. *Continued on next page.*

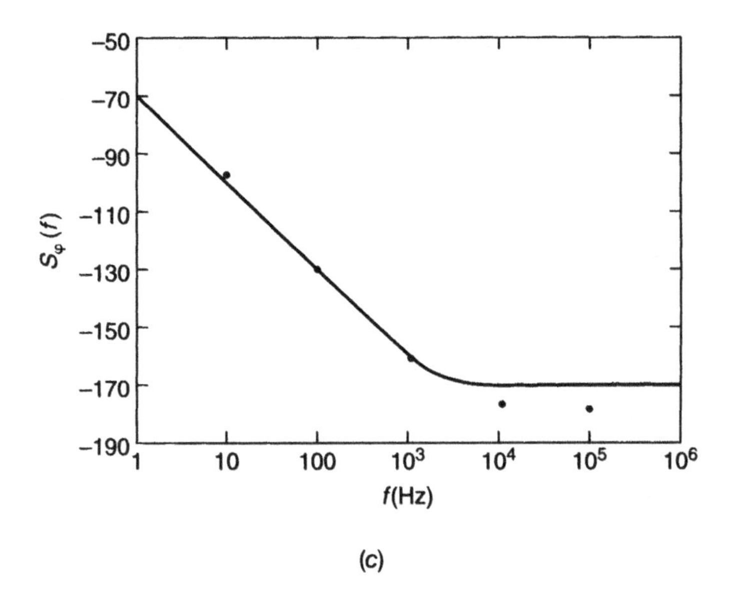

(c)

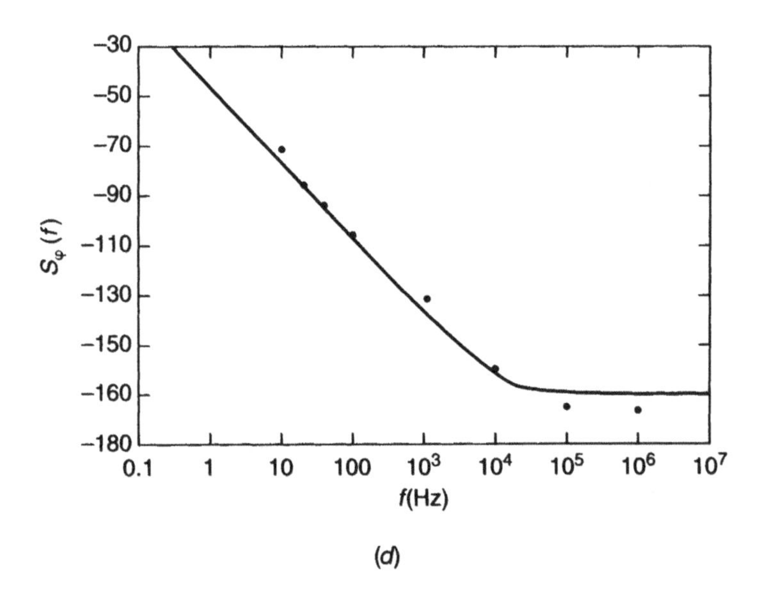

(d)

Fig. 3.1 (*cont.*) Phase noise characteristics of precise quartz crystal oscillators: (*c*) 100 MHz and (*d*) 500 MHz.

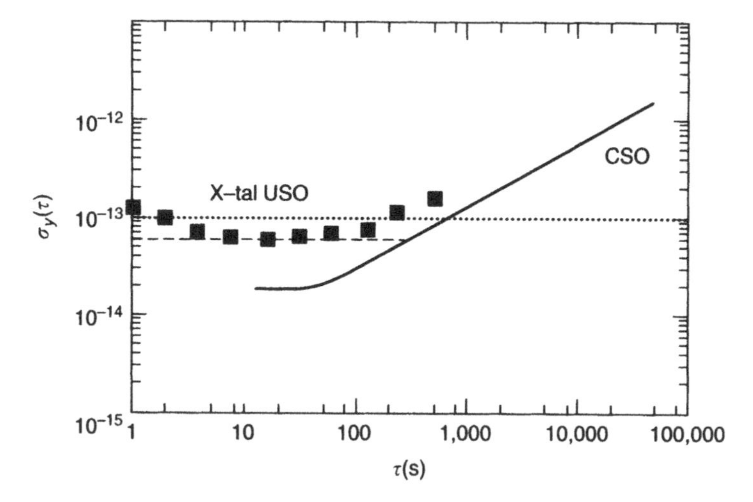

Fig. 3.2 Time domain frequency stability of a 5-MHz ultrastable quartz crystal oscillator measured with the CSO reference [3.5]. (Copyright © IEEE. Reproduced with permission.)

since the flicker noise term due to losses in the resonator, a_{R1}, starts to be masked by the term, a_E, generated in the maintaining electronics (see the cooled sapphire oscillators below). Nevertheless, they are praised for their lower white-phase noise floor, which is much lower compared with that after multiplication of the best crystal oscillator (cf. Fig. 3.3). The reason is because the overall phase noise, in accordance with (3.1), still holds and a larger power, P_o, is used. Another possibility is application of the phase-lock systems that make it possible to retain both of the low-phase noises that are close to the carrier and in the white noise range (see Chapter 6 for more information).

3.1.2.1 Microwave Self-Excited Oscillators

In its simplest form, these oscillators may recall the well-known Leeson model, which is investigated in Section 2.4. The corresponding block diagram is shown in Fig. 3.4. Note that the additional bandpass filter takes care of oscillation in the intended mode of the microwave resonator. The output phase noise can be computed with the assistance of (2.38):

$$\phi_{\text{out}} = \phi_a + \frac{1}{s\tau}(\phi_a + \dot{\phi}_{\text{res}}\tau) + \Delta\phi \qquad 3.9$$

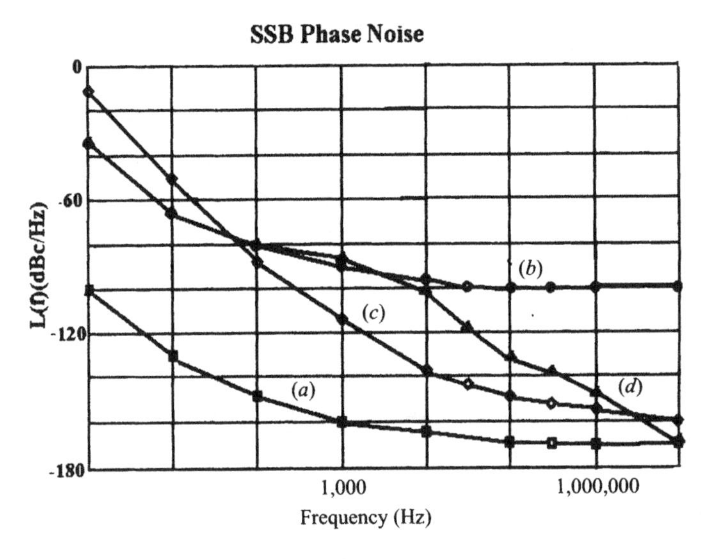

Fig. 3.3 Phase noise of the 10-MHz crystal oscillator (*a*), compared with its multiplied noise at 20 GHz (*b*), with the phase noise of cooled sapphire 20-GHz microwave oscillator (*c*), and with a phase-locked loop (PLL) system noise of 20-GHz DRO (dielectric resonator oscillator) and a 100-MHz crystal oscillator (*d*). (Adapted from [3.6]. Copyright © IEEE.)

The oscillation condition requires that the mean value of frequency fluctuations is zero (the term in parentheses) and the overall phase is equal to $2\pi N$ ($N = 0, 1, 2, \ldots$). To this end, a phase shifter φ is included. The corresponding output phase noise is

$$S_{\phi,\mathrm{out}}(f) = \left(\frac{f_o}{2Qf}\right)^2 [S_{\phi,\mathrm{res}}(f) + S_{\phi_a}(f)] + S_{\phi_a}(f) \qquad 3.10$$

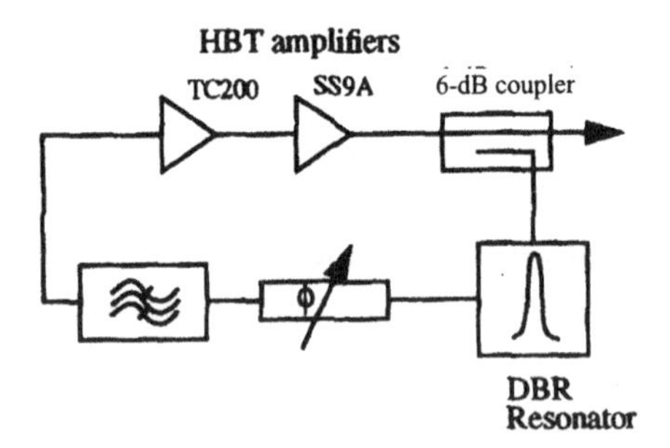

Fig. 3.4 Block diagram of the self-excited oscillator (Leeson model).

In the instance at which the noise generated in the resonator is small compared with the electronic noise originating in the loop, we get

$$S_{\phi,\text{out}}(f) \approx \left(\frac{f_o}{2Qf}\right)^2 \left[\frac{a_{a,e}}{f} + a_{a,o}\right] + \frac{a_{a,e}}{f} + a_{a,o} \qquad 3.11$$

EXAMPLE 3.1

Let us examine phase-noise properties of a precise microwave oscillator [3.7] with

$$f_o = 9\text{ GHz} \qquad Q_{\text{unloaded}} = 700{,}000 \qquad a_{a,e} = 10^{-11.5}$$

From the measured characteristic in [3.7, Fig.7], we read numerical values presented in Table 3.1.

After their plot, we draw asymptotes for establishing the polynomial PSD (see Fig. 3.5) and find the following numerical values of the PSD characteristic:

$$S_{\phi}(f) \approx \frac{10^{-2.6}}{f^4} + \frac{10^{-3.6}}{f^3} + \frac{10^{-10}}{f^2} + \frac{10^{-11.5}}{f} + 10^{-16} \qquad 3.12$$

In Fig. 3.5*a*, we evaluate the asymptotic slopes and in Fig. 3.5*b* we verify or correct the noise coefficients with the assistance of the frequency noise characteristic (cf. Example 2.5). Finally, with the assistance of (3.5), we find the noise constants to be approximately

$$a_{R2} = 10^{-11} \qquad a_{R1} = 10^{-12} \qquad a_{a,e} = 10^{-15} \qquad a_{a,0} = 10^{-16}$$

The expected Allan variance is plotted in Fig. 3.5*c*.

3.1.2.2 Stabilized Microwave Local Oscillators

The block diagram of the stabilized microwave local oscillator (STALO) is displayed in Fig. 3.6. It is a feedback system correcting

Table 3.1 Phase noise properties of a precise microwave oscillator with $f_o = 9$ GHz[a]

f (Hz)	1	3	10	30	100	400	700	1000	4000	10^4	10^5
$S(f)$ (dB)	−25	−45	−60	−77	−95	−112	−120	−125	−142	−151	−160

[a] $Q_{\text{unloaded}} = 700{,}000$; $a_{a,e} = 10^{-11.5}$ [3.7].

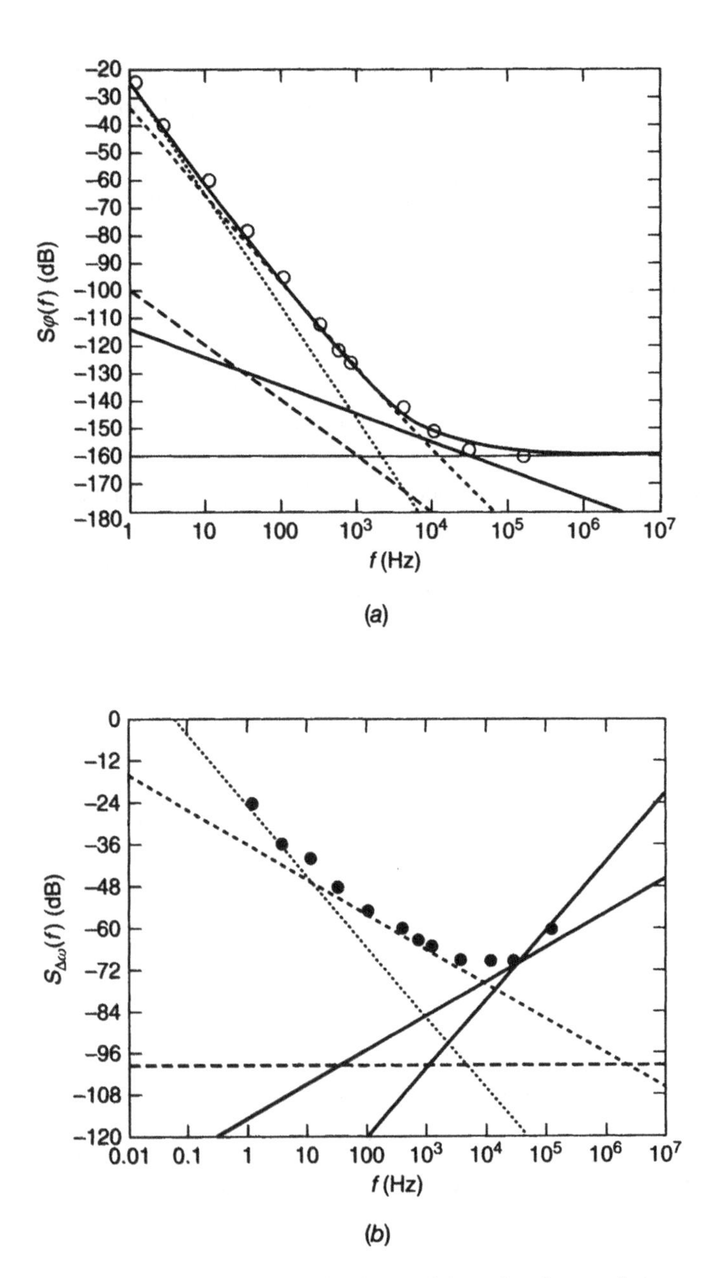

Fig. 3.5 Evaluation of the asymptotic slopes of the noise characteristic computed in Example 3.1: (*a*) evaluation of the asymptotic slopes from the published phase noise characteristic (Adapted from [3.7]). (*b*) Verification or correction of the noise coefficients with the assistance of the modified fractional frequency noise characteristic.

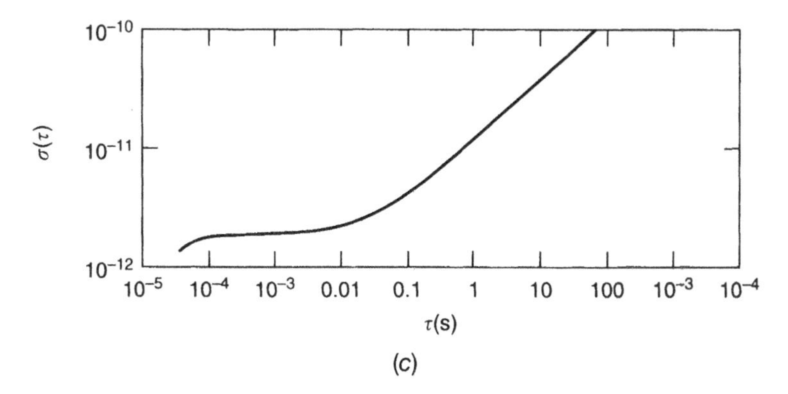

Fig. 3.5 (*cont.*) Evaluation of the asymptotic slopes of the noise characteristic computed in Example 3.1: (*c*) The corresponding Allan variance, which will be discussed in Chapter 5.

the oscillator phase noise φ_{osc} with the assistance of the signal passed through a stable resonator. By using PLL theory [e.g., 3.8], we get for the phase detector (PD) output

$$
\begin{aligned}
K_d(s) &= K_d[\varphi_{out}(s) - F_{res}(s)][\varphi_{out}(s) + \varphi_{res}(s)] + \varphi_{PD}(s) = \\
&K_d\left[\varphi_{out}(s) - \frac{1}{1+s\tau}[\varphi_{out}(s) + \varphi_{res}(s)] + \varphi_{PD}(s)\right] \approx \\
&K_d\left[\frac{\varphi_{out}(s)\, s\tau - \varphi_{res}(s)}{1+s\tau} + \varphi_{PD}(s)\right]
\end{aligned}
\tag{3.13}
$$

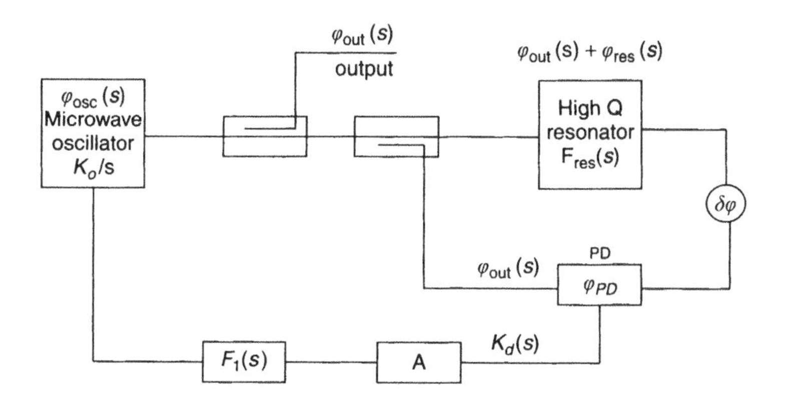

Fig. 3.6 Block diagram of the microwave STALO.

where $\varphi_{\mathrm{res}}(s)$ is an additive phase noise generated in the resonator itself and filtered together with the output noise $\varphi_{\mathrm{out}}(s)$ by the corresponding transfer function, $F_{\mathrm{res}}(s)$:

$$F_{\mathrm{res}}(s)=\frac{1}{1+s\dfrac{2Q_{\mathrm{res}}}{\omega_{\mathrm{res}}}}=\frac{1}{1+s\tau} \qquad \tau=\frac{2Q_{res}}{\omega_{\mathrm{res}}} \tag{3.14}$$

Note the phase shifter, $\delta\varphi$, which takes care of the necessary 90° shift between PD input signals. Proceeding further in accordance with PLL theory, we encounter the amplifier gain K_A; the loop filter, $F_L(s)$, between the phase detector and the oscillator (if any); and the oscillator gain K_o/s, which form the *open-loop gain* $G(s)$:

$$G(s)=\frac{K_d(s)}{1+s\tau}K_AF_L(s)\frac{K_o}{s}\approx\frac{K}{(1+s\tau)s} \tag{3.15}$$

The preliminary output phase is

$$\varphi_{\mathrm{out}}(s)=\varphi_{\mathrm{osc}}(s)-G(s)[\varphi_{\mathrm{out}}(s)s\tau-\varphi_{\mathrm{res}}(s)+(1+s\tau)\varphi_{\mathrm{PD}}] \tag{3.16}$$

After rearranging the above relation, we obtain

$$\varphi_{\mathrm{out}}(s)=\frac{G(s)[\varphi_{\mathrm{res}}(s)-(1+s\tau)\varphi_{\mathrm{PD}}(s)]+\varphi_{\mathrm{osc}}(s)}{1+G(s)s\tau} \tag{3.17}$$

In the case in which the gain $G(s)$ is large compared to *one* (for small Fourier frequencies), the approximation reveals

$$\varphi_{\mathrm{out}}(s)\approx\frac{[\varphi_{\mathrm{res}}(s)-\varphi_{\mathrm{PD}}(s)]}{s\tau}+\frac{\varphi_{\mathrm{osc}}(s)}{G(s)s\tau} \tag{3.18}$$

and the PSD of the output phase is (after introducing for the time delay $\tau=\omega_o/2Q_{\mathrm{res}}$)

$$S_{\varphi,o}(f)\approx\left(\frac{f_o}{2Q_{\mathrm{res}}f}\right)^2\left[S_{\varphi,\mathrm{res}}(f)+S_{\varphi,\mathrm{PD}}(f)+\frac{S_{\varphi,\mathrm{osc}}(f)}{K^2}\right] \tag{3.19}$$

Note that PDS of the STALO close to the carrier is controlled by the resonator phase fluctuations (cf. 3.1) if the gain, K, is large compared to *one*.

$$S_{\varphi,\text{res}} = \frac{a_{R2}}{f^2} + \frac{a_{R1}}{f} + a_o \tag{3.20}$$

3.1.2.3 Interferometric Stabilization of Microwave Oscillators

In instances where the noise contributions of the open-loop VCO is large compared with the intrinsic noise of the resonator, we would need to increase the gain, K, substantially, particularly with the assistance of an amplifier in the feedback path beyond a reasonable value. Another solution may be provided by an interferometric system decreasing the VCO noise level. The principle is depicted in Fig. 3.7

To a summation amplifier, we directly feed the output signal $[v_1(t)]$, and to the subtracting input we feed the same signal $[v_2\ (t)]$, passed though the high-Q resonator but in the opposite phase. The corresponding low-frequency output of the summation amplifier is

$$\begin{gathered} A[V_1\cos(\omega_o t + \phi_{\text{out}}) - V_2\cos(\omega_o t + (\phi_{\text{out}} + \phi_{\text{res}})F_{\text{res}})] \approx \\ A[(V_1 - V_2)\cos(\omega_o t) - (V_1 - V_2)[\phi_{\text{out}} - (\phi_{\text{out}} - F_{\text{res}}(\phi_{\text{out}} + \phi_{\text{res}})]\sin(\omega_o t)] \end{gathered} \tag{3.21}$$

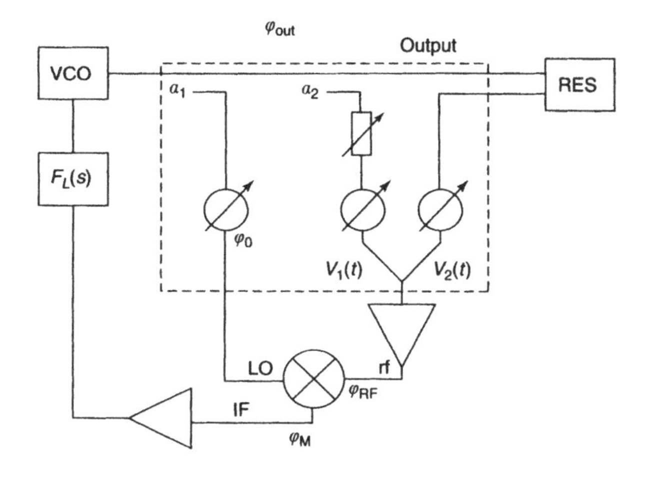

Fig. 3.7 Interferometric reduction of the output noise level (adapted from [3.9]).

After introduction of the resonator filtering function F_{res} (3.14) and following phase detection in the mixer, we arrive at the output in the Fourier transform:

$$M(s) \approx K_d \left[AV_1 \frac{s\tau\phi_{\text{out}}(s) + \phi_{\text{res}}(s)(1+s\tau)}{1+s\tau} + \phi_M(s) \right] \quad 3.22$$

And after closing the effective PLL we get

$$\phi_{\text{out}}(s) = \phi_{\text{osc}}(s) \quad 3.23$$

$$-\frac{K_o}{s} F_L(s) \left[K_d K_{AV} \frac{s\tau\phi_{\text{out}}(s) + \phi_{\text{res}}(s)(1+s\tau)}{1+s\tau} + K_d \phi_M(s) \right]$$

Finally, we get the effective output phase

$$\phi_{\text{out}}(s) = \frac{\phi_{\text{res}}(s) \dfrac{\frac{K_o}{s} F_L(s) K_d K_{AV}}{1+s\tau} + \phi_M(s) \dfrac{K_o F_L(s) K_d}{s}}{1 + \dfrac{\frac{K_o}{s} F_L(s) K_d K_{AV} s\tau}{1+s\tau}} \quad 3.24$$

$$+ \frac{\phi_{\text{osc}}}{1 + \dfrac{\frac{K_o}{s} F_L(s) K_d K_{AV} s\tau}{1+s\tau}}$$

In the vicinity of the carrier $s\tau \ll 1$ and for $F_L \approx 1$,

$$\phi_o(s) \approx \frac{\phi_{\text{res}}(s) + \dfrac{\phi_{IF}(s)}{K_{AV}}}{s\tau} + \frac{\phi_M / s}{1 + \tau K F_L(s)} \quad 3.25$$

Note again the similarity to relation (2.40). Consequently, we may write for the output phase noise PSD

$$S_{\phi,\text{out}}(f) = \left(\frac{f_o}{2Qf} \right)^2 \left[S_{\phi,\text{res}}(f) + \frac{S_{\phi,M}(f)}{K_{AV}^2} \right] + \frac{S_{\phi,\text{osc}}(f)}{|1 + \tau K F_L(s)|^2} \quad 3.26$$

$$K = K_o K_d K_{AV}$$

After comparing (3.26) with that for the STALO output phase noise (3.19), we see that the interferometric stabilization results in reduction of the spurious phase noises due to the reduction of the carrier signal. After comparing (3.26) to (3.4), we arrive at

$$S_{\phi,\mathrm{out}}(f) = \left(\frac{f_o}{2Qf}\right)^2 \left[\frac{a_{R2,res}}{f^2} + \frac{a_{R1,\mathrm{res}}}{f} + \frac{a_E}{f}\right] + S_{\phi,\mathrm{add}}(f) \qquad 3.27$$

The result is that the noise close to the carrier is that of the resonator and associated circuitry, whereas the contribution of VCO and other spurious sources is substantially reduced.

3.1.2.4 Microwave Oscillators Stabilized with Ferrite Circulators

Another possibility for the carrier reduction provides application of a ferrite circulator that works on the same principle as the interferometer (see Fig. 3.8).

The circulator consists of a ferrite cylinder with three ports at 120°, biased on an axial direct current (DC) field that introduces a Larmor precession. The voltage transfer function between planes *A-A* and *C-C* is

$$F_V(\omega) \approx \Gamma(\omega) = \frac{e_{r'}(\omega)}{e_{i'}(\omega)} \qquad 3.28$$

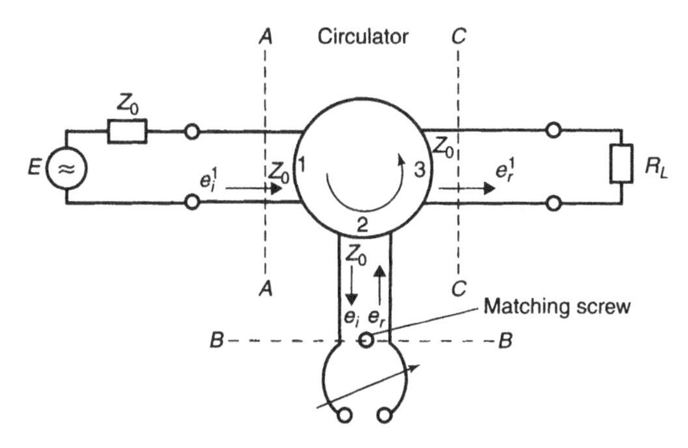

Fig. 3.8 Ferrite circulator as a carrier suppression network [3.10]. (Copyright © IEEE. Reproduced with permission.)

where e_r and e_i are voltages at the output and input ports of the circulator, respectively. Since the circulator insertion loss is usually small, < 0.2 dB from port to port, the transfer function is equal to the reflection coefficient at the *B-B* plane:

$$F_V(\omega) \approx \Gamma(\omega) = \frac{e_r(\omega)}{e_i(\omega)} \tag{3.29}$$

To arrive at the transfer function $\Gamma(\omega)$, we use the equivalent circuit in Fig. 3.9*a*. The Thevenin theorem reveals the effective voltage (Fig. 3.9*b*) and, finally, in Fig. 3.9*c*, the Norton current source (Y is the total circuit admittance):

$$Y_c = \frac{1}{nZ_o} + j\left(\omega C - \frac{1}{\omega L}\right) \approx \frac{1 + jQ_U(2\Delta\omega/\omega_o)}{nZ_o} \tag{3.30}$$

where $Y_c(\omega)$ is the cavity admittance, $Q_U = nZ_o/\omega_o$, the unloaded cavity Q, and $\omega_o = 1/\sqrt{LC}$ the cavity resonant frequency.

Finally, with the input admittance and the transformed load admittance, $Y_G(\omega) = (n^2 R_G)^{-1}$, we arrive at the reflection coefficient:

$$\Gamma(\omega) = \frac{Y_G - Y_c(\omega)}{Y_G + Y_c(\omega)} = \frac{\frac{nZ_o}{n^2 R_G} - 1 - jQ_U\frac{2\Delta\omega}{\omega_o}}{\frac{nZ_o}{n^2 R_G} + 1 + jQ_U\frac{2\Delta\omega}{\omega_o}} = \frac{\beta - 1 - jQ_U\frac{2\Delta\omega}{\omega_o}}{\beta + 1 + jQ_U\frac{2\Delta\omega}{\omega_o}} \approx \frac{ST}{2} \qquad \beta = \frac{Z_o}{nR_G} \approx 1 \tag{3.31}$$

Its plot is shown in Fig. 3.10.

Note that the circulator isolates the input generator from the output load and that the typical insertion loss between points 1 and 3 in Fig. 3.8 is ~ 0.3 dB. When maximum cavity absorption (maximum power transfer) occurs, the result is close to the carrier and the amplitude of the output signal is proportional to $|\Gamma(\omega)|$ and the phase fluctuations are changed into frequency fluctuations and resonator frequency varia-

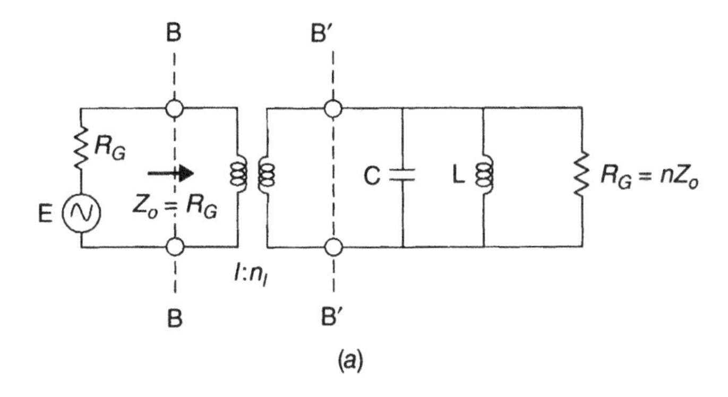

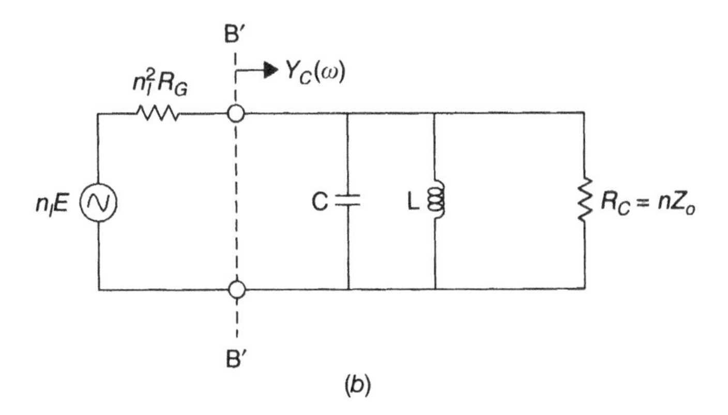

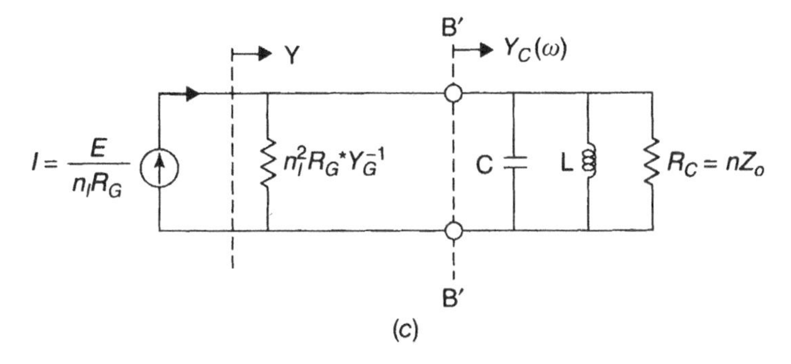

Fig. 3.9 Equivalent input and output circuit of the circulator: (*a*) equivalent circuit, (*b*) the Thevenin theorem of the effective input voltage, (*c*) the effective Norton current source. The parameter $Y_c(\omega)$ is the cavity admittance and $Y(\omega)$ is the total circuit admittance (adapted from [3.10]).

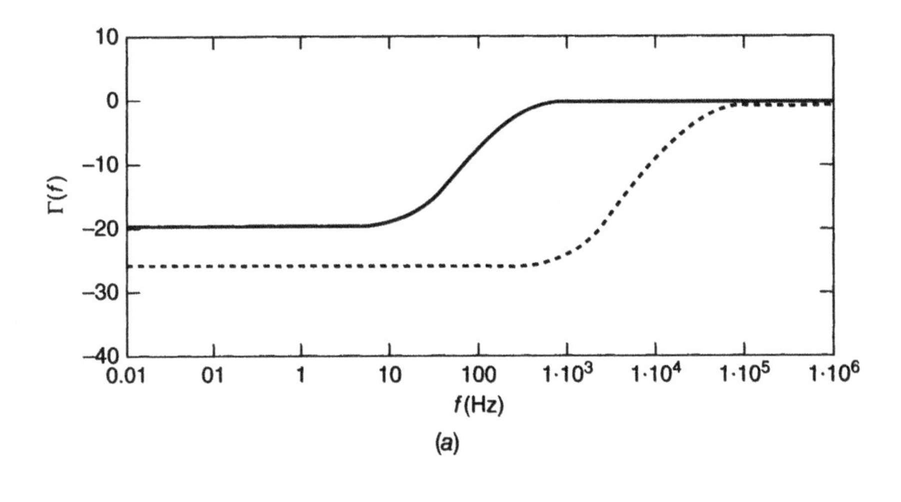

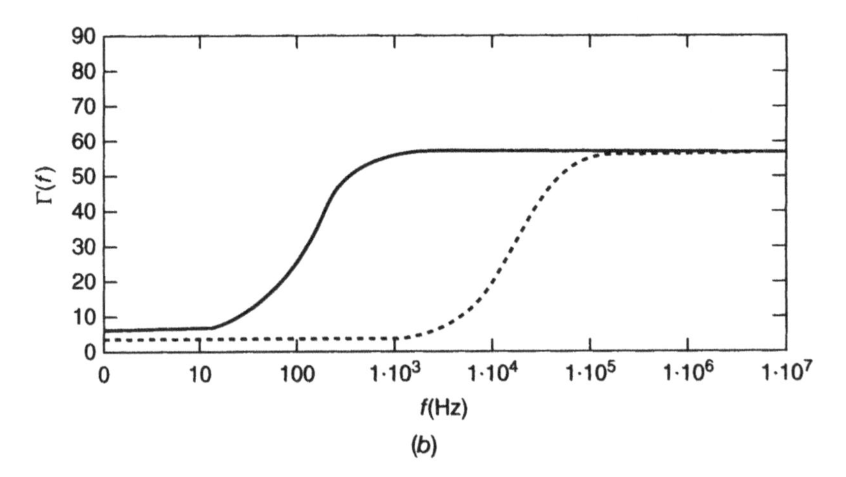

Fig. 3.10 Carrier suppression by circulator versus Fourier frequency for two cavities: $Q_{load} = Q_U/2$ (*a*) carrier suppression for $Q_{load} = 1 \times 10^8$ (full line) and $Q_{load} = 75 \times 10^4$ (points), (*b*) $Q_{load} = 1 \times 10^8$ (full line) corresponding phases.

tions. Evidently (3.25–3.29) are also valid for microwave oscillators stabilized with ferrite circulators.

Recently, combination of interferometric and circulator carrier suppression was reported [3.9, 3.11], with the result that carrier suppression may be as large as 90 dB. The principle is explained with the assistance of Fig. 3.11. The phase noise close to the carrier is determined by the noise generated in the resonator only.

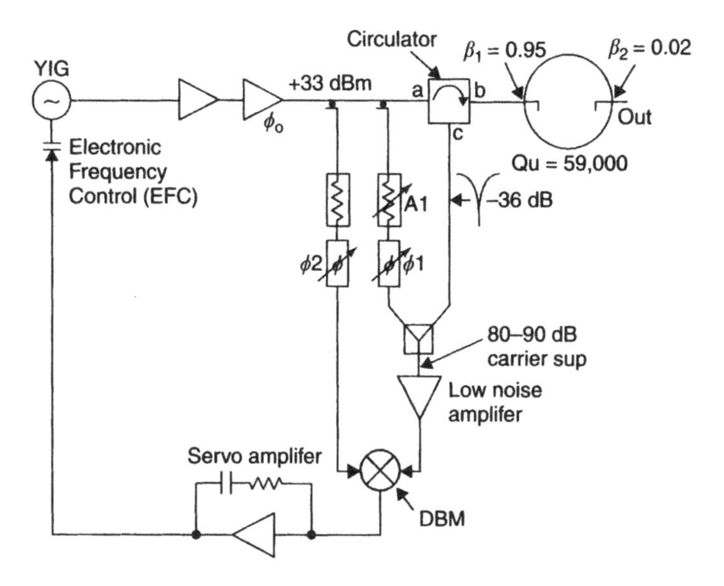

Fig. 3.11 Combination of simultaneous interferometric and circulator carrier suppression [3.9]. (Copyright © IEEE. Reproduced with permission.)

3.1.2.5 Pound Stabilization

Another possibility for stabilizing microwave oscillators is based on the arrangement suggested by Pound [3.12]. Again the ferrite circulator is used for correcting the VCO. However, the efficiency is increased with an auxiliary frequency modulation of the carrier (see Fig. 3.12).

To the input port (1) of the circulator, we supply the frequency modulated carrier

$$e_1(t) = E_1\left[\sin\left(\omega_o t + \phi(t) + m_f \sin(\nu t)\right)\right] \qquad 3.32$$

where $\omega_o = \omega_{res} + \Delta\omega$, $\varphi(t)$ is the phase instability close to the carrier, and $m_f \sin(\nu t)$ is the signal generating a large auxiliary frequency modulation. Expansion of (3.32) proceeds with the assistance of Bessel functions, $J_n(m_f)$,

$$\begin{aligned} e_1(t) = E_1[&J_o(m_f)\sin[\omega_o t + \phi(t)] \pm J_1(m_f)\sin[\omega_o t + \phi(t) \pm \nu t] \pm \\ &J_2(m_f)\sin[\omega_o t + \phi(t) \pm 2\nu t] \pm J_3(m_f)\sin[\omega_o t + \phi(t) \pm 3\nu t] + \cdots] \end{aligned} \qquad 3.33$$

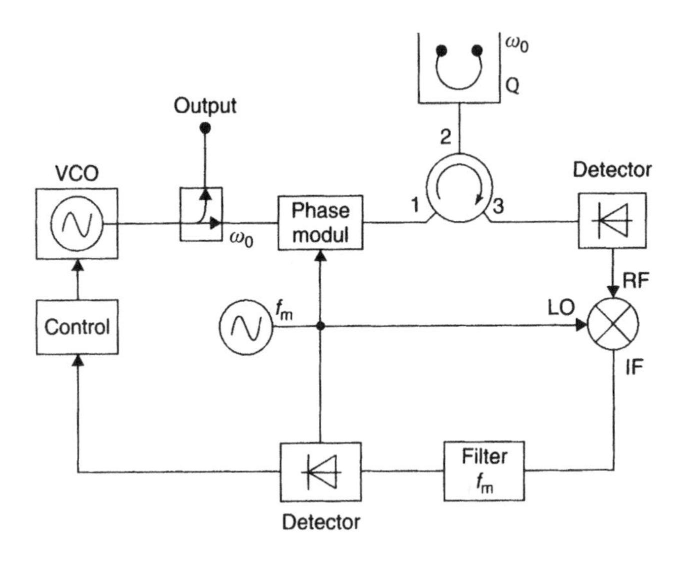

Fig. 3.12 Block diagram of the Pound stabilized oscillator.

and for the output signal, $e_3(t)$, we have

$$e_3(t) \approx E_1[J_o(m_f)\Gamma(\omega_o)\sin(\omega_o t)) \pm \\ J_1(m_f)\Gamma(\omega_o \pm \nu)\sin(\omega_o + \phi(t) \pm \nu t) + \cdots] \qquad 3.34$$

which is composed of the carrier reduced in accordance with (3.33). Further, it exhibits the delayed phase, the added frequency noise generated in the resonator, and the reflected and effectively unattenuated modulation side bands. Operation of the Pound circuit can be understood as conversion of the phase modulation (PM) to the amplitude modulation (AM). The process may be explained with the assistance of Figs. 3.13 and 3.14 (cf. also the discussion in [3.13]).

EXAMPLE 3.2

Inspection of Fig. 3.13 reveals for a nearly zero amplitude of the carrier the modulation index $m \approx 2.4$, which is an important contribution of the sidebands J_1 (2.4) ≈ 0.518, J_2 (2.4) ≈ 0.433, J_3 (2.4) ≈ 0.198, and nearly negligible values for all other modulation sidebands. Another possibility is the choice of the modulation index $m \approx 1$. In this case, due to the attenuation of the carrier, the output signal $e_3(t)$ will be amplitude modulated as in Fig. 3.14. Then, after HF detection, we filter out the modulation sig-

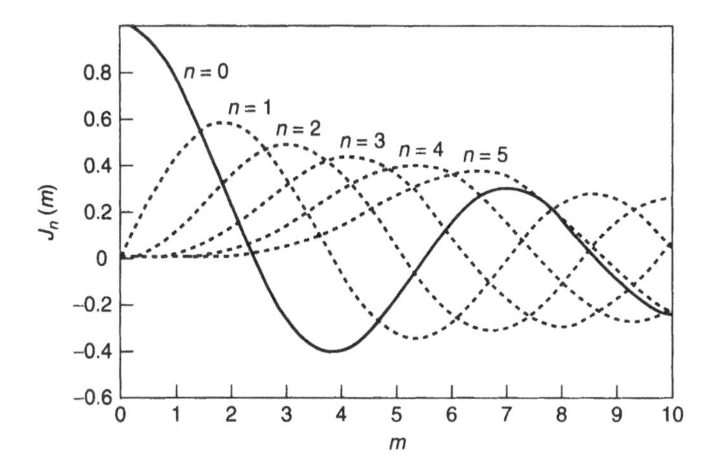

Fig. 3.13 Bessel functions.

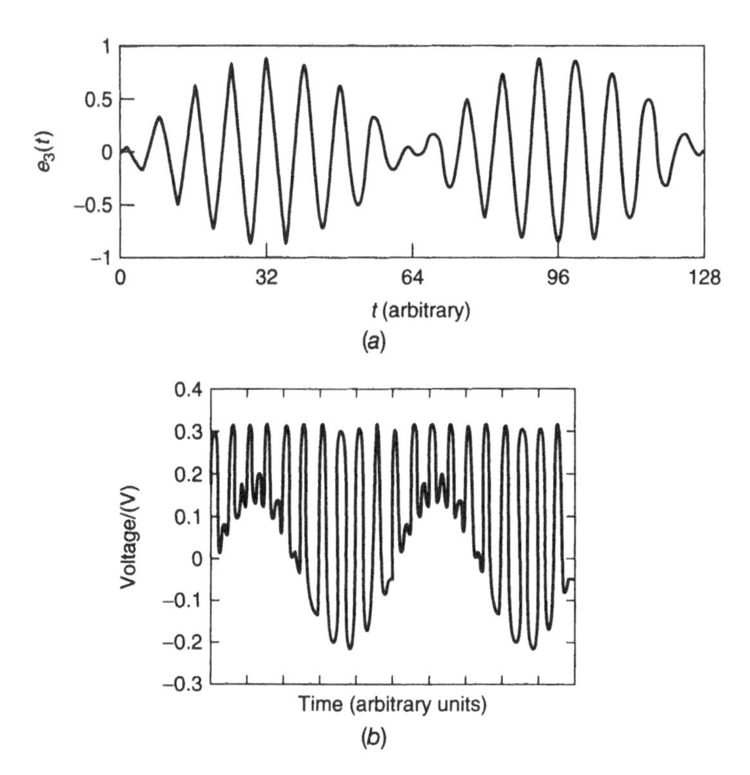

Fig. 3.14 (*a*) Amplitude modulated output signal $e_3(t)$, and (*b*) sample of the detected output of the Pound discriminator [3.14]. (Copyright © IEEE. Reproduced with permission.)

nal, which contains the phase shift introduced by the resonator that will be extracted with the assistance of the synchronous detection and used further as the feedback signal. In cases in which a larger modulation index is used, say $m = 2.4$, one must filter out the higher harmonics of the modulation signal (cf. Fig. 3.15 where a block diagram of an actual Pound stabilized low noise receiver with a cryogenic sapphire oscillator is reproduced).

3.1.3 Cryogenic Stabilized Oscillators

Improvements of both short- and long-term stability of precision oscillators could be obtained by cooling resonators to very low temperatures, thus reducing acoustic losses. Generally, the cryogenic temperatures are imperative in cases in which augmentation of Q by one or more orders is desired. However, this is the source of several difficulties, such as separation of the maintaining electronics network from the deep cooled resonator in liquid gasses, and so on. (Recently, application of the tunnel diode, which does not freeze, was reported for amplitude detection [3.15]). In this connection, Table 3.2 summarizes the boiling temperature of several gases.

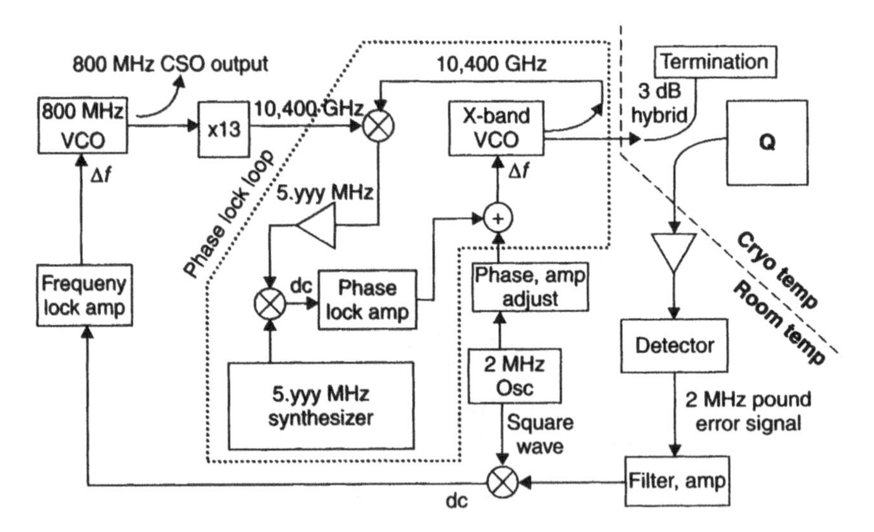

Fig. 3.15 Schematic diagram of the "low-noise" receiver with a cryogenic sapphire oscillator [3.15].

Table 3.2 The Boiling temperature of several gases

Gas	Boiling temperature (K)
Carbon dioxide	192.5
Krypton	120.1
Air	78.6
Nitrogen	77.1 (57 freezing temp.)
Hydrogen	20.1
Helium	4.1

3.1.3.1 Quartz Crystal Oscillators

To the author's knowledge, only a few experiments were performed with quartz crystal resonators cooled to cryogenic temperatures. Here, we illustrate the expected quality factor (inversion of acoustic losses) according to the measurements by Robichon et al. [3.16] (his results are reproduced in Fig. 3.16). Similar results about material losses, $tg(\delta)$, were reported by Halliburton and Koehler [3.17], and the corresponding characteristic is similar to the one reproduced in Fig. 3.16. Inspection of both figures teaches us that the expected increase of the quality factor, Q, of quartz crystal resonators, even when cooled to liquid helium temperature, provides only about one order of improvement. This value is not too impressive. In such a situation, one may expect that the $1/f$ noise generated by processes in the resonator will be comparable with the $1/f$ electronic contribution and the random walk of frequency ($1/f^2$) processes. The only real advantage would be reduced aging.

3.1.3.2 Oscillators with Ceramic Resonators

Typical commercial dielectric resonators (DR) are made from ceramic materials with large permittivity and quality factors ranging from several thousands to a few hundred thousand at room temperature that increases only three to four times after cooling too a few kelvin [3.18]. The most popular are sapphire resonators (cylinders or rings), machined from a low defect Al_2O_3 monocrystals in which microwave resonance modes are excited with Q-factors in the range of 10^5 at room temperature.

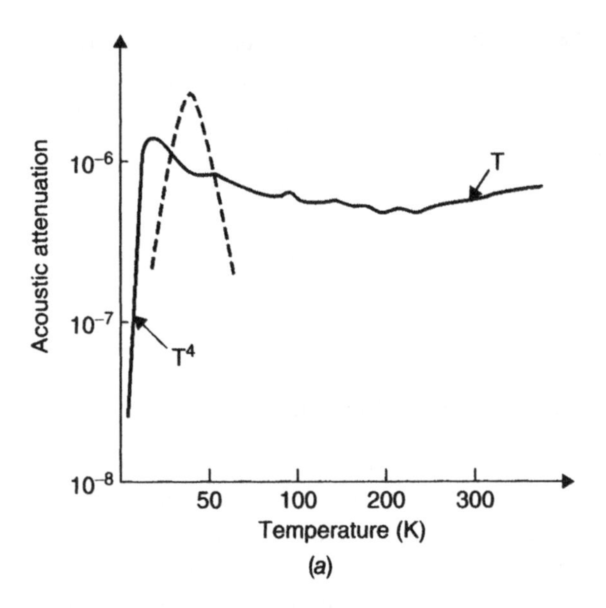

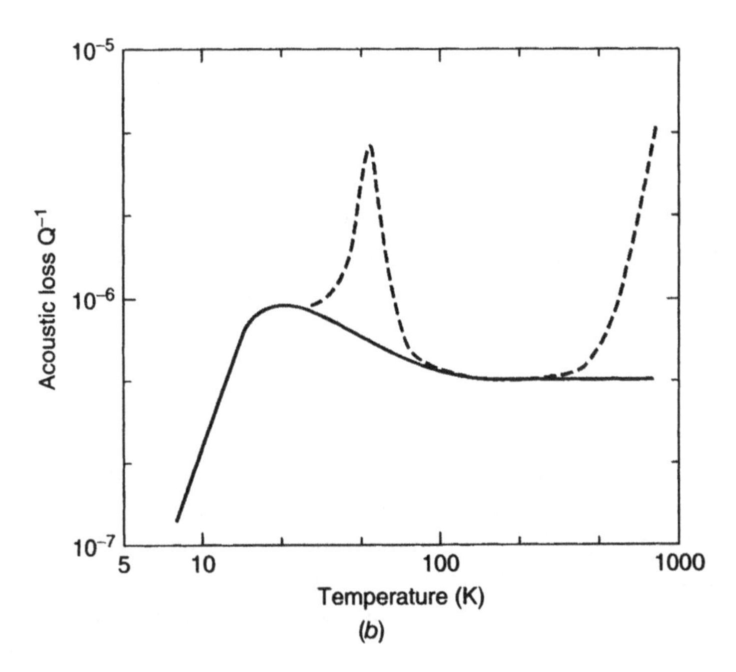

Fig. 3.16 Acoustic losses of the cooled quartz resonator. (*a*) Linear increase in temperature. (*b*) Logarithmic increase in temperature.

3.1.3.3 Cooled Sapphire Oscillators

Cooled sapphire resonators are preferred over quartz crystal resonators since the cooled single-crystal sapphire resonators offer the highest short-term stability of any secondary frequency sources designed to date. Due to the high Debye temperature of 1047 K, a reasonable increase of the Q-factor is observed for temperatures just < 100 K and another augmentation of the effective Q, by more than one order, for the liquid helium bath (see Fig. 3.17) (compared with cooled crystal and ceramic resonators).

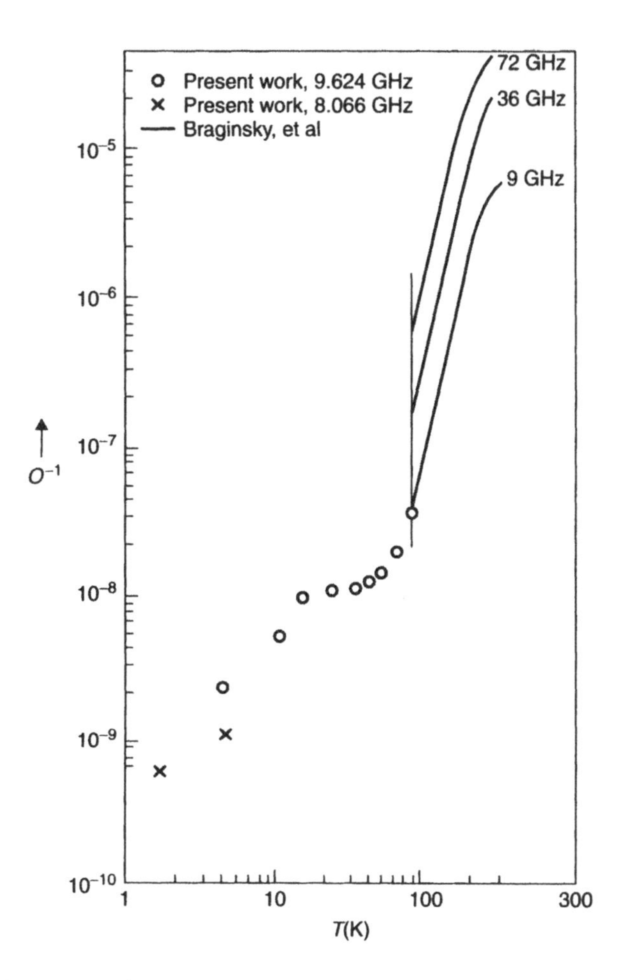

Fig. 3.17 Acoustic losses of the cooled sapphire resonator (adapted from [3.19]).

The problem is the nonexistence of the piezoelectric effect with the consequence that the sapphire oscillators depend on the properties of dielectric resonators. For cryogenic applications, TE_{01x}-mode resonator arrangements that have rather large dimensions compared to the effective wavelengths (cf. Fig. 18) are popular. However, the eventual radiation losses may reduce Q for the device. A remedy provides use of the whispering-gallery modes that confine the electromagnetic energy to the resonator body (cf. Fig. 3.18*c* and [3.19]). However, in these cases we encounter problems with identification of the resonant frequency because of an infinite number of TE_{01x} modes.

The temperature dependence causes another inconvenience since pure sapphire does not exhibit a turnover temperature [3.20]. However, there are always residual impurities that are responsible for an eventual turnover temperature peak (e.g., for the 10 K systems). An example is reproduced in Fig. 3.19.

Different compensation techniques were devised at higher temperatures (additional doping, auxiliary dielectric tuning posts, etc.). Evidently, cryogenic sapphire oscillators are suitable to provide secondary frequency standards in microwave ranges as required by a new generation of passive atomic frequency standards for space and other applications. In Fig. 3.20, we reproduce a cryogenic insert.

3.1.3.4 Frequency Stability of Sapphire Oscillators

Frequency stability of sapphire oscillators evaluated from the effective Q of the resonator may be treated from different points of view (the frequency domain, time domain, practical applications domain, etc.). If we revert to the time domain we may consider the short-term stability. This stability is closely related to the effective Q of the resonator; the medium term stability, with emphasis on good isolation from the environment (vibration, temperature, pressure, etc.), and long or very long term stability characterized by aging of resonators, electronic circuits, and so on.

The short-term stability is generally limited by flicker noise. For room temperature, the Qf_o may dominate (cf. 3.2, 3.5, and 3.7) and the expected minimum of the Allan variance would be (5.27)

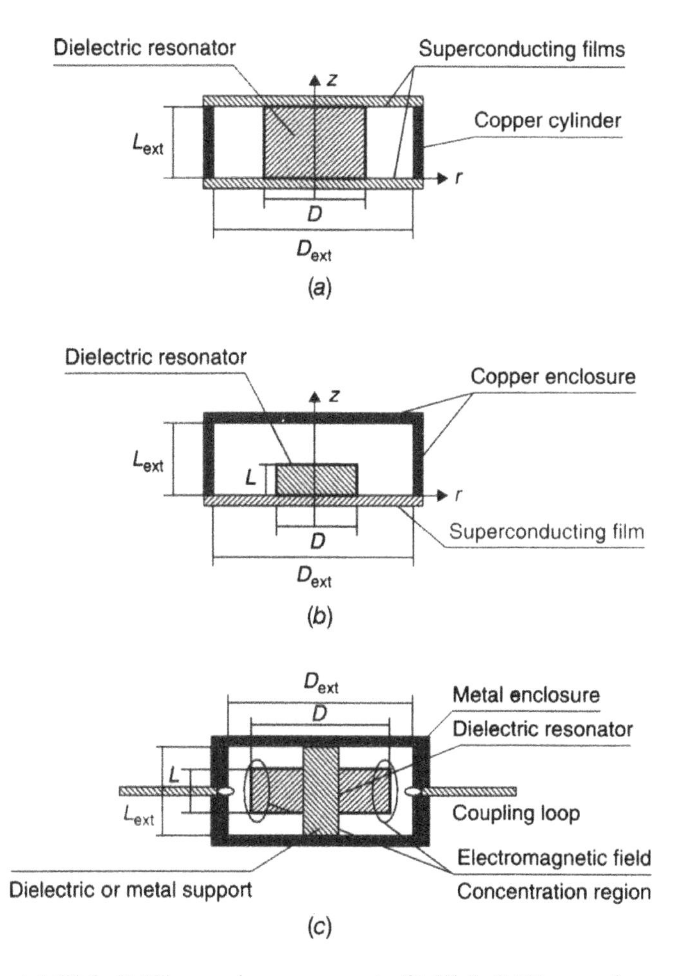

Fig. 3.18 (*a*) High-Q TE_{011}-mode arrangement, (*b*) High-Q TE_{01x}-mode arrangement, and (*c*) whispering-gallery mode arrangement (adapted from [3.18]).

$$\sigma(\tau) = \sqrt{h_{-1} 2 lg(2)} \approx \frac{1}{2Q}\sqrt{\frac{1}{Q f_o}} \qquad 3.35$$

(see Example 3.3). Equation (3.35) is also applicable in stabilized oscillator systems discussed in previous paragraphs. In Table 3.3, we summarized noise properties of several cryo-cooled sapphire frequency generators and in Fig. 3.21 we reproduce two Allan variance measurements.

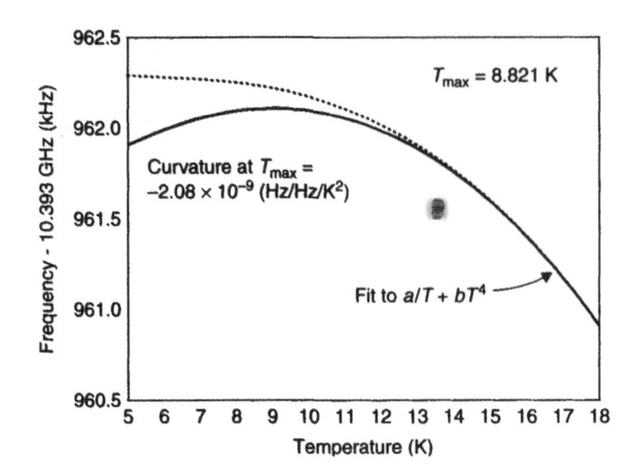

Fig. 3.19 Temperature turnover characteristic for a cryo-cooled sapphire resonator [3.21]. (Copyright © IEEE. Reproduced with permission.)

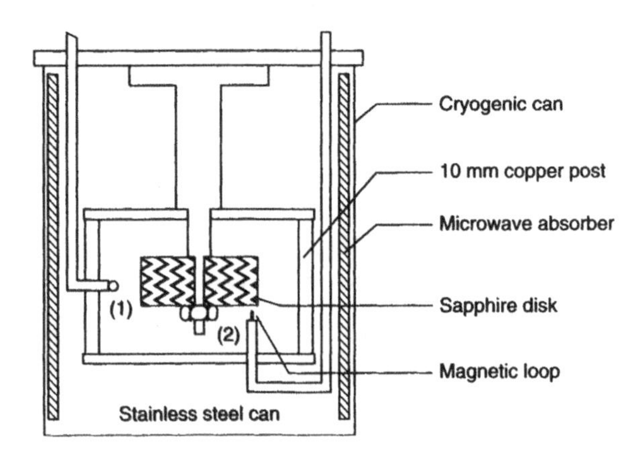

Fig. 3.20 An example of a cryogenic insert arrangement.

Table 3.3 Properties of sapphire oscillators

f_o (GHz)	Q	h_{-2}	h_{-1}	$\Phi(\vartheta)_{min}$	h_o	$T(K)$	Reference
11.9	2×10^9	-331	-296	3×10^{-15}	-270	≈7° [He]	[3.23]
12	2×10^8	-305	-278	2×10^{-14}	-270	≈6° [He]	[3.5, 3.24]
16	1.4×10^8	-290	-277	3×10^{-14}	-270	≈40°	[3.25]
9/12.6	3×10^7	-306	-294	3.5×10^{-14}	-295	≈50° [N]	[3.26]
11.2	1.5×10^9	-334	-307	5×10^{-16}	-296	≈7° [He]	[3.27]
9	2×10^5		-266	6×10^{-14}		300° [room temperature]	[3.28]

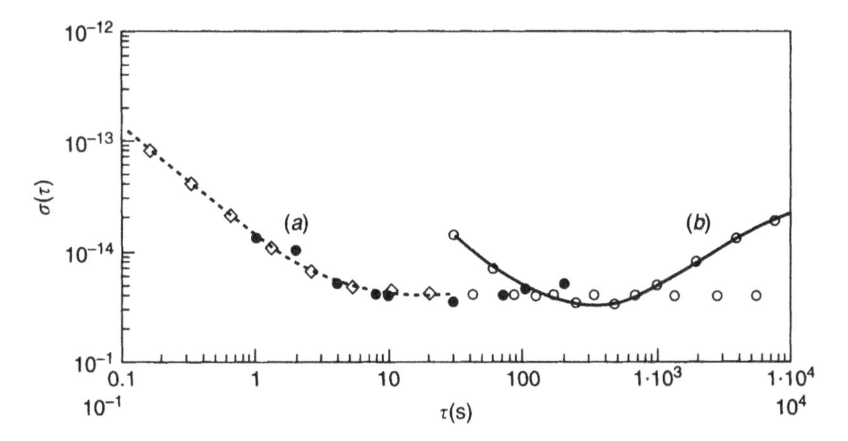

Fig. 3.21 Allan variance measurements of two cryo-cooled sapphire frequency generators. (*a*) adapted from [3.21] and (*b*) adapted from [3.23].

EXAMPLE 3.3

Here, we will only discuss the short-term stability of the stabilized sapphire oscillator at room temperature. In accordance with the theory for the origin of the flicker noise in crystal resonators and in accordance with relation (3.7), we evaluate the h_{-1} noise coefficient and the corresponding Allan variance. By assuming the carrier frequency f_o = 10 GHz and the resonator $Q \approx 2 \times 10^5$ we get for the Qf_o product $Qf_o = 2*10^{15}$ and, finally, from relation (3.5) $h_{-1} \approx 10^{-26}$ and $\sigma(\tau) \approx 10^{-13}$. Actual measurement revealed nearly the same results (see the last line in Table 3.3). Note that similar quantities would be found for a 10-MHz quartz crystal oscillator (cf. Fig. 3.2). On the other hand, Q's of the cooled sapphire resonators are so large that the noise introduced by maintaining electronics dominates, but it is still lower than with the best crystal oscillators, due to the sophisticated noise reduction methods. Recently, it was also due to using low-noise SiGe amplifiers.

The white frequency noise constant h_o due to the resonator itself is

$$h_o \approx \frac{1}{(2Q)^2} \cdot \frac{2kT}{P_{res}} \tag{3.36}$$

where P_{res} is the energy stored in the resonator itself. For room temperature, $Q \approx 2 \times 10^5$ and $P_{res} \approx 0$ (dBm), we get $h_o \approx 10^{-28}$ (cf. Table 3.3).

3.1.4 Opto-electronic Oscillators

Research on opto-electronic devices started some 20 years ago. Since then, substantial improvements in the level of spurious signals and of noise properties have been demonstrated, particularly in oscillator applications. The latter approach has the unique features of providing spectrally pure signals up to 80 GHz [3.30, 3.31].

3.1.4.1 Basic Arrangement

The basic scheme of an opto-electronic oscillator is illustrated in Fig. 3.22.

In its simplest form, these oscillators may recall the well-known Leeson model investigated in Chapter 2. They include a CW laser, an electro-optic modulator, a photodetector at the end of a single-mode optical fiber with the necessary RF amplification, coarse RF filtering, and RF coupling in the feedback branch. They allow a high tunability and almost no limitations on the range of possible oscillation frequencies, due to high mode density generated as long as the oscillating condition is met, that is, for all wavelengths λ satisfying

$$\left(N+\frac{1}{2}\right)\lambda = L \quad \lambda = \frac{c}{nf_o} \qquad f_o \approx \frac{Nc}{\mathrm{Ln}} \qquad 3.37$$

where N is an integer, c is the speed of light, n is the effective refractive index of the fiber, and L is its length in meters (m). By considering the distributed length element R, L, and C (cf. Fig. 3.23), we can approxi-

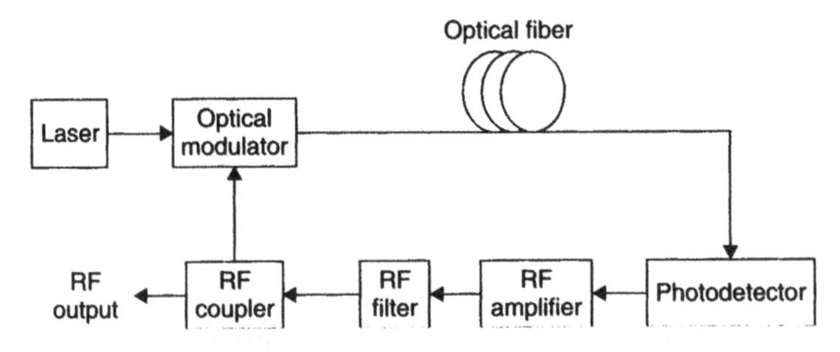

Fig. 3.22 The basic scheme of an opto-electronic oscillator [3.31]. (Copyright © IEEE. Reproduced with permission.)

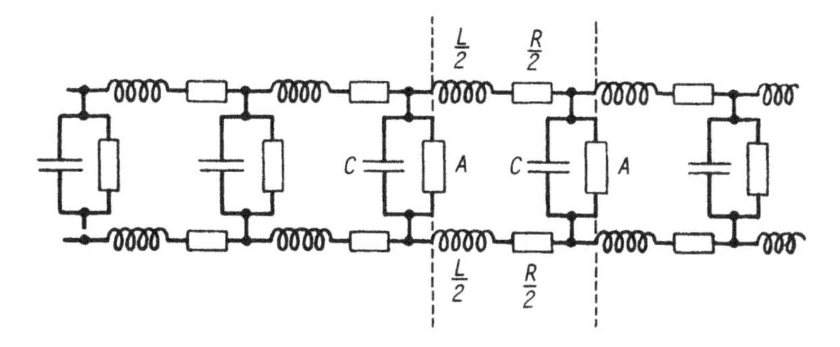

Fig. 3.23 Distributed length elements R, L, and C of the transmission line.

mate the RF propagation in the optical fiber with the assistance of the telegraph equation (by neglecting the leakage A):

$$LC\frac{\delta^2 v}{dt^2}+RC\frac{\delta v}{dt}=\frac{\delta v^2}{dx^2} \quad 3.38$$

For a periodic signal $v(t) = Ve^{j(\omega t+x)}$, its solution is

$$V=V_{+}e^{\gamma x}+V_{-}e^{-\gamma x}$$
$$\gamma=\sqrt{j^2\omega_o^2LC+j\omega_oRC}\approx j\sqrt{1-j\omega_oRC}\approx \quad 3.39$$
$$\frac{\omega_oRC}{2}+j=\beta+j\alpha$$

where β is the damping constant and α is the phase constant. The resonating transmission line exhibits the effective quality factor Q:

$$Q=\frac{L}{2\beta\lambda} \quad 3.40$$

Another approach for the computation of the Q-factor provides the propagation time τ_d:

$$\tau_d=\frac{nL}{c} \quad 3.41$$

from which, with the assistance of (3.37), the quality factor of the opto-electronic oscillator (OEO) is

$$Q = \frac{\omega_o \tau_d}{2} = \pi f_o \frac{nL}{c} = \pi N \qquad 3.42$$

Note that the quality factor, Q, is proportional to the multiplication factor, N, of the effective resonant frequency. This difficulty presents neighboring modes at distances Δf apart:

$$\Delta f = \frac{c}{nL} \qquad 3.43$$

For illustration, we present a numerical example.

EXAMPLE 3.4

Find the Q of the optical fiber of length $L = 2$ km with the refractive index $n = 1.5$. From (3.41), we get for propagation delay

$$\tau_d \approx \frac{1.5}{3 \times 10^8} \times 2 \times 10^3 = 10^{-5} \text{ (s)}$$

From (3.42), for $f_o \approx 10$ GHz the Q factor

$$Q \approx \pi \times f_o \times 10^{-5} \approx 3 \times 10^5$$

and the mode difference

$$\Delta f = 100 \text{ kHz}$$

3.1.4.2 Noise Properties of OEOs

Noise in OEOs is subjected to similar analysis as in other oscillators discussed in previous sections. By assuming the validity of the relation (3.1), first we evaluate the integration factor with the assistance of (3.42):

$$\left(\frac{f_o}{2Q}\right)^2 = \left(\frac{f_o \cdot c}{2 \cdot \pi f_o nL}\right)^2 = \left(\frac{c}{2\pi nL}\right)^2 \qquad 3.44$$

Note that it is independent of the carrier frequency but is inversely proportional to the square of the length of the fiber, and one may begin investigations of the phase noise properties with discussion of the relation (3.4). Close to the carrier, we expect either flicker frequency noise (FFN) with the slope of the phase noise characteristic $1/f^3$ or the random walk of frequency (RWF) with the slope of the phase noise characteristic $1/f^4$ or $1/f^5$. To this end, we have summarised some data in Table 3.4.

1. *FFN Noise:* Its sources may be either acoustic losses in the resonator or in the maintaining electronics. By using the heuristic (1.111), we would get for the flicker noise constant, a_{R1}, of the OEO (in accordance with the discussions in *Example 3.4*):

$$a_{R1} \approx (Qf_o)^{-1} \approx 1 \times 10^{-15} \tag{3.45}$$

However, from the data summarized in Table 3.4 one may conclude that the contribution of the electronics noise at such high frequencies is much more important because of the combination

$$a_{R1} + a_E \approx 1 \times 10^{-11\ 1} \tag{3.46}$$

Evidently, in the present state of OEOs the electronics noise constant, a_E, predominates (cf. 3.4)

2. *RWF Noise:* The proportionality to f^{-4} follows from the assumption that the resonator is subjected to the RW dimensions–fluctuations changing the resonant frequency, in most instances by fluctuations of the fiber longitude or of its refractive index:

$$\Delta f_o = -\frac{f_o}{2}\left(\frac{\Delta L}{L} + \frac{\Delta n}{n}\right)\Delta T \tag{3.47}$$

Table 3.4 Properties of OEOs

f_o (GHz)	h_{-2} (dB)	h_{-1} (dB)	h_o (dB)	Q	L (m) Qf_o	Reference
10	-200	-220			2000	[3.30]
10	-200	-223				[3.31]
10.57	-200	-210	- 280	0.2×10^5	1200	[3.32]
10.4		-212		0.3×10^5	2000	[3.33]
10	-190	-220		0.6×10^5	4000	[2.34]
10	.	-235		0.1×10^6	6000	[3.35]

Authors in [3.31] state that the temperature dependence of the effective refraction index is dominant for the frequency instability. Some information noise coefficients provide data summarized in Table 3.4.

By comparing OEO phase noise with that of the multiplied 10-MHz quartz crystal oscillator, we see a substantial improvement at Fourier frequencies > 1 kHz and a practical equivalence with the free-running sapphire oscillator (cf. noise characteristics in Fig. 3.24). The difficulty is the mentioned high density of the generated neighboring modes, of the thermal dependence of the fiber length, of its refractive index, and of the RF filter.

3.1.4.3 More Loops Arrangement

In addition to the investigated noises, one must consider the spurious signals. The problem of the large number of spurious modes [cf. (3.43)] is discussed in some depth by Eliyahu and Maleki [3.30, 3.33, 3.34]. For alleviating the drawback, they suggest an interferometric suppression of

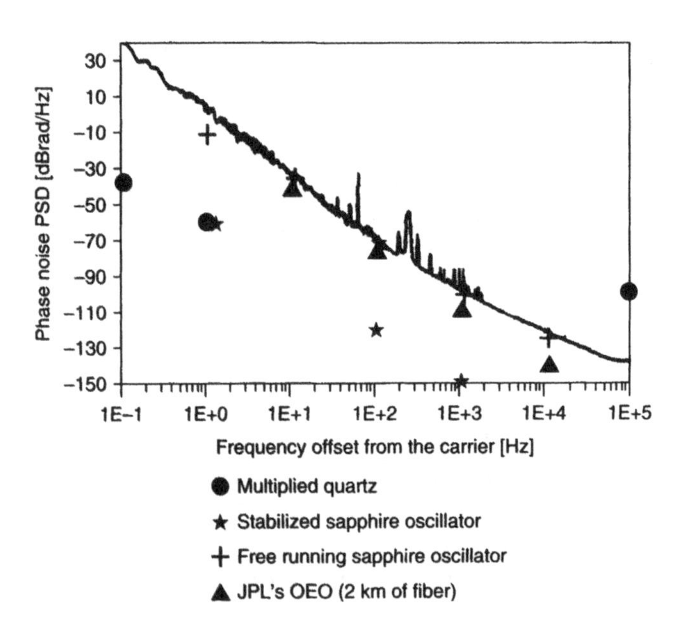

Fig. 3.24 Phase noise of an experimental opto-electronic device is comparable with that of the multiplied 10-MHz quartz crystal oscillator with stabilized sapphire oscillators [3.32]. (Copyright © IEEE. Reproduced with permission.)

some modes with the assistance of the dual (or threefold) optical loop OEO reproduced in Fig. 3.25. The corresponding suppression is illustrated in Fig. 3.26 and the corresponding phase noise characteristic of the single loop is reprinted in Fig. 3.27. Inspection of this figure reveals the predominance of the random walk coefficient a_{R2} at low Fourier frequencies.

3.1.4.4 Synchronized Opto-Electronic Oscillators

Recently, Zhou and Blasche [3.35] suggested application of the principle of synchronized oscillators for suppression of the spurious modes (cf. Fig. 3.28). The slave short fiber of the OEO is synchronized with the low noise signal of the long loop. The problem is discussed in depth in Chapter 6, Section 6.10, where for the PSD of the output signal was found (6.162):

$$S_{\phi,\mathrm{out}}(f) = S_{\phi,\mathrm{long}}(f)\frac{K^2}{K^2+f^2} + S_{\phi,\mathrm{short}}(f)\frac{f^2}{K^2+f^2} \tag{3.48}$$

the constant K is defined as

$$K = \frac{V_{\mathrm{syn}}\omega_o}{2QV_o} = \frac{\omega_o}{2Q}\sqrt{\frac{P_{\mathrm{syn}}}{P_o}} \tag{3.49}$$

An example of the output phase noise PSD is shown in Fig. 3.29. Inspection of (3.48) reveals that for small Fourier frequencies, $K < f$, the output noise is controlled by that of the long loop and vice versa.

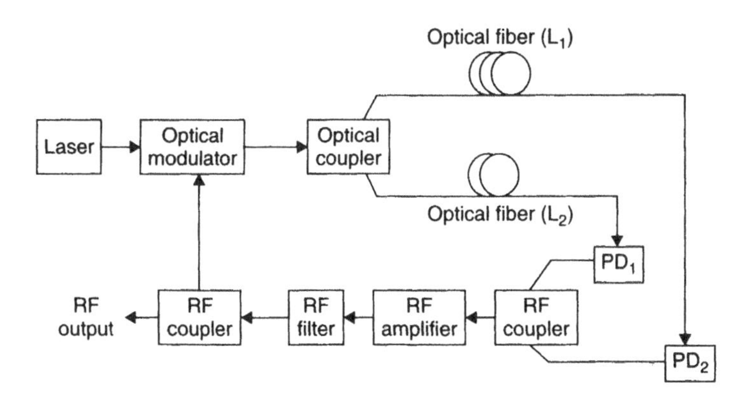

Fig. 3.25 Opto-electronic oscillator with dual optical loop [© 3.34].

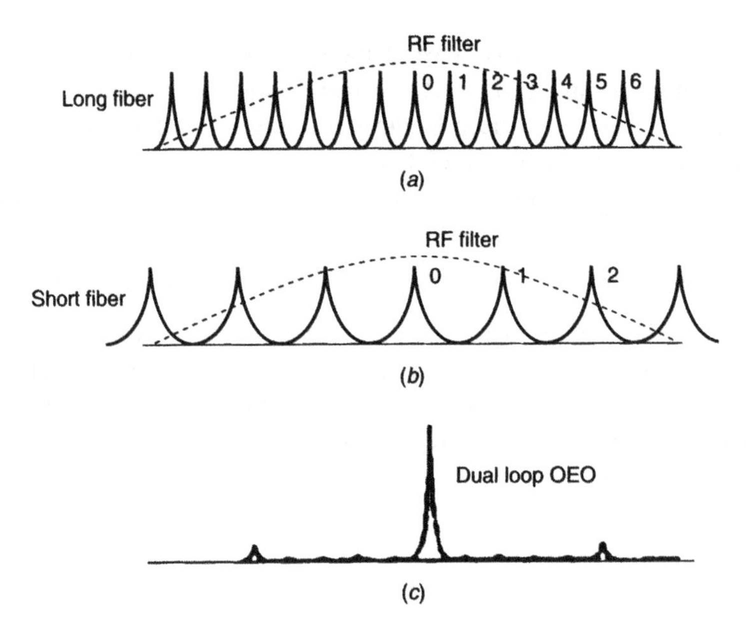

Fig. 3.26 (*a*) Illustration of the modes in a single loop of an OEO, (*b*) interferometric suppression, and (*c*) the output.

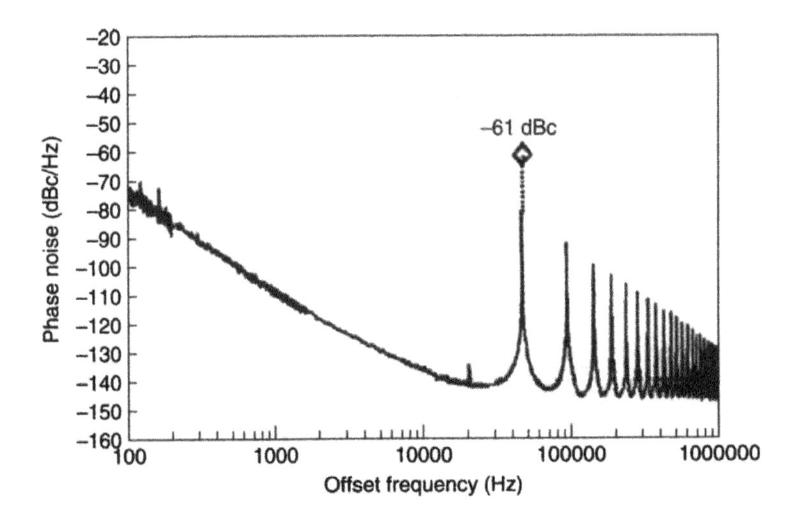

Fig. 3.27 The SSB phase noise of a single-loop OEO measured by the heterodyne method [3.34]. The solid curve describes phase noise measurment in units of dBc/Hz. The dotted line describes highest spurious measured in units of dBc. (Copyright © IEEE. Reproduced with permission.)

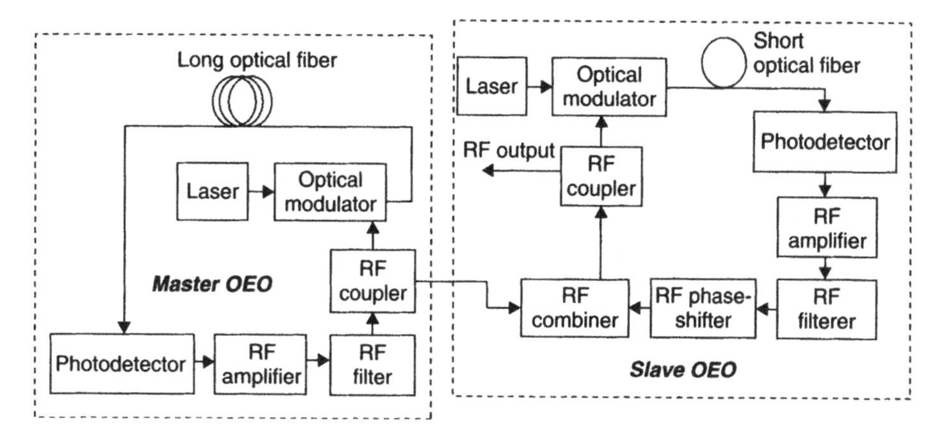

Fig. 3.28 Injection-locked dual OEO with ultralow-phase noise [3.35]. (Copyright © IEEE. Reproduced with permission.)

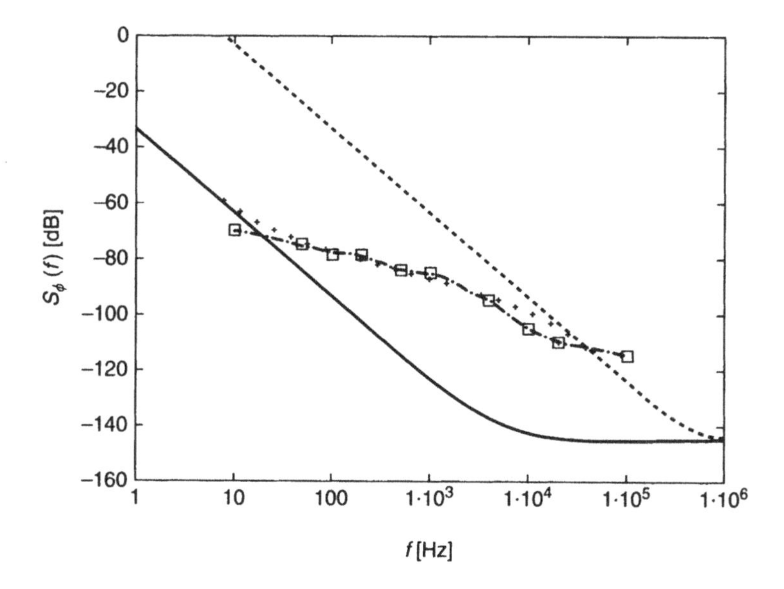

Fig. 3.29 Output phase noise characteristic of injection-synchronized dual optical loop of an OEO. The thin full line is the PSD of the OEO with the long fiber, the dashed line is that of the short fiber OEO, the thick dashed line has been computed with the assistance of (3.48), and rectangles are taken from [3.35].

3.2 PRACTICAL OSCILLATORS

First the wavelength was in meters (m), but very soon their carrier frequencies were in kHz, MHz, and GHz, generated both by transmitter exciters and receiver local oscillators. Their reliability and precision was indispensable for reliable transfer of information. Shortly, the role was dominated by LC oscillators with the resonant circuits built from an inductor and a capacitor. Since the early 1930s, we encountered more precise quartz oscillators (which were able to demonstrate irregularities in the Earth's rotation). At the same time, we witnessed efforts to reduce frequency instability and phase fluctuations of the generated carrier frequencies both on the transmitter and receiver sites with some sort of frequency synthesis which, with present applications, is widespread.

3.2.1 LC Oscillators

The basic problems of noise in oscillator design have been discussed in the previous chapters together with the expected noise properties (Section 2.3). The main feature is the phase noise close to the carrier, which is generally inversely proportional to the square of the quality factor, Q, of the resonant circuit, irrespective of the oscillator type. To prove the validity of this feature, we plotted h_{-1} and h_o constants for several earlier LC oscillators in Fig. 2.4. However, due to the progress of frequency synthesis, VCOs based on LC oscillators lost their importance in lower RF wave ranges, but this is not the case with microwave systems.

3.2.2 Microwave LC Oscillators

Microwave oscillators are at the heart of modern communications systems, providing clock signals for microprocessors and transmission channels to wireless base stations, radar, satellite links, and so on. The ever growing importance of telecommunications in microwave ranges has introduced the need for small, cheap, and low-power RF components and even efforts for design of single-chip transceivers fitted with frequency synthesizers for generation of the local oscillator signals with corresponding noise qualities. To meet the latter conditions, we witness efforts for construction of the LC voltage controlled oscillators

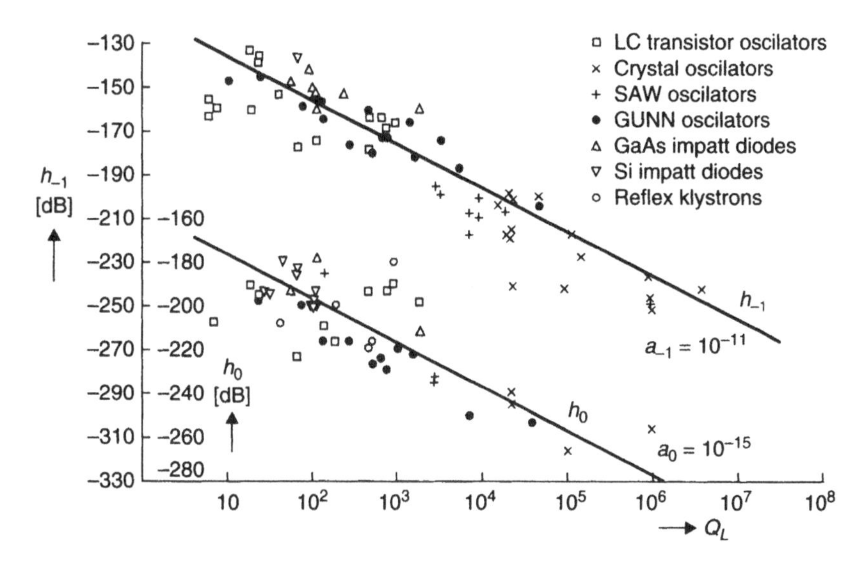

Fig. 3.30 The earlier measurements of the noise constant h_{-1} and h_o for different oscillators as functions of load Q_L [2.18].

(VCO) on IC chips. Listed here are several techniques [3.36–3.67] for the carrier frequency generation:

1. Divide by two via a frequency divider following the (VCO) oscillator. This arrangement requires the smallest area.
2. Two VCOs forced to run in quadrature.
3. Application of phase shifters, particularly for operation in frequency synthesis.
4. Other techniques [flat-coil inductors suitable for integration on chips (Fig. 3.31).

A typical block diagram of the monolithic VCO is depicted in Fig. 3.32.

3.2.3 Noise in Microwave LC Oscillators

Noise properties of microwave LC oscillators are the same as discussed in Section 3.1. Equations for the phase (3.1) and fractional frequency noise (3.4), respectively, are valid:

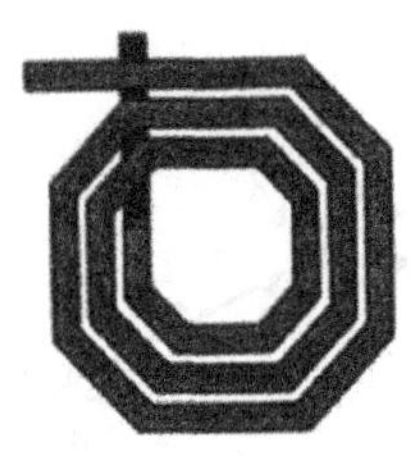

Fig. 3.31 An example of a flat-coil inductor suitable for integration on chips [© 3.36]. (Copyright © IEEE. Reproduced with permission.)

$$S_y(f) = S_\phi(f)\left(\frac{f}{f_o}\right)^2 = \frac{h_{-1}}{f} + h_o + h_1 f + h_2 f^2 \qquad 3.50$$

Interdependence between h_i-constants and phase noise constants is shown in Table 3.5. Note that the lowest output phase noise requires resonant circuits with the quality factor Q as high as possible, low resonator noises a_{R2} and a_{R1}, as well as the electronic contributions a_E, and the low thermal noise in the maintaining amplifiers. However, modern microwave communications require construction of LC oscillators (VCOs) in the IC form with flat coils suitable for integration on chips (cf. Fig. 3.31). Their properties are generally low quality factor Q, from 3 to 25 (exceptionally higher [3.37]), and inductances in nH. A basic block diagram of the microwave LC oscillators in the differential arrangement is illustrated in Fig. 3.32. All important noise sources are the same as encountered in other oscillators, namely, those of the resonant circuit. Transistors and diodes used, however, are subjected to the changes due to the differential disposition.

3.2.3.1 White Noise Constant

Estimation of the white noise constant starts again with (3.5):

$$h_0 \approx \frac{2kT}{4Q^2 P_o} \approx \frac{10^{-20}}{4 \cdot 10^2 \cdot 10^{-3}} \approx -196 \quad \text{(dB)} \qquad 3.51$$

which is in agreement with experimental data summarized in Table 3.6 as well as with Example 3.5.

EXAMPLE 3.5

In accordance with the noise in amplifiers and mixers discussed in Chapter 4, we find that the white current noise is generated in

Table 3.5 h_i-Constants as a function of the phase noise constants

h_{-2}	h_{-1}	h_0	h_1	h_2
$a_{R2}/(2Q)^2$	$(a_{R1}+a_E)/(2Q)^2$	$2kT(1+F)/(2Q)^2P_{osc}$	a_E/f_o^2	$2kT(1+F)/(f_o^2P_{osc})$

the biasing circuit (cf. 4.87), in the switching transistors (cf. 4.92), and in the LC output circuit:

$$\frac{<i_n^2>}{B_W}=4kT\gamma g_{m3}+\frac{16kT\gamma}{\pi}\cdot\frac{I_B}{V_{LO}}+\frac{4kT}{R_L} \quad 3.52$$

The corresponding PSD of the white phase noise is

$$S_\phi(f)=\left[4kT\gamma g_{m3}+\frac{16kT\gamma}{\pi}\cdot\frac{I_B}{V_{LO}}+\frac{4kT}{R_L}\right]\frac{R_L^2}{V_{LC}^2/2} \quad 3.53$$

By approximating in (4.37), the current I_D with $I_B \approx V_{DE}/R_L \approx 2 \times (1/R_L) \approx 2 \times (1/Q^2R_S)$ (A) and by putting $V_{GS}-V_t \approx 0.2$ (V), we obtain for the transconductance of the biasing stage

$$g_{m3}\approx\frac{2I_B}{V_{GS}-V_T}\approx\frac{20}{R_L}>\frac{8}{\pi R_L}>\frac{1}{R_L} \quad 3.54$$

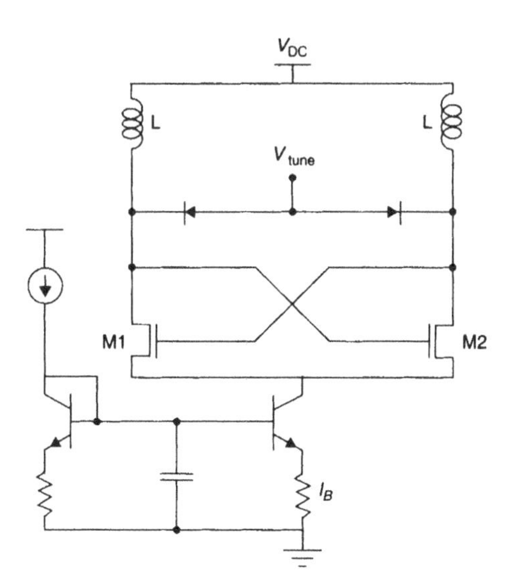

Fig. 3.32 Basic block diagram of the LC microwave oscillator in the differential arrangement (NMOS topology [3.36]). (Copyright © IEEE. Reproduced with permission.)

Table 3.6 Noise properties of several recent LC microwave VCOs

f_o (GHz)	Q	h_{-1} (dB)	h_o (dB)	FOM (dB)	$a_{R1}+a_E$ (dB)	a_o (dB)	Reference	log (h_{-1}/h_{-2})
1.2	13	−150	−202	192	−121	−179	[3.37]	5.2
1	10	−146	−195	183	−119	−175	[3.38]	4.9
1.3	4.5	−133	−186	175	−111	−173	[3.39]	5.3
1.57	20	−154	−202	187	−105	−176	[3.40]	4.8
1.74	4	−137	−185	174	−119	−173	[3.41]	4.8
1.8	10	−140	−195	187	−114	−159	[3.49]	5.5
2	11	−155	−196	180	−128	−175	[3.42]	4.1
2.27	6	−145	−197	184	−124	−181	[3.43]	5.2
3.98	13	−127	−187	185	−99	−169	[3.44]	6.0
4.9	4	−139	−189	168	−121	−183	[3.45]	5.0
4.88	9	−146	−199	185	−121	−180	[3.46]	5.3
5.35	7	−137	−191	182	−114	−175	[3.47]	5.4
5.32	10	−145	−200	188	−119	−180	[3.48]	5.5
4.5	11.5	−143	−195	185	−115		[3.50]	5.2
4.6	12	−149	−193	181	−121	−166	[3.51]	5.4
9.93	20	−165	−202	190	−133	−168	[3.52]	3.7
15.2	30	−128	−184	179	−92	−148	[3.53]	5.6
24.2	6.5	−124	−183	174	−111	−160	[3.54]	5.9
		142.4±10.5	193.4±6.5	182±6.3	−116±.9.9	−171.7±9		4.8±0.12
265	20	−124	−188	182	−104	−156	[3.68]	6.4

Consequently, we realize that g_{m3} is the largest contributor to the white noise coefficient h_o in (3.52). By assuming further that in LC microwave oscillators with Q for 5–10, the series resistance is close to $R_S \approx 1$ [Ω] ($R_L = 25$–100 Ω) and we get for the noise coefficient h_o

$$h_o \approx 4kT\gamma \frac{20}{R_L} \frac{R_L^2}{4Q^2} \cdot \frac{2}{V_{LC}^2} \approx 4kT\gamma \cdot 5R_S \cdot 2 \approx 10^{-19.2} \qquad 3.55$$

3.2.3.2 Flicker Noise Constant

Reverting to the relations (3.4) and (3.5), we get for the flicker noise constant h_{-1}

$$h_{-1} = \frac{a_{RI} + a_E}{(2Q)^2} = \frac{(Qf_o)^{-1} + a_E}{(2Q)^2} \qquad 3.56$$

In Example 3.6, we estimate its size:

EXAMPLE 3.6

Let us choose $f_o \approx 5$ GHz, $Q \approx 10$, $a_E \approx 10^{-12}$ and apply (1.111), then

$$h_{-1} \approx \frac{10^{-9.7-1} + 10^{-12}}{10^{2.6}} \approx \frac{10^{-10.7}}{10^{2.6}} = 10^{-13.3} \qquad 3.57$$

In the past 10 years, noise properties of differential microwave LC oscillators were investigated by many authors from various points of view. The published experimental data are fairly consistent and prove the randomly chosen data in Table 3.6 for the range from 1 to 25 GHz. In addition, analysis reveals some important facts. From the available data, we have computed the noise coefficients h_{-1} with the mean value,

$$h_{-1} \approx -142 \pm 10 \qquad \text{(dB)} \qquad 3.58$$

which is about one order smaller compared with the estimation in Example 3.6. This discrepancy may be explained with the rather crude estimation of (3.57), with low-frequency feedback (introduced by the current source in the differential arrangement in Fig. 3.32, which suppresses the flicker noise component generated in the LC resonator itself [3.37]), and other not yet specified processes. Note that the mean intersection frequency, $f_c \approx h_{-1}/h_{-2}$, recalled in the ninth column in the Table 3.6 is ~ 100 kHz .

3.2.3.3 Random Walk Noise Constant

We will estimate the value of the random walk noise constant with the assistance of (3.5) and the discussion in Section 2.1.3:

$$h_{-2} \approx \frac{1}{(2 \times 10)^2} \frac{10^{-20.3}}{10^{-4\pm1}} \approx 10^{-19\pm1} \qquad 3.59$$

3.2.3.4 Expected Output Phase Noise of the Microwave LC Oscillators

By summarizing all of the above results, we may expect the following relation for the output phase noise for differential LC microwave oscillators:

$$S_\phi(f)=\left(\frac{f_o}{f}\right)^2 S_y(f)=\left(\frac{f_o}{f}\right)^2\left[\frac{h_{-1}}{f}+h_o+h_1f+h_2f^2\right] \quad 3.60$$

Next, we examine the phase noise parameters,

$$S_\phi(f)\approx\frac{a_{-3}}{f^3}+\frac{a_{-2}}{f^2}+\frac{a_{-1}}{f}+a_o=\left(\frac{f_o}{f}\right)^2\left[\frac{h_{-1}}{f}+h_o+h_1f+h_2f^2\right] \quad 3.61$$

with the following numerical mean values for VCO in the gigahertz ranges $f_o \approx 3$ GHz:

$$S_\phi(f)=10^{17}\cdot\left[\frac{10^{-19\pm1}}{f^4}+\frac{10^{-14\pm1}}{f^3}+\frac{10^{-19\pm1}}{f}\right]+\frac{10^{-12\pm1}}{f}+10^{-15\pm1}=$$
$$\frac{10^{-2\pm1}}{f^4}+\frac{10^{3\pm1}}{f^3}+\frac{10^{-2\pm1}}{f^2}+\frac{10^{-12\pm1}}{f}+10^{-15\pm1} \quad 3.62$$

A typical noise characteristic of LC microwave oscillators close to the carrier is shown in Figure 3.33.

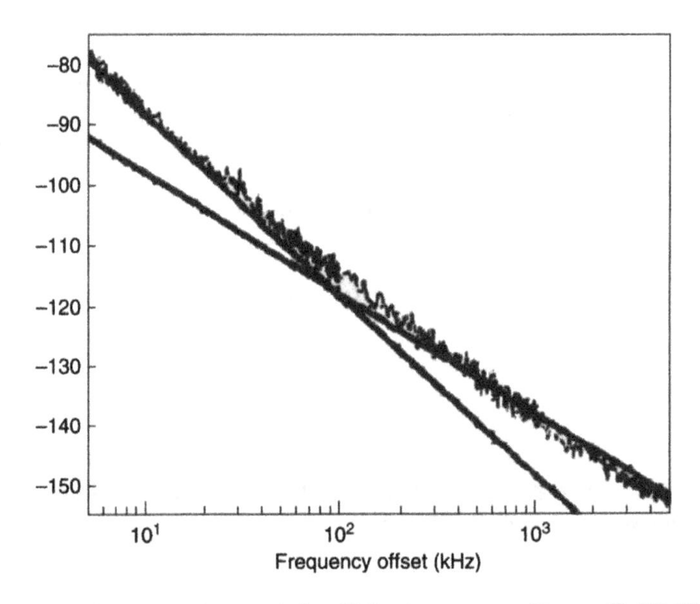

Fig. 3.33 A typical noise characteristic of LC microwave oscillators. [3.40] (Copyright © IEEE. Reproduced with permission.)

3.2.4 Varactor Tuning of Microwave LC Oscillators

By assuming the LC resonant circuit with the capacity fully or at least partially formed by a varactor, we get for the change of the oscillator frequency due to the varactor tuning voltage

$$\Delta\omega \approx -\frac{\omega_o}{2}\cdot\frac{1}{C_o}\cdot\Delta C \approx \frac{\omega_o}{2}\cdot\frac{1}{C_o}\cdot\frac{\delta C}{\delta V}\cdot\Delta v \qquad 3.63$$

With the assistance of the relation for the varactor capacity

$$C_v = C_{(v=0)}\left(\frac{\Phi}{v+\Phi}\right)^{\gamma_{var}} \qquad 3.64$$

and its derivation with respect to the voltage

$$\frac{dC_v}{dv} = C_{(v=0)}\left(\frac{\Phi}{v+\Phi}\right)^{\gamma_{var}}(v+\Phi)^{-1}(-\gamma_{var})dv \qquad 3.65$$

we get for the frequency change

$$\Delta\omega \approx -\frac{\omega_o}{2}(v+\Phi)^{-1}(-\gamma_{var})dv \approx K_o\Delta v \qquad 3.66$$

where we have introduced the VCO gain due to the vararctor voltage, that is, K_o (Hz/V). Evaluation of (3.66) reveals for an approximate value of K_o [cf. 3.55]

$$K_o \approx \omega_o\frac{1}{V_{DD}/2}\cdot\frac{1}{3} \approx 4\frac{f_o}{V_{DD}} \approx 2f_o \qquad [V_{DD}=2(V)] \qquad 3.67$$

Note that, $K_o \approx f_o$. Spurious voltages that may cause modulation of varactors in LC oscillating circuits are fluctuations of the DC voltage, parasitic couplings generated on the chip itself, vibrations, and so on. In addition, we must consider the influence of the noise current generated in the biasing stage (cf. Fig. 3.32) and, eventually, the noise voltages generated in the resonant circuit itself, which are of the $1/f$ and white noise type. The corresponding fractional frequency is

$$<\left(\frac{\Delta\omega}{\omega_o}\right)^2> \approx \left(\frac{\gamma_{var}}{v+\Phi}\right)^2 <e_n>^2 \qquad 3.68$$

By reverting to the results in Section 3.2.3, we expect the PSD of the additive fractional frequency to be

$$S_{y,\text{add}}(f) \approx \frac{1}{\gamma_{\text{var}}^2}\left(\frac{h_{-1}}{f} + h_o\right) \qquad 3.69$$

[see (3.55 and 3.57)]. A further reduction of the noise generated in the varactor may be achieved by combination with of the circuit capacity (see Fig. 3.34).

3.2.5 Figure of Merit of Microwave Oscillators

For evaluation of the quality of the microwave VCOs, the fractional frequency noise at the Fourier frequency, $\Delta f = 600$ kHz, the overall input power is taken into account. Its definition, used by many authors, is

$$\text{FOM} = 10 \times \log\left[\left(\frac{f_o}{\Delta f}\right)^2 \frac{10^{-3}}{\mathcal{L}(\Delta f) P_{\text{DC}}}\right] \qquad 3.70$$

and called figure of merit (FOM) with PSD (Δf) in the side band, $\Delta f =$ 600 kHz. The P_{DC} parameter is the DC input power in W. From (3.62), we conclude that $\mathcal{L}(\Delta f)$ generally would be in the white frequency noise range. In that case, we can simplify the above relation, with the assistance of (3.4), into

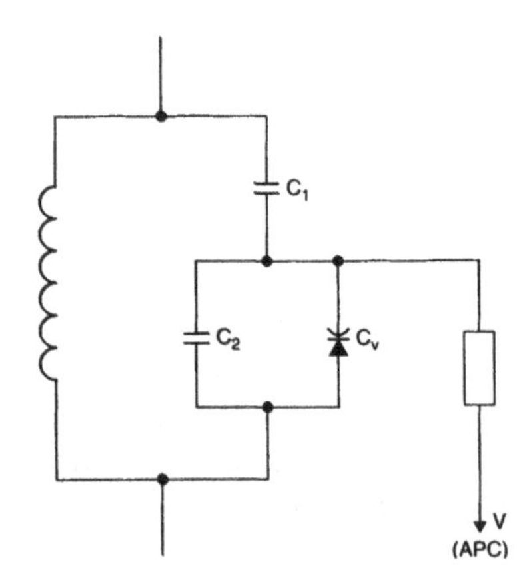

Fig. 3.34 A general arrangement of the LC resonant circuit with varactor tuning.

$$\mathrm{FOM} = 10 \times \log\left[\frac{2\times10^{-3}}{h_o P_{\mathrm{DC}}}\right] \qquad 3.71$$

Finally, with the above relation and values introduced in Example 3.5, we arrive at a numerical estimate (V_{DD} =2 V, I_B = 15 mA)

$$\mathrm{FOM} = 192 + 3 - 10\log(V_{DD} I_B) \approx 182 \qquad 3.72$$

Referring to Table 3.6, where we compare several recently published results, we find for the mean value FOM = 182. On the other hand, recent evaluation presented by Linten et al. [3.53] provides a bit larger value, namely, FOM = 186.7 or even FOM = 190 [3.66, 3.67].

3.3 PRACTICAL RC OSCILLATORS

Another type of integrated microwave generators provide simple RC oscillators either in the relaxation or the time delay form (ring oscillators). The advantage is a simple and straightforward integration with a much smaller area required on the IC chip compared to LC oscillators. However, this leads to increased phase noise, due to the nonexistence of the storing element (circuit Q) [cf. 3.56], which in some instances causes limitations of applications in various modern microwave communications systems.

3.3.1 Relaxation Oscillators

Earlier RC oscillators were based on the relaxation circuits mostly on the emitter-coupled multivibrator with a floating timing capacitor. The major drawback is a poor frequency stability (~ 1%), a rather low frequency range from 1 to 100 kHz, and, occasionally, to several megahertz (e.g. [3.68]), and rather large output noise.

A schematic design is reproduced in Fig. 3.35(a). The switching process is due to charging and discharging the capacity C with the assistance of the resistor R. Consequently, the output wave form has the shape of two e-functions with the period $T_o = 2\tau = RC$ [Fig. 3.35(b)].

To estimate the output phase noise, we must first evaluate the time jitter due to uncertainties of individual voltage crossings. Similarly, as with the noise in mixers (4.81) and dividers, we write

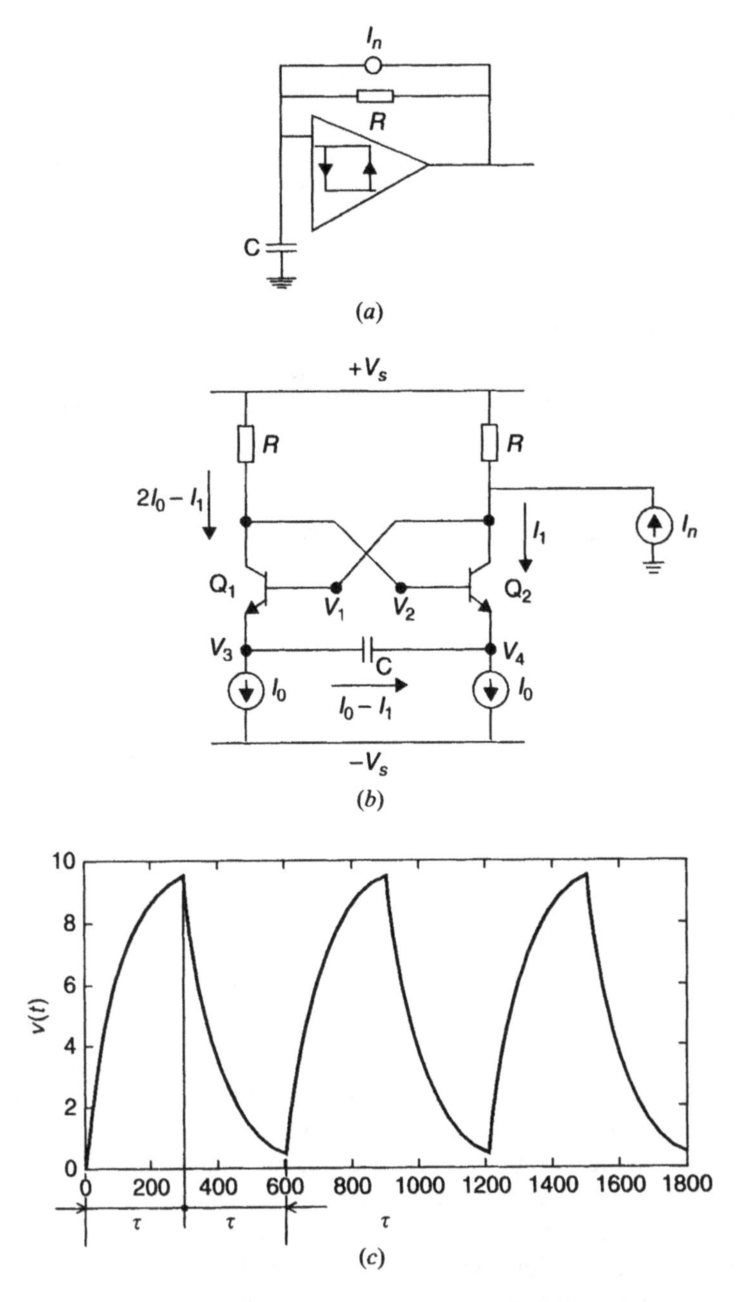

Fig. 3.35 Relaxation oscillator circuit: (*a*) a simplified block diagram of a Schmitt trigger in an RC feedback loop, (*b*) the corresponding schematic design, and (*c*) the output wave form in form of two e-functions with the period $T_o = 2\tau$.

$$\Delta_{T_i} = \frac{v(t_i) - v(iT)}{\dot{v}(iT)} \qquad 3.73$$

Here the numerator presents the spurious noise voltage at the instants of the level crossings, whereas the denominator is the corresponding voltage slope at the moments of mutual intersection. By taking into account that the reference level is in the middle of the output wave, we may approximate, without any appreciable error, the slope of the tangent as

$$\frac{dv(t)}{dt} = \frac{V_{pp}}{T_o / 2} \qquad \left(f_o = \frac{1}{T_o} \right) \qquad 3.74$$

and the corresponding time uncertainty due to the noise voltage e_n as

$$\Delta T = \frac{e_n}{2V_{pp}} T_o = \frac{e_n}{V_{pp}} \tau \qquad 3.75$$

where τ is the time constant, $\tau = RC$, in the relaxation system. The effective phase noise is easily computed as

$$< \left(\frac{\Delta T}{T_o} \right)^2 > = < \left(\frac{e_n}{2V_{pp}} \right)^2 > = S_{\phi,V}(f) B_W = S_y(f) B_W \qquad 3.76$$

However, the output phase noise of an oscillator is subjected to integration (cf. 3.1 and 3.4). Consequently, we get

$$S_{\phi,\mathrm{osc}} = S_{\phi,V} \left(\frac{f_o}{f} \right)^2 \qquad 3.77$$

where the noise voltage is approximately

$$S_{\phi,V}(f) \approx \frac{4kTR}{(2V_{pp})^2} \approx \frac{2kT}{P_o} \qquad 3.78$$

EXAMPLE 3.7

By putting $V_{pp} = 1$ (V) and $V_{pp}/R = 0.025$ (mA) we get for the PSD of the white fractional frequency noise

$$S_{y,\mathrm{RC}} \approx \frac{10^{-20.1}}{1 \times 25 \times 10^{-6} / 2} \approx 10^{-15} = h_o \qquad 3.79$$

By considering the intersection frequency to be $\sim f_c \times 10^5$ Hz, as in Example 3.6, we may estimate the value of the $1/f$ factional frequency to be

$$S_{y,\mathrm{RC}}(f) \approx \frac{10^{-10}}{f} = \frac{h_{-1}}{f} \qquad 3.80$$

and, finally, we arrive at the expected output phase noise of the regenerative oscillator

$$S_\phi(f) = \left(\frac{f_o}{f}\right)^2 S_y(f) \approx f_o^2\left(\frac{10^{-10}}{f^3} + \frac{10^{-15}}{f^2}\right) + S_{\phi,\mathrm{add}} \qquad 3.81$$

This crude noise evaluation is in a good agreement with much more profound calculations performed, for example, by Abidi and Meyer [3.56] or by Nizhnik et al. [3.57]. Note that $Q_{\mathrm{eff}} = 1$.

3.3.2 Ring Oscillators

Another type of *RC* oscillators, particularly in microwave ranges, brings about the *ring oscillator system* built up by a chain of invertors. The corresponding block diagram is depicted in Fig. 3.36. In contradistinction to the above discussed relaxation oscillators, the maximum carrier frequency can be much higher and tuning ranges can be larger. In addition, these circuits are used to build up high-speed blocks in digital and optical communications systems, due to their simplicity, compact size, and ease of integration, even in tens of gigahertz wave ranges.

A ring oscillator is formed by a loop of an odd number, N, of inverting amplifiers with approximately the same gains $A(j\omega)$ (see Fig. 3.37). This feedback system will oscillate whenever the overall gain $G(s)$ is equal to -1. This condition expressed in complex form is

$$|G(j\omega)|\, e^{j\psi} = -1 \qquad 3.82$$

that is, for a chain of N stages,

$$|G(j\omega)| = |A(j\omega)|^N \quad \text{and} \quad N\psi = (2k+1)\pi \;\; (k = \pm 1, \pm 2, \ldots) \qquad 3.83$$

From (3.83) one concludes that generally N should be odd. The oscillating frequency is given by charging and discharging times:

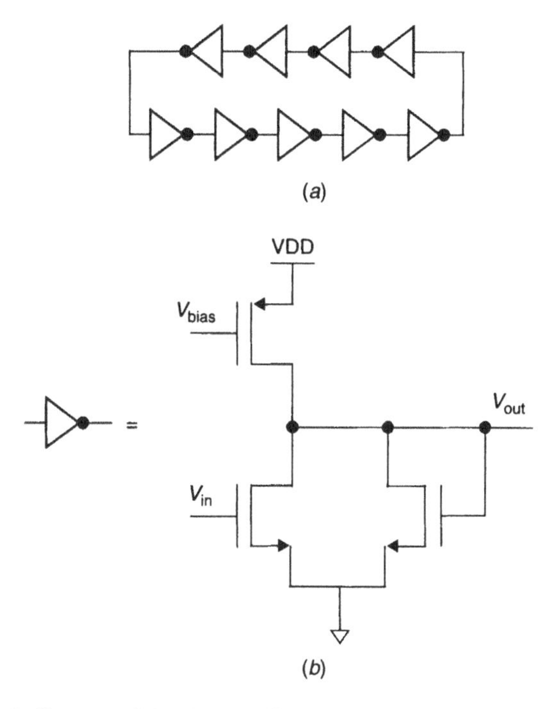

Fig. 3.36 Block diagram of the ring oscillator. (*a*) The corresponding chain of inverters and (*b*) the inverting amplifier [3.56]. (Copyright © IEEE. Reproduced with permission.)

$$f_o = \frac{1}{N(\tau_{ch} + \tau_{dis,ch})} \approx \frac{1}{2N\tau} \quad 3.84$$

where τ is an approximate time delay generated by charging and discharging the steering capacity (in Fig. 3.36) formed by pnp field-effect transistors (FET) and N is the corresponding number of invertors.

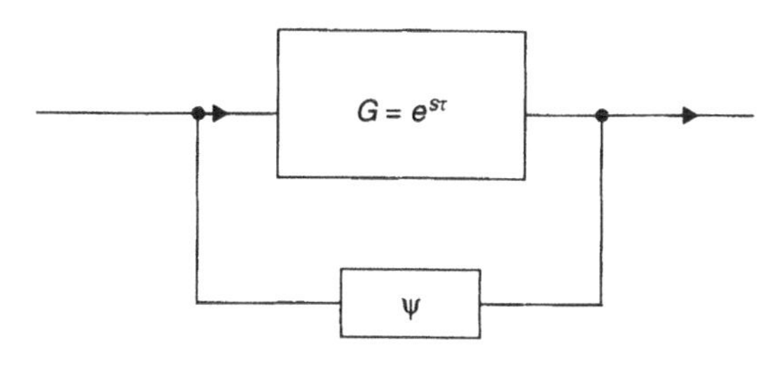

Fig. 3.37.

3.3.2.1 Noise in Ring Oscillators

To evaluate the expected time jitter of the $2N$ voltage crossings (with application of the central limit theorem) we start with (3.76):

$$< \Delta T^2 > \approx 2N\left\langle \frac{e_n^2}{V_{pp}^2} \right\rangle T_o^2 \qquad 3.85$$

The PSD of the corresponding fractional frequency noise is

$$<\left(\frac{\Delta T}{T_o}\right)^2 > \approx \frac{1}{2N}\left\langle \frac{e_n^2}{V_o^2} \right\rangle \qquad 3.86$$

which is increased by $(f_o/f)^2$ in the case in which the system oscillates:

$$S_{\phi,\mathrm{osc}}(f) \approx \left(\frac{f_o}{f}\right)^2 \frac{1}{2N} \cdot \frac{e_n^2}{V_o^2} \approx \left(\frac{f_o}{f}\right)^2 \frac{4kTR}{2NV_o^2} \approx$$
$$\left(\frac{f_o}{f}\right)^2 \frac{2kT}{P_{\mathrm{osc}}} \approx \left(\frac{f_o}{f}\right)^2 \frac{kT}{NP_{\mathrm{invertor}}} \approx \left(\frac{f_o}{f}\right)^2 h_o \qquad 3.87$$

By assuming the smallest number of invertors in the chain, that is, $N = 3$, we get for the power of individual inverters, $P_{\mathrm{invertor}} \approx 0.1$ mW,

$$h_o \approx \frac{2kT}{2NP_{\mathrm{inverter}}} \approx \frac{kT}{NP_{\mathrm{inverter}}} \approx \frac{10^{-20.4}}{10^{-4}} \approx 10^{-16.4} \qquad 3.88$$

which is in good agreement with practical values summarized in the third column in Table 3.7. The flicker frequency noise will be introduced by the electronics noise current and its value may be expected to be the same as in (3.80). In Table 3.7, second column, only a few h_{-1} values are mentioned since time jitter below ~ 10 kHz is generally not referred to.

3.3.2.2 Figure of Merit of Ring Oscillators

Computation of the FOM follows from (3.70) and (3.71):

$$\mathrm{FOM} = 10 \times \log\left(\frac{2\times 10^{-3}}{h_o P_{\mathrm{DC}}}\right) = 10 \times \log\left(\frac{2\times 10^{-3} \times NP_{\mathrm{inverter}}}{kT \times P_{\mathrm{DC}}}\right) \qquad 3.89$$

By introducing $P_{\text{invertor}} = 0.1$ (mW) and $P_{\text{DC}} = 2*0.02$ (W) we have for $N = 3$

$$\text{FOM} \approx 10 \times \log\left(\frac{2 \times 3 \times 10^{-3-4}}{kT \times 0.04}\right) \approx 156 \;\; (\text{dB}) \qquad 3.90$$

which is again in good agreement with practical values summarized in Table 3.7, fifth column, but compared to LC microwave LC oscillators it is ~ 25 dB smaller. There are two reasons for the lower value: the absence of the resonator Q and nearly *2N-times* larger input power.

EXAMPLE 3.8

Comparison of FOM of the LC and ring microwave oscillators:

1. Reduction due to the absence of Q — $20 \log (2 \times Q)$
2. Increase of the input power — $10 \log (2N)$

For $Q \approx 5$ and $N = 3$, the difference is $-20 \log (2 \times 5) + 10 \log (6) \approx 28$ dB

For an LC oscillator with $Q \approx 5$ we get from (3.70) $\text{FOM}_{\text{LC}} \approx 181$ dB

The actual difference due to (3.88) is 181 – 156 = 25 dB

Table 3.7 Properties of several ring oscillators

f_o (GHz)	h_{-1}	h_o	h_2	FOM	K_o (MHz/V)	References
0.106/9	–111	–158				[3.58]
0.261/5	–128	–175				[3.58]
0.232	–116	–166			220	[3.59]
0.838/4	–100	–160	Fig.13		615	[3.60]
0.3908/4		–160		140	280	[3.61]
0.475/3	–112	–167.5	Fig. 17c	154	290	[3.62]
2		–170				[3.63]
5.8/3		–175				[3.64]
1.81/9		–175				[3.64]
3.87/9	–106	–160			Sun 05	[3.60]
1.33/3		–152	Fig. 11			[3.60]
11.5/2		–169		150	16%	[3.65]
32/1		–175		156		[3.66]
	–112±9	–168±8.5				

A similar calculation was recently provided by Abidi (3.56) who found the ring oscillator is not suitable for use in integrated devices.

REFERENCES

3.1. V.F. Kroupa, Theory of 1/*f* Noise: A New Approach, *Phys. Lett. A, 336*(2–3) (2005), p. 126,

3.2. F.L. Walls and J. Vig, Fundamental Limits on the Frequency Stability of Crystal Oscillators, *IEEE Tr. UFFC, 42*(4), (July. 1995), pp. 576–589.

3.3. Ch.A. Adams, J.A. Kusters, and C.M. Sousa, Improved Quality Factor for Quartz Oscillators, in *Proceedings of the 1996 IEEE International Frequency Control Symposium,* pp. 699–705.

3.4. S. Galliou, M. Mourrey, F. Marionnet, R.J. Besson, and P. Guillemot, An Oscillator for Space, in *Proceedings of the 2003 IEEE International Frequency Control Symposium,* pp. 430–434.

3.5. P-Y. Bourgeois, Y. Kersale, N. Bazin, M.Chaubet, and V. Giordano, A Cryogenic Open Cavity Sapphire Reference Oscillator with Low Spurious Mode Density, *IEEE Tr. UFFC, 51*(10) (Oct., 2004), pp. 1232–1239.

3.6. S.A. Vitusevich, K. Schieber, I.S. Ghosh, N. Klein, and M. Spinnler, Design and Characterisation of an All-Cryogenic Low Phase-Noise Sapphire K-Band Oscillator For Satellite Communications, *IEEE Tr., MTT-51* (Jan., 2003), pp. 163–169.

3.7. C.A. Flory and H.L. Ko, Microwave Oscillators Incorporating High Performance Distributed Bragg Reflector Microwave Resonators, in *Proceedings of the 1997 IEEE International Frequency Control Symposium,* pp. 994–999.

3.8. V.F. Kroupa *Phase Lock Loops and Frequency Synthesis,* Hoboken, NJ: Wiley, (2003).

3.9. A.S. Gupta, D.A. Howe, C. Nelson, A. Hati, F.L. Walls, and JF. Nava, High Spectral Purity Microwave Oscillator: Design Using Conventional Air-Dielectric Cavity, *IEEE Tr. UFFC, 51*(10) (Oct. 2004), pp 1225–1230.

3.10. J.G. Ondriar, A Microwave System for Measurement of AM and FM Noise Spectra, *IEEE Tr., MTT-16* (Sept. 1968), pp. 767–781.

3.12. E.I. Ivanov and M.E. Tabor, Application of Interferometric Signal Processing to Phase-Noise Reduction in Microwave Oscillators, *IEEE Tr., MTT-46*(10), (1998), pp. 1537.

3.12. R.V. Pound, Electronic Frequency Stabilisation of Microwave Oscillators, Rev. Sci. Instr. 17(11), (Nov. 1946), pp. 490–503.

3.13. G.J. Dick, Pound Circuit-Induced Frequency Sensitivities in Ultra-Stable Cryogenic Oscillators, in *Proceedings of the 2004 IEEE International Frequency Control Symposium,* pp. 3224–8.

3.14. R. Basu, R.T Wang, and J. Dick, Novel Design of an All-Cryogenic RF Pound Circuit, in *Proceedings of the 2005 IEEE International Frequency Control Symposium,* pp. 562–5688.

3.15. A.N. Luiten, A.G. Mann, and D.G. Blair, Cryogenic Sapphire Microwave Resonator-Oscillator with Exceptional Stability, *Electron. Lett., 30*(5), (March 1994), pp. 417–419.

3.16. G. Robichon, J. Groslambert, and J.J. Gagnepain, Frequency Stability of Quartz Crystal at Very Low Temperatures: Preliminary Results, in *Proceedings of 38th Frequency Control Symposium,* pp. 201–205 (1984).

3.17. L.E. Halliburton and D.R. Koehler, Properties of Piezoelectric Materials, in *Precision Frequency Control*, Academic Press, Orlando, eds. E.A. Gerber and A. Ballato 1985, pp. 1–45.

3.18. J. Krupka and J. Mazierska, ASingle Crystal Dielectric Resonators for Low-Temperature Electronics Applications, *IEEE Tr., MTT-48* (July, 2000), pp. 1270–1273.

3.19. G.J. Dick and D.M. Dtrayer, Measurement and Analysis of Cryogenic Sapphire Dielectric Resonators and DRO's, in *Proceedings of 41st Frequency Control Symposium* (1987), pp. 87–89.

3.20. A.N. Luiten, Frequency Measurement and Control: Advanced Techniques and Future Trends, Berlin–Heidelberg: Springer (2001).

3.21. G.J. Dick and R.T. Wang, Stability and Phase Noise Tests of Two Cryo-Cooled Sapphire Oscillators, *IEEE Tr. UFFC, 47*(5), (Sept., 2000), pp. 1098–1100.

3.22. P-Y. Bourgeois, Y. Kersale, N. Bazin, M.Chaubet, and V. Giordano, Cryogenic Opened Cavity Sapphire Resonator for Ultra-Stable Oscillator, *El. Lett., 15* (May 2003), *39*(10), pp. 780–781.

3.23. CR. Locke, S. Munro, ME. Tabor, EN. Ivanov, and G. Santarelli, Constructing the Next Generation Cryogenic Sapphire Oscillator, in *Proceedings of the 2003 IEEE International Frequency Control Symposium,* pp. 350–354.

3.24. P.Y. Bourgeois, Y. Kersale, N. Bazin, J.G. Harnet, M. Chaubet, and V. Giordano, Progress in the Building Of Sapphire Helium Clock at LPMO Ultra Low Drift Cryogenic Sapphire Microwave Oscillator, in *Proceedings of the 2003 IEEE International Frequency Control Symposium,* pp. 355–359.

3.25. R.T. Wang and G.J. Dick, High Stability 40 Kelvin Cryo-Cooled Sapphire Oscillator, in *Proceedings of the 2003 IEEE International Frequency Control Symposium,* pp. 371–375.

3.26. J.D. Anstie, J.G. Hartnett, M.E. Tabor, E.N. Ivanov, and P.L. Stanwix, Evaluation of UWA Solid Nitrogen Dual Mode Sapphire Oscillator Julia, in *Proceedings of the 2004 IEEE International Frequency Control Symposium,* p. 3869.

3.27. C. McNeilage, J.H. Searls, M.E. Tabor, P.R. Stockwell, D.M. Green, and M. Mossammaparast, A Review of Sapphire Whispering Gallery-Mode Oscillators Including Technical Progress and Future Potential of the Technology, in *Proceedings of the 2004 IEEE International Frequency Control Symposium,* pp. 210–218.

2.28. J.G. Hartnett. CR. Locke, E.N. Ivanov, M.E. Tabor, and P.L. Santwix, Cryogenic Sapphire Oscillator with Exceptionally High Long-Term Frequency Stability, in *Proceedings of the 2007 IEEE International Frequency Control Symposium,* pp. 1028–1031.

3.30. X.S. Yao and L. Maleki, Multiloop Optoelectronic Oscilator, *IEEE J. Quantum Electronics, 36* (Jan. 2000), p. 79–84

3.31. D. Eliyahu, K. Sariri, M. Kamran, and M. Tokhmakhian, Improving Short and Long Term Stability of the Opto-Elctronic Oscillator, in *Proceedings of the 2002 IEEE International Frequency Control Symposium,* pp. 580–583.

3.32. S. Roemisch, J. Kitching, E. Ferre-Pikal, L. Hollberg, and F.L. Walls, Performance Evaluation of an Optoelectronic Oscillator, *IEEE Tr. UFFC, 47*(5), (Sept. 2000), pp. 1159–1165.

3.33. S. Huang, L. Maleki, and T. Lee, A 10 GHz Optoelectronic Oscillator With Continuous Frequency Tunability And Low Phase Noise, in *Proceedings of the 2001 IEEE International Frequency Control Symposium,* pp. 720–727.

3.34. D. Eliyahu and L. Maleki, Low Phase Noise and Spurious Level in Multi-Loop Opto-Electronic Oscillators, in *Proceedings of the 2003 IEEE International Frequency Control Symposium,* pp. 405–410.

3.35. W.Zhou and G. Blasche, Injection-Locked Dual Opto-Electronic Oscillator with Ultra-Low Phase Noise and Ultra-Low Spurious Level, *IEEE Tr., MTT-53* (March, 2005), pp. 929–933.

3.36. J. Craninckx and M.J.S. Steyaert, A 1.8-GHz Low-Phase-Noise CMOS VCO Using Optimized Hollow Spiral Inductors, *IEEE J. Solid-State Circuits, 32* (May 1997), pp. 736–744.

3.37. E. Hegazi et al., A Filtering Technique to Lower LC Oscillator Phase Noise, *IEEE J. Solid-State Circuits, 36* (Dec. 2001), pp. 1921–1930.

3.38. E.Ch. Park et al., Fully Integrated Low Phase-Noise VCO's with on Chip MEMS Inductors, *IEEE Tr. MTT, 51* (Jan. 2003), pp. 289–296.

3.39. F. Svelto, S. Deantoni, and R. Castello, A 1.3 GHz Low-Phase Noise Fully Tunable CMOS LC VCO, *IEEE J. of Solid-State Circuits, 35* (March. 2000), pp. 356–661.

3.40. P. Vancorenland et. al., A 1.57-GHz Fully Integrated Very Low-Phase-Noise Quadrature VCO, *IEEE J. Solid-State Circuits, 37* (May, 2002), pp. 653–661.

3.41. F. Herzel et al., Phase Noise in a Differential CMOS Voltage-Controlled-Oscillator for RF Applications, *IEEE Tr. on Circuits and Systems II, 47*(1), (Jan. 2000), pp. 11–15.

3.42. B. De Muer, M. Borremans, M. Stayaert, and G. Li Puma, A 2GHz Low-Phase Noise Integrated LC–VCO Set with Flicker-Noise Up-conversion Minimization, *IEEE J. Solid-State Circuits, 35* (July. 2000), pp. 1034–1038.

3.43. P.Andreani et al., On the Phase-Noise and Phase-Error Performance of Multiphase LC CMOS VCOs, *IEEE J. Solid-State Circuits, 39* (Nov. 2004), pp. 1883–1893.

3.44. J. Maget, M. Tieboout, and R. Kraus, MOS Varactors with B and p- Type Gates and their Influence on an LC–VCO in Digital CMOS, *IEEE J. Solid-State Circuits, 38* (July, 2003), pp. 1139–1147.

3.45. J. van der Tang et al., Analysis and Design of an Optimally Coupled 5-GHz Quadrature LC Oscillator, *IEEE J. Solid-State Circuits, 37* (May. 2002), pp. 657–661.

3.46. S.L.L. Gierkink, A Low-Phase-Noise 5 GHz CMOS Quadrature VCO Using Superharmonic Coupling, *IEEE J. Solid-State Circuits, 38*(7), (July 2003), pp. 1148–1153

3.47. Ch-M. Hung, B.A. Floyd, N. Park, and Keneth K.O., Fully Integrated 5.35-Ghz CMOS VCO's and Prescalers, *IEEE Tr. MTT, 49* (Jan. 2001), p. 17.

3.48. A. Jerng, and Ch.G. Sodini., The Impact Of Device Type and Sizing on Phase Noise Mechanisms, *IEEE J. on Solid State Circuits* (Feb. 2005), p.360.
3.49. P. Vaananen, N. Mikkola, and P. Helio, VCO Design with On-Chip Calibration System, *IEEE Tr. MTT, 53* (Oct. 2006), p. 2157.
3.50. Lee S-H. Et al., Low-Phase Noise Hartley Differential CMOS Voltage Controlled Oscillator, *MW Com Let.*, Feb. 2007, p. 145.
3.51. H. Jacobson, S.Gevorgian, M. Mkhtari, C. Hedenas, B. Hansson, T. Lewin, H. Berg, W. Rabe, and A. Schuppen, Low-Phase-Noise Low-Power IC Vcos for 5-8- GHz Wireless Applications, *IEEE Tr. MTT, 48* (Dec. 2000), pp. 2533–1538.
3.52. Maxim, A., Multi-Rate 9.953-12.5-GHz 0.2 μm, SiGe BiCMOS LC Oscillator Using a Resistor-Tuned Varactor and a Supply Pushing Cancellation Circuit, *J. of Solid State Circuits, 41* (April 2006), pp. 918–934.
3.53. D. Linten et al., Low Power Voltage-Controlled Oscillators in 90-Nm CMOS Using High-Quality Thin-film Post Processed Inductors, *IEEE J. Solid-State Circuits, 40* (Sept., 2005), pp. 1922–1930
3.54. A.W.L. Ng, G.C.T., Leung, K-C. Kwok, L.L.K. Leung, and C. Luong, A 1-V 24-GHz 17.5 mV Phase Locked Loop in a 0.18-μm, CMOS Process, *IEEE J. Solid State Circuits, 41* (June, 2006), pp. 1236–1244.
3.55. E. Hegazi and A.A. Abidi, Varactor Characteristics, Oscillator Tuning Curves, and AM–PM Conversion, *IEEE J. Solid-State Circuits, 38* (June, 2003), pp. 1033–1039.
3.56. A.A. Abidi and R.G. Meyer, Noise in Relaxation Oscillators, *IEEE J. Solid-State Circuits, 18* (Dec. 1983), pp. 794–802.
3.57. R. Navaid, T.H. Lee, and R.W. Dutton, Minimum Achievable Phase Noise of RC Oscillators, *IEEE J. Solid-State Circuits, 40* (March, 2005), pp. 630–637.
3.58. Y. Ou, N. Barton, R. Fetche, N. Seshan, T. Fiez, U-K Moon, and K. Mayaram, Phase Noise Simulation and Estimation Methods: A Comparative Study, *IEEE Tr. on Circuits and Systems II, 49*(9) (Sept. 2002), pp. 635–638.
3.59. A. Hajimiri et al. A General Theory of Phase Noise in Electrical Oscillators, *IEEE J. Solid-State Circuits, 33*(2) (February 1998), pp. 179–194.
3.60. AQ.A. Abidi, Phase Noise and Jitter in CMOS Ring Oscillators, *IEEE J. Solid-State Circuits, 41* (Aug., 2006), pp. 1803–1816.
3.61. Seog-Jun Lee, et al., A Fully Integrated Low-Noise 1-GHz Frequency Synthesizer Design for Mobile Communication Application, *IEEE J. Solid-State Circuits, 32*(5) (May 1997), pp. 760–765.
3.62. M. Thamsirianunt et al., CMOS VCO's for PLL Frequency Synthesis in GHz Digital Mobile Radio Communications, *IEEE J. Solid-State Circuits, 32* (Oct.. 1997), pp. 1511–1524.
3.63. C. Vaucher and D. Kasperkovitz, A Wide-Band Tuning System for Fully Integrated Satellite Receivers, *IEEE J. Solid-State Circuits, 33*(7) (July 1998), pp. 987–997.
3.64. Y.A. Eken and J.P. Uyemura, A 5.9-Ghz Voltage-Controlled Ring Oscillator in 0.18-μm CMOS, *IEEE J. Solid-State Circuits, 39*(1) (Jan. 2004), pp. 230–233.
3.65. J. van der Tang, van de Ven, D. Kasperkovitz, and A. van Reormund, A 9.8-11.5 GHz Quadrature Ring Oscillator for Optical Receivers, *IEEE J. Solid-State Circuits, 37* (March. 2002), pp. 438–442.

3.66. W-M.L. Kuo J.D. Cressler, Y-J.E. Chen, and A.J. Joseph, An Inductorless K-band SiGe Ring Oscilator, *IEEE MW and Wireless Comp. Lett.* (Oct. 2005), pp. 682–684.

3.67. O. Nizhnik R.L. Pokharel, H. Kanaya, and K. Yoshida, Low Noise Wide Tuning Range Oscillator for Multi-Standard Receiver, *IEEE MW and Wireless Comp. Lett.* (July, 2009), pp. 470–472.

3.68. Y.-S. Tiao, M.-L. Sheu, Full Range Voltage-Controlled Ring Oscillator in 0.18 μm CMOS for Low Voltage Operation, *El. Lett., 7,* Jan. 2010, pp. 30–32.

3.69. B. Razavi, AA 300-Ghz Fundamental Oscillator in 65-nm CMOS Technology, *IEEE J. Solid-State Circuits, 46* (April 2011), pp. 894–903.

CHAPTER 4

Noise of Building Elements

Practically all physical systems are generators of fluctuations around the desired output signal, which are mostly designated as noise. The problem was discussed in many papers and books. Here we mention one, very comprehensive paper, prepared years ago by Gupta [4.1].

4.1 RESISTORS

The thermal noise generated in resistors is white noise and was discussed in Section 1.1.1. However, in many instances we also encounter, at low Fourier frequencies, the flicker noise component (cf. Section 1.5.2.2).

4.1.1 Resistors: 1/*f* Noise

In Chapter 1, we started with the finding by Johnson [4.2] that at very low Fourier frequencies the shot noise in vacuum tubes did not follow the white noise law any more. He introduced for this additive noise the name flicker noise, and the name is still used. Subsequent observations proved its validity for a much larger set of physical phenomena on one hand and its prevalent appearance at very low Fourier frequencies on the other. Some years later, Bernamont [4.3] suggested a law for its PSD:

$$S_n(f) \approx \frac{1}{f^{\alpha}} \qquad 4.1.$$

where the magnitude of α was in the vicinity of one.

Reverting to resistors, Van der Ziel [4.4] states that practically all resistors, including wire wound, semiconductor films, thin metal films, and others, show $1/f$ noise, particularly at very high current densities. This finding is in agreement with our conclusions in Section 1.5.2.2, where we arrived at the hypothesis that $1/f$ fluctuations are caused by periodic losses of energy (cf. 1.108 and Fig. 1.14).

4.2 INDUCTANCES

Let us consider an inductance in series with a resistance R at a temperature T. Estimate a total mean-square noise current $<I^2>$ in the inductor. The classical statistical mechanics state that a system that is in equilibrium (with constant temperature) contains an average energy [e.g., 4.5]

$$P_{\text{av}} = \frac{1}{2} kT \quad \text{(Ws)} \qquad 4.2$$

Consequently, the energy stored in the inductor is (Fig. 4.1*a*)

$$\frac{1}{2} L < I^2 > = \frac{1}{2} kT \quad \text{(Ws)} \qquad 4.3$$

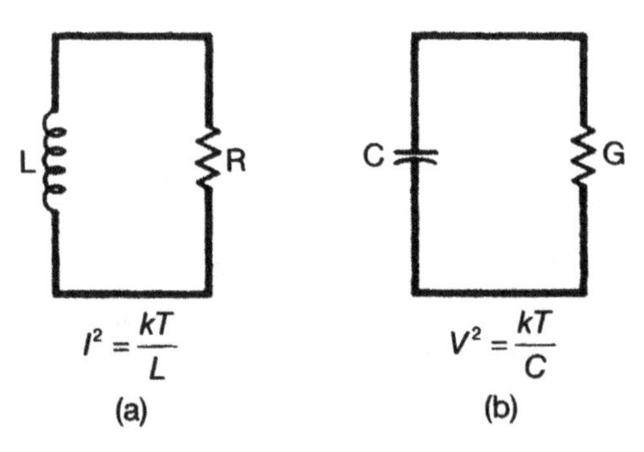

Fig. 4.1 The total Johnson noise current (voltage) squared in an inductance in series with a resistance (*a*) and across a capacitance in shunt with a conductance (*b*) [4.5]. (Copyright © IEEE. Reproduced with permission.)

The corresponding noise current is

$$< I^2 > = \frac{kT}{L} \tag{4.4}$$

This must be true for all values of the series resistance R. Since the noise current is composed of various frequency components, more low-frequency current must flow in instances where R is small, and vice versa, since the total mean-square current should be the same in both cases.

4.2.1 Inductances: White Noise

By considering the current noise

$$i_n = \frac{e_n}{R + j\omega L} \tag{4.5}$$

and by introducing the PSD of the noise power, we get for lower frequencies at room temperature, $T = 300$ (K),

$$S_n(f) = \frac{4hf}{e^{hf/kT} - 1} \approx 4kT \quad \left(f_H \ll \frac{kT}{h} \approx 6.34 \times 10^{13} \right) \tag{4.6}$$

After integration over the range from zero to $f_B < f_H$, we arrive at the well-known equation for the square of the noise voltage:

$$< e_n^2 > = \int_0^{f_B} S_{in}(f) R df = 4kTf_B \cdot R \tag{4.7}$$

and for the noise current

$$< i_n^2 > \approx \frac{4kTR \cdot f_B}{R^2 + (\omega L)^2} \tag{4.8}$$

Note that h is the Planck constant, $h = 6.625 \times 10^{-35}$ (Js), and k is the Boltzmann constant, $k = 1.380 \times 10^{-23}$ (J/K).

4.2.2 Inductance: 1/*f* Noise

To the author's knowledge, the publications about this topic at radio frequency (RF) are scarce. Here, we refer to one of them [4.6]. From this work, we infer that the phase modulation (PM) noise in the air core coils is nearly indistinguishable from the background noise. On the other hand, inductances with ferrite and other windings exhibit the additional $1/f$ noise. They may be explained by losses in the bulk of the material, in accordance with discussions in (1.108) (cf. Fig. 4.2).

4.3 CAPACITANCE

The capacitors' noise property is closely related to their Q factor, which is frequency dependent and dominated at low frequencies by losses due to the surface leakage through parallel resistance R_p. But at high frequencies, the capacitor Q is lowered due to the losses in the dielectric body itself, which is represented by the series resistance R_s (Fig. 4.3*a*). Consequently,

$$Q = \frac{Im(Y)}{Re(Y)} \approx \frac{\omega C R_P}{1 + \omega^2 C^2 R_S^2 + \omega^2 C^2 R_S R_P} \approx \frac{\omega C R_P}{1 + \omega^2 C^2 R_S R_P} \quad 4.9$$

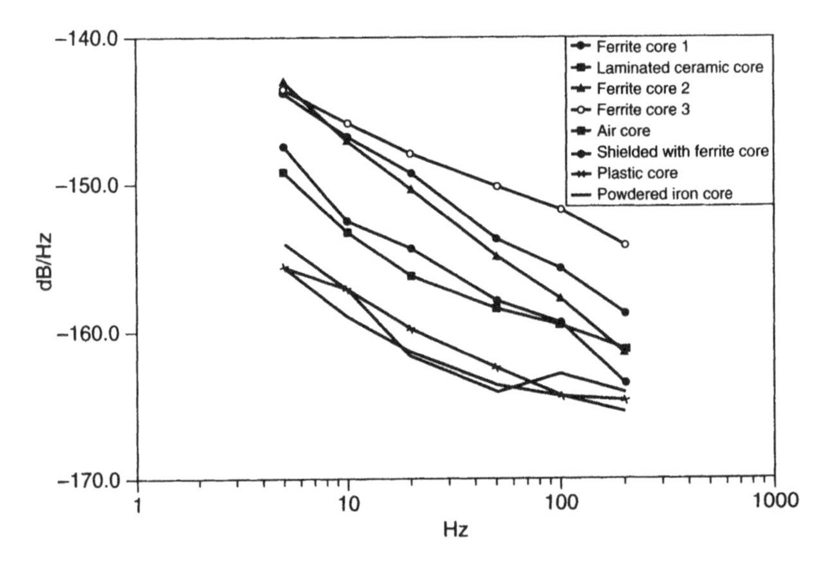

Fig. 4.2 Flicker phase noise at different inductances [4.6]. (Copyright © IEEE. Reproduced with permission.)

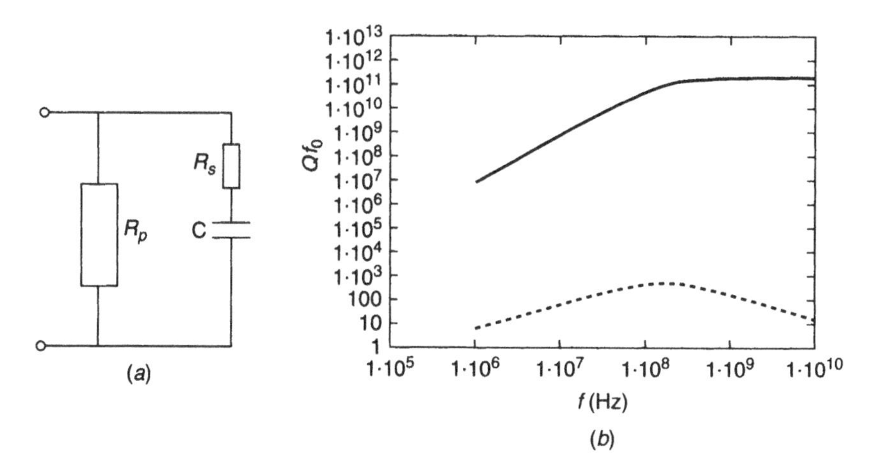

Fig. 4.3 (*a*) Losses in the capacitor: R_p represents losses due to the surface leakage and resistance R_s are losses in the dielectric body itself. (*b*) The corresponding Q-factor (dashed) and the Qf_o relation (solid line).

As an example, we reproduce the Q-factor of a varactor (Fig. 4.3b). Note that the peak of Q is

$$Q_{\max} = \frac{1}{\omega C}\sqrt{\frac{1}{R_P R_S}} \tag{4.10}$$

4.3.1 Capacitance: White Noise

The situation is similar to that in Section 4.2.1. Again, the classical statistical mechanics is satisfied [cf. (4.2)], and the noise energy stored in the capacitor is (cf. Fig. 4.1*b*)

$$\frac{1}{2}C < V^2 > = \frac{1}{2}kT \qquad \text{(Ws)} \tag{4.11}$$

The corresponding noise voltage is

$$< V^2 > = \frac{kT}{C} \tag{4.12}$$

However, for the noise current at lower frequencies we arrive at (4.13) with the assistance of (4.6) and circuit equations

$$<i_n^2> \approx \frac{4kTR \cdot f_B}{R^2 + (\omega C)^{-2}} \tag{4.13}$$

The noise voltage is computed in Example 4.1:

EXAMPLE 4.1

Evaluate the variance of the noise voltage for a capacitive load. With the assistance of Fig. 4.1*b* and (4.13), we have

$$\begin{gathered} <e_n^2> = \int_0^{f_h} \frac{S_\varphi(f)}{R^2 + (\omega C)^{-2}} \frac{1}{(\omega C)^2} df = \\ 4kT \int_0^{f_h} \frac{1}{1 + (\omega \mathrm{RC})^2} df = \frac{2k\mathrm{TR}}{\pi\tau} \int_0^{2\pi f_h \tau} \frac{1}{1 + x^2} dx \end{gathered} \tag{4.14}$$

where $x = \omega\mathrm{RC} = 2\pi f\tau$. After integration, we get for the noise voltage a constant value, that is,

$$<e_{cn}^2> = \frac{2kTR}{\pi\tau} \tan^{-1}(2\pi f_h \tau) \approx \frac{kTR}{\tau} = \frac{kT}{C} \qquad \tau = RC \tag{4.15}$$

4.3.2 Capacitance: 1/*f* Noise

Compared with inductances, the situation was much more often analysed, particularly due to the investigations of the 1/*f* noise in quartz crystal resonators and resonators with other dielectrics where the intrinsic losses with delays forming a memory system are believed to be at the origin of the 1/*f* noise (cf. Chapter 2). Experimental observations revealed that quality factor multiplied by the resonant frequency is a material constant:

$$Q \cdot f_o = \text{const} \tag{4.16}$$

from which the PSD of the 1/*f* noise is [cf. (1.111)]

$$S_\phi(f) = \frac{(Qf_o)^{-1}}{2f} = \frac{tg(\delta)}{f} \tag{4.17}$$

The Qf_o constants for several dielectric materials are summarized in Table 4.1. In addition, $1/f$ noise characteristics of capacitors with different dielectrics are reproduced in Fig. 4.4.

EXAMPLE 4.2

We present a simple evaluation of the Qf_o product using a rather heuristic approach. For carrier frequencies $f_o > f_{o,Q\max}$, the quality factor is from (4.9) approximately

$$Q \approx \frac{1}{\omega_o C R_s} \qquad 4.18$$

In this case, the effective capacity and resistance of the dielectric resonator can be approximated as follows:

$$C = \frac{A}{d}\varepsilon\varepsilon_o \qquad Y = \frac{1}{R_s} = \frac{A}{d\rho} \qquad 4.19$$

where A is the approximate area of the electrodes, d is the mutual distance, ε is the dielectric constant, and ρ is the resistivity. For Qf_o or the inverse loss factor, that is, $tg\ \delta$, we get (4.20) after introducing the above relations into (4.17) and multiplying the result with the carrier frequency

$$Q \cdot f_o = \frac{1}{2\pi \cdot \varepsilon\varepsilon_o \cdot \rho} = \text{const} \qquad 4.20$$

4.4 SEMICONDUCTORS

When discussing frequency stability problems, we must consider properties of semiconductor diodes and transistors that we encounter in amplifiers, oscillators, frequency mixers, and synthesizers, in output

Table 4.1 The Qf_o constants for several dielectric materials

Material	Qf_o	Temperature	Dielectric Constant	Reference
ε_o			$8.859 * 10^{-18}$ [As(Vm)$^{-1}$]	
Quartz	10^{13}		4.4	[3.17]
Sapphire	10^{15}	Room temp	90	[3.28]
Sapphire	10^{19}	10 K	90	[3.28]
Varactors	10^{12}			

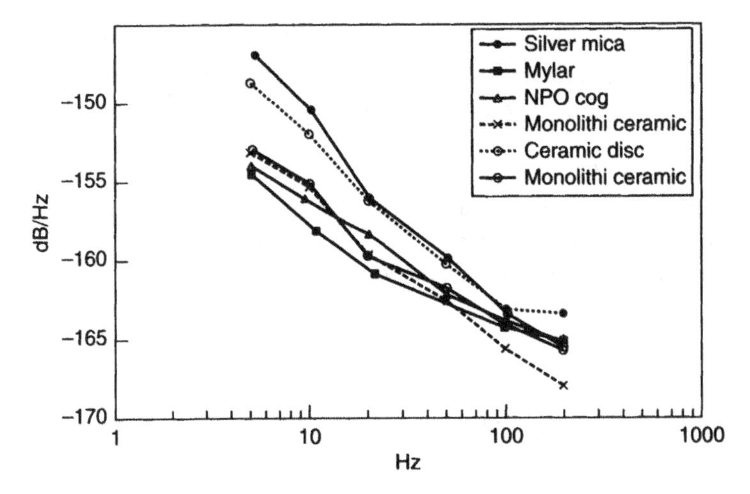

Fig. 4.4 Flicker phase noise in capacitors with different dielectric material ([4.6]). (Copyright © IEEE. Reproduced with permission.)

and distributing circuits, and so on. There are devices with either *n*- or *p*-type conducting channels: bipolar junction transistors (BJT), heterojunction bipolar transistors (HBT), field-effect transistors with junction gates (JFET), metal oxide semiconductors (MOSFET), Schottky barrier gates (MESFET), complementary metal oxide semiconductors (CMOS), modulation-doped (MODFET or HMT) based on GaAs junction, and others [4.4, 4.7].

Semiconductor devices operate as amplifiers or rectifiers with an effective channel between a source and a drain that presents a modulated resistor that is the origin of the thermal noise. At the same time, the current flow is formed by carriers crossing barriers, independently and at random. Consequently, the shot noise is generated. In addition, the ever-present $1/f$ or flicker noise is encountered.

4.4.1 Shot Noise in Diodes

As discussed in Chapter 1, Section 1.4.2.3, the PSD of the shot noise is white. Here, we follow Van der Ziel's reasoning [4.4]. Let us consider an *n-type* Schottky barrier diode with the following characteristics between the current I and the voltage V:

$$I = I_o[e^{\lambda V} - 1] \qquad \lambda = \frac{q}{kT} \approx 40\ (\mathrm{V}^{-1}) \tag{4.21}$$

where λ is a constant relaying i–v semiconductor exponential law characteristics. Note that the current I is composed of a nearly constant backward current I_o and a much larger forward component. The same characteristic holds for the n + –p junctions as well. Since the diode current consists of carriers crossing barriers at random, we have for the PSD of the shot noise current

$$S_I(f) = 2qI_o e^{\lambda V} + 2qI_o = 2q(I + 2I_o) \tag{4.22}$$

With the assistance of the low-frequency conductance

$$g_o = \frac{dI_o}{dV} = I_o \lambda e^{\lambda V} = \frac{q(I + I_o)}{kT} \tag{4.23}$$

we can also express the PSD of the shot noise current as

$$S_I(f) = 2kTg_o \frac{I + 2I_o}{I + I_o} \qquad [I \gg I_o] \tag{4.24}$$

At zero bias ($V = 0$, $I = 0$), the noise is approximately equal to the thermal noise of the conductance itself having the PSD

$$S_I(f) = 4kTg_{o0} \qquad (g_{o0} = I_o q / kT) \tag{4.25}$$

By comparing the latest result with (4.24), we find that the high-frequency shot noise is equal to one-half of the above thermal noise for $I \gg I_o$, that is,

$$S_I(f) = 2kTg_{o0} \tag{4.26}$$

After introducing the internal resistance of the device ($R_o = 1/g_o$) the open-circuit noise voltage PSD is

$$S_V(f) = \frac{S_I(f)}{(g_o)^2} \approx 2k\mathrm{TR}_o \frac{I + 2I_o}{I + I_o} \approx 2k\mathrm{TR}_o \qquad [I \gg I_o] \qquad 4.27$$

4.4.2 Shot Noise in Bipolar Transistors

We will continue our investigations with the assistance of [4.4], the Ebbers–Moll transistor model [4.7, 4.8] for emitter and collector currents:

$$\begin{aligned} I_E &= I_{\mathrm{ES}}[e^{\Lambda v_{be}} - 1] - \alpha_R I_{\mathrm{CS}}[e^{-\Lambda v_{be}} - 1] \\ I_C &= \alpha_F I_{\mathrm{ES}}[e^{\Lambda v_{be}} - 1] - \alpha_R I_{\mathrm{CS}}[e^{-\Lambda v_{be}} - 1] \end{aligned} \qquad 4.28$$

where v_{bv} is the emitter base voltage, α_F and α_R are the forward and reverse short-circuit current gains, and I_{ES} and I_{CS} are emitter and collector saturation currents mutually related through the reciprocity relation

$$\alpha_F I_{ES} = \alpha_R I_{CS} \qquad \alpha_F, \alpha_R \approx 1 \qquad 4.29$$

All the currents in (4.28) show the full shot noise in accordance with (4.22), that is,

$$S_{I_E} = 2qI_E + 2qI_{\mathrm{ES}} \quad \text{and} \quad S_{I_C} = 2qI_C + \alpha_F 2qI_{\mathrm{ES}} \qquad 4.30$$

Note that currents I_{BE} and I_{BC} are very small and can be neglected unless one is operating devices at extremely low currents [cf. (4.28)]. However, there is difficulty with the mutual correlation between the above noises (cf. Fig. 4.5*a*). The arrangement in Fig. 4.5*b* removes this drawback by introducing the voltage noise with the PSD,

$$S_e = i_l^2 R_{\mathrm{eo}}^2 = 2kT R_{e0o} \frac{I_B + 2I_{\mathrm{BE}}}{I_B + I_{\mathrm{BE}}} \qquad 4.31$$

originating in the input conductance at the emitter:

$$g_{\mathrm{eo}} = \frac{1}{R_{\mathrm{eo}}} = \frac{dI_E}{dv_{\mathrm{be}}} = (q/kT) I_{\mathrm{ES}} e^{\Lambda v_{be}} = (q/kT)(I_E + I_{\mathrm{ES}}) \qquad 4.32$$

The situation with the common emitter connection is displayed in Fig. 4.6*a*; for low frequencies, the transconductance at the collector is given by

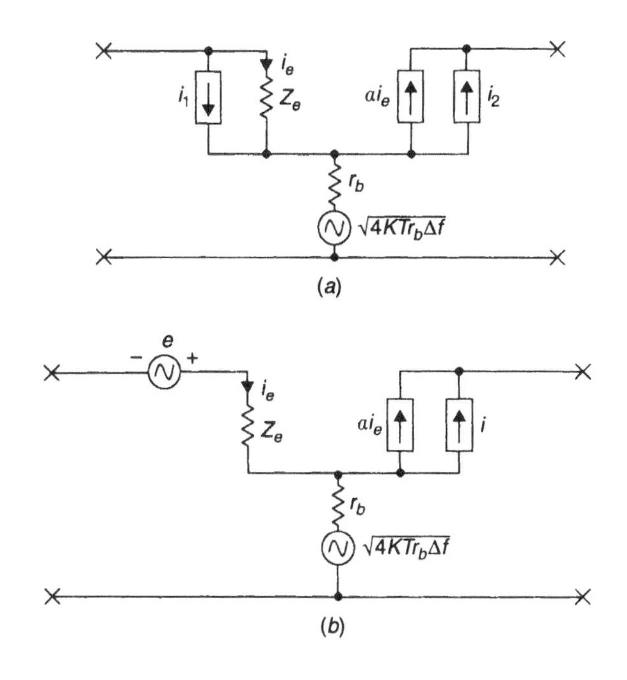

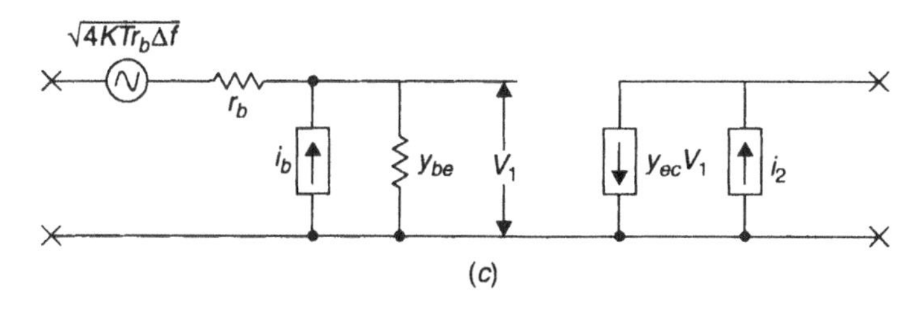

Fig. 4.5 Shot noise sources in transistors. (*a*) Common base configuration with two strongly correlated current noise generators, (*b*) common base configuration with two nearly uncorrelated current noise generators, (*c*) equivalent circuit ([4.4, 4.9]).

$$g_{\text{mo}} = \frac{dI_C}{dv_{\text{be}}} = \alpha_F(q/kT)I_{\text{ES}}e^{\lambda v_{be}} + \alpha_R(q/kT)I_{\text{CS}}e^{-\lambda v_{be}} \approx \alpha_R(q/kT)I_C \approx (q/kT)I_C \qquad 4.33$$

where g_{mo} is the transconductance and V_{be} is the base-emitter voltage. Nevertheless, here we must also consider the shot noise of the base current

$$S_{I_B} = 2qI_E(1-\alpha_F) \approx 2qI_B \qquad 4.34$$

The difficulty might be correlation between the base and collector noises

$$S_{\mathrm{corr}} = <\bar{i}_{\mathrm{bn}} \cdot i_{\mathrm{cn}}> = <(\bar{i}_{\mathrm{en}} - i_{\mathrm{cn}}) \cdot i_{\mathrm{cn}}> = 2kT(Y_{\mathrm{ce}} - g_{\mathrm{mo}}) \qquad 4.35$$

However, it is generally so small that without any appreciable error we can discard it and use, for high frequencies, the equivalent circuit presented in Fig. 4.6*b* [4.4]. Another phenomenon one must take into consideration, namely, the lifetime of the carriers, is with the consequence of a larger transconductance, which for a *long* junction is approximately

$$g' = g_0 Re(1 + j\omega\tau)^{1/2} \approx g_0\left[1 + \frac{1}{2}(\omega\tau)^2\right] \qquad 4.36$$

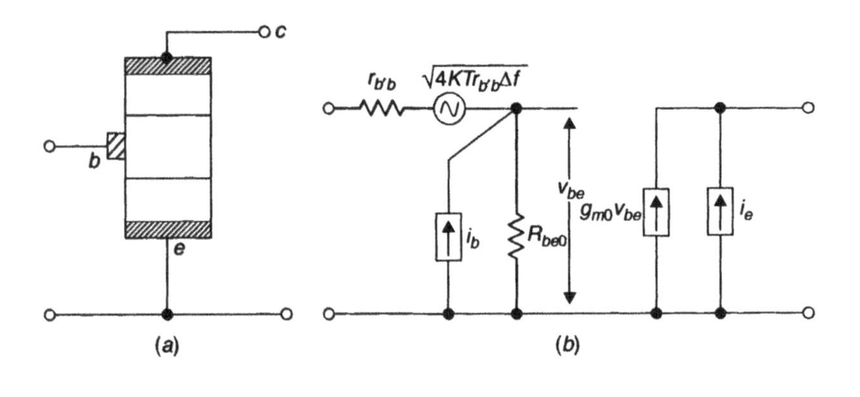

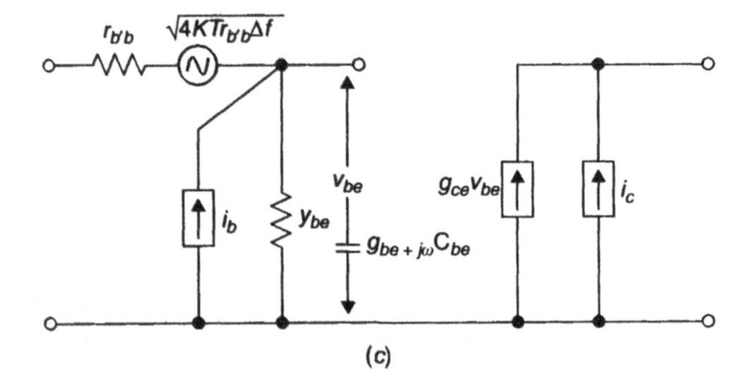

Fig. 4.6 (*a*) Equivalent circuit of the common emitter configuration. (*b*) For low-carrier frequencies. (*c*) For high-carrier frequencies [4.4].

4.4.3 White Noise in Field-Effect Transistors

In contradistinction to the above discussed amplifiers, all field-effect transistor (FET) systems exhibit low input admittance, highly linear (nearly ideal) tranconductance, and lately, microwave applications [e.g., 4.10, 4.11]. The drain current of the MOSFET is given by

$$I_D = K\frac{(V_{GS}-V_T)^2}{1+\theta(V_{GS}-V_T)} \approx K(V_{GS}-V_T)^2 \qquad K = \frac{1}{2}\cdot\frac{\mu W C_{ox}}{L} \tag{4.37}$$

where μ is the mobility, W is the channel width, C_{ox} is the capacitance per unit area, L is the channel length, and θ is the normal field degeneration factor. Further, V_D is the drain voltage, V_G is the gate voltage, and V_T is the turn-on or threshold voltage [4.4, 4.7]. The differentiation reveals the transconductance g_m:

$$g_m = K\frac{(V_{GS}-V_T)[2+\theta(V_{GS}-V_T)]}{[1+\theta(V_{GS}-V_T)]^2} \approx$$
$$2K(V_{GS}-V_T) = \frac{2I_D}{V_{GS}-V_T} \qquad (\text{for } \theta \ll 1) \tag{4.38}$$

The drain conductance for zero drain bias, g_{do}, is given by

$$g_{do} = \frac{\mu W C_{ox}}{L}(V_G - V_T) \qquad g_m = \frac{\mu W C_{ox}}{L} V_D \tag{4.39}$$

The corresponding thermal noise is expressed as

$$S_I(f) = 4kT\gamma_{sat} g_{do} \tag{4.40}$$

where $\gamma_{sat} = 2/3$ for a fully closed channel, $\gamma = 1/2$ for a fully open channel, and $\gamma = 1$ is for $V_D = 0$ [more detail is found in 4.4].

4.4.4 The Flicker (1/*f*) Noise in Semiconductors

Today, the origin of low-frequency $1/f$ noise in bipolar transistors is still not fully understood. There are several theories about its origin: the recombination–generation process, fluctuation of the diffusion constant [4.4, 4.7, 4.12], temperature dependence [4.13], bulk material losses [4.14], and so on. Despite these difficulties, many attempts were made to find empirical models, particularly for the simulation computer programs (SPICE) and so on. In this case, the low-frequency current noise PSD is evaluated as a function of the base current, I_b,

$$S_{nb}(f) \approx 2qI_b + KF\frac{I_b^{AF}}{f} \tag{4.41}$$

where the first term on the right-hand side (rhs) is the shot noise due to the base current, I_b, and KF and AF bring up empirical constants. The difficulty with the above relation is the value of the exponent AF in the second rhs term. Some authors suggest $AF = 1$ and others $AF = 2$ [e.g., 4.15, 4.16]. The latter estimation is supported by the physical dimension of the left-hand side (lhs) (A^2) and by the corresponding phase noise.

The flicker noise model of the MOSFET is similar. In accordance with the Berkeley User's Manual [4.9],

$$S_{id}(f) = \frac{KF \cdot I_{ds}^{AF}}{C_{ox} \cdot L_{eff}^2 \cdot f} \tag{4.42}$$

where KF is again the flicker noise coefficient, AF is the flicker noise exponent ($AF = 1$), I_{ds} is the saturation current, and L_{eff} is the effective channel length. Note that low-frequency noise increases with shrinking of the active area [4.17, 4.18].

Another model, particularly in a system working in higher RF and MW ranges (carrier frequencies $f_o > f_{o,Q\max}$), may start with the general $1/f$ noise (1.109) when the quality factor is expressed with the assistance of (4.20). Then

$$S_\phi(f) = \frac{1}{f} 2\pi\varepsilon\varepsilon_o\rho = \frac{1}{f}\frac{\varepsilon\rho}{18\times 10^9} \approx \frac{1}{f}10^{-12\pm 1} \tag{4.43}$$

$$[\varepsilon = 9;\ \rho = 2\times 10^{-3\pm 1}(\Omega\text{m})]$$

Finally, a general approximation suggested for the overall phase noise PSD, has been put forth in [4.19]:

$$S_\phi(f) = \frac{S_{nb}(f)}{I_b^2} = \frac{2q}{I_b}\left(1 + KF\frac{I_b}{2qf}\right) = \frac{2q}{I_b}\left(1 + \frac{f_c}{f}\right) \tag{4.44}$$

where f_c is the intersection frequency between flicker and white noise characteristics. It may be as low as 10 Hz, but more often it is found in the low kilohertz ranges. With the assistance of (4.43) and thermal noise we get approximately (cf. Table 4.2)

$$f_c \approx 10^{-12\pm 1}\frac{P_o}{2kT} \approx 10^{8\pm 1}P_o \quad (f_c \approx 10^5 \text{ for } P_o \approx 10^{-3}) \qquad 4.45$$

4.5 AMPLIFIERS

By discussing frequency stability problems, we encounter amplifiers in oscillators, frequency synthesizers, output circuits, distributing nets, measurement devices, and others. The appreciated features are voltage or power gain, low *noise figure,* and stability, particularly of feedback systems.

Silicium (Si) bipolar junction transistors are suitable for wireless RF applications, whereas GaAs HBTs are valued for their superior flicker noise properties, short base transit time, and so on [4.19].

The JFETs are praised for their high input and out impedances, good linearity, low noise, and high cut-off frequencies f_T, making it possible to work in microwave ranges. Nowadays, field-effect transistors, particularly CMOS systems, are predominant in IC applications, in mobile communications, in synchronized data transmissions, and so on.

4.5.1 Linear Two Ports Representation

In the last 50 years, many papers and books were written about design and applications of transistor amplifiers. Here, we mention solutions based on the linear two-port models (Fig. 4.7). We start with the y parameters preferred in RF systems (Fig. 4.7*a*):

$$\begin{aligned} i_1 &= y_{11}V_1 + y_{12}V_2 \\ i_2 &= y_{21}V_1 + y_{22}V_2 \end{aligned} \qquad 4.46$$

where y_{11} represent the short-circuit input admittance ($V_2 = 0$), y_{12} the short-circuit reversed transfer admittance ($y_{21} = g_m + j\omega C_{\text{in}}$), y_{21} the transconductance, and y_{22}, the short-circuit output admittance. For mi-

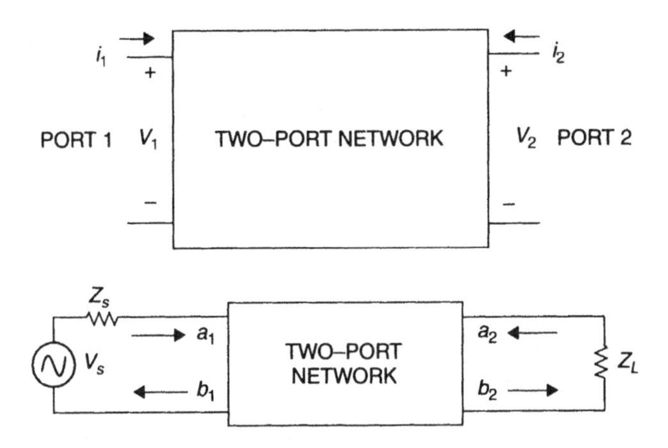

Fig. 4.7 Two-port models. Top: *y* parameters; bottom: *s* parameters (adapted from [4.20]).

crowave frequencies > 1 GHz, the *s* parameters, that is, the generalized scattering parameters based on the wave functions, are predominantly used (Fig. 7.4*b*).

The wave functions, the independent variables a_i ($i = 1,2$), and dependent variables b_i ($i = 1,2$) are defined [4.20] as

$$a_i = \frac{V_i + I_i Z_o}{2\sqrt{Z_o}} \qquad b_i = \frac{V_i - I_i Z_o}{2\sqrt{Z_o}} \tag{4.47}$$

and equations describing the two-port network are

$$\begin{aligned} b_1 &\approx s_{11}a_1 + s_{12}a_2 \\ b_2 &\approx s_{21}a_1 + s_{22}a_2 \end{aligned} \tag{4.48}$$

where a_1 and a_2 are incident voltage waves and b_1 and b_2 are reflected voltage waves. The *s*-factors ($s_{m,n}$) are defined as

$$s_{m,n} = S_{m,n}e^{\phi_{m,n}} = \frac{b_m}{a_n}\bigg|_{a_m=0} \qquad s_{m,m} = S_{m,m}e^{\phi_{m,m}} = \frac{b_m}{a_m}\bigg|_{a_n=0} \tag{4.49}$$

where s_{11} and s_{22} are input and output reflection factors and s_{12} and s_{21} are reverse and forward transfer factors. In the ideal case, where no reflection and reverse transfer takes place, then

$$s_{21} = -\frac{y_{21}}{2} \quad 4.50$$

4.5.2 Feedback in Amplifiers

In amplifiers, we encounter feedback connections, wanted or unwanted, between input and output ports. The positive outcomes are the increased linearity or bandwidth. On the other hand, we may encounter desired or undesired oscillations, modified transconductance, and other factors.

EXAMPLE 4.3

As an example, we investigate the effective transconductance of the bipolar transistor amplifier shown in Fig. 4.8. After introduction into (4.28), we have

$$i_c(t) \approx I_c e^{\lambda v_{be}(t)} = I_c e^{\lambda\left[v_s(t) - i_c(t)\cdot\left(Z_e + \frac{R_s + r_{bb}}{\beta}\right)\right]} \quad 4.51$$

Differentiation reveals the transconductance

$$\frac{di_c(t)}{dv_s(t)} = g_m = \frac{I_c\lambda}{1 + I_c\lambda\left(Z_e + \frac{R_s + r_{bb}}{\beta}\right)} \approx \frac{1}{Z_e}\left[\text{for } Z_e \gg \frac{R_s + r_{bb}}{\beta}\right] \quad 4.52$$

in a much more general case (Fig. 4.8), we arrive at

$$g_m = \frac{I_c\lambda}{1 + I_c\lambda\left[Z_e + \frac{R_s Z_L}{Z_M - R_s} + \frac{1}{\beta}\left(\frac{R_s Z_M}{Z_M - R_s} + r_{bb}\right)\right]} \quad 4.53$$

Note that for $Z_M \to \infty$ the above relation converges into (4.52) as expected.

4.5.3 Voltage and Power Gains

As an example, we have chosen the simplified circuit arrangement of the common emitter configuration drawn in Fig. 4.8. The amplifier voltage gain is ($Z_M \to \infty$)

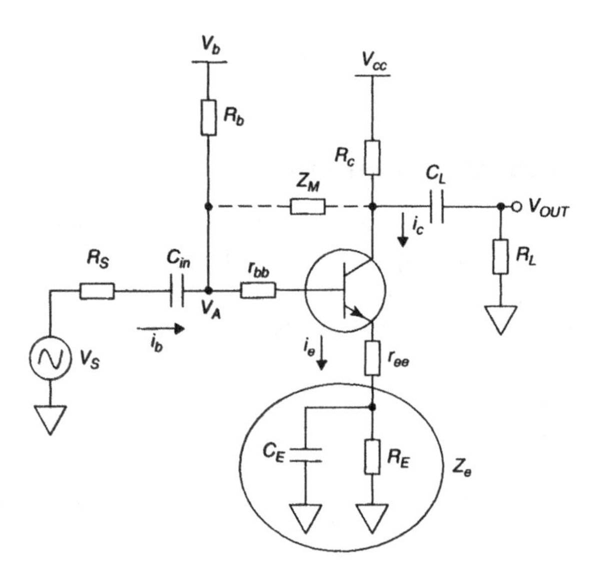

Fig. 4.8 The simplified circuit arrangement of the common emitter configuration.

$$G_v = \frac{V_{\text{out}}}{V_S} = \frac{i_c Z_L}{V_S} = g_m \frac{v_G}{V_S} Z_L = \frac{g_m Z_{\text{in}}}{R_S + Z_{\text{in}}} Z_L = y_{21} Z_L \approx g_m R_L \qquad 4.54$$

and the corresponding power gain is

$$G_p = Re\,|\,y_{21}^2 Z_L Z_S\,| \approx g_m^2 R_L R_S \qquad 4.55$$

4.5.4 Maximum Operating Frequency

At very high frequencies field-effect transistors (MOSFET and CMOS) exhibit a rather small input impedance $Z_i = 1/j\omega C_{gs}$ (cf. Fig. 4.9). By reverting in such an instant to the relation for the voltage gain (4.54), we find that the gain may be lost when the product $g_m Z_i$ is reduced to < 1. The result is that the maximum operating frequency can hardly exceed

$$\omega_m = \frac{g_m}{C_{gs}} \approx \mu \frac{V_D}{L^2} \qquad 4.56$$

This drawback can be alleviated by increasing the input and other impedances; for example, to parallel resonances with auxiliary inductances.

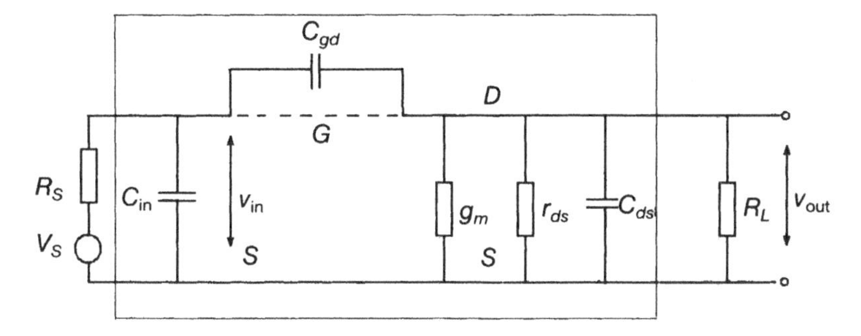

Fig. 4.9 Simplified block diagram of the field-effect transistors (MOSFET and CMOS).

4.5.5 Noise Figure

The noise figure is an important property characterizing amplifiers, mixers, and so on, and is equal to the ratio of the noise accompanying the output signal to the noise accompanying the input signal in the white noise region. In this connection, we encounter thermal noises generated in the associated network, noises provided by the active semiconductor layers, and current shot noises [4.4, 4.21]. First, we will investigate the common base circuit arrangement. The overall input noise voltage, in accordance with the block diagram in Fig. 4.5*b*, is (in 1-Hz band pass)

$$e^2_{n,\text{input}} = 4kTR_s + 4kTr_{bb} + e^2_{n,\text{uncor}} + i^2_{n,c} \,|\, Z_s + r_{bb} + Z_e + Z_{\text{corr}}|^2 \qquad 4.57$$

where $e^2_{n,\text{uncorr}}$ is the thermal noise (cf. 4.31) and $e_{n,c}{}^2$ is the collector noise current (cf. 4.35):

$$e^2_{n,\text{uncorr}} = 4kTR_n \qquad i^2_{n,c} = \alpha^2 4kT g_m \approx 4kT g_m \qquad 4.58$$

The noise contribution generated in the transistor transconductance is white and defined by

$$<(e_{n,i})^2> = \frac{<i^2_{nd}>}{g^2_{fs}} = 4kTB_W R_n \qquad R_n = \frac{\gamma}{g_{fs}} + R_E + r_{ee} \qquad 4.59$$

where g_{fs} is the common source JFET forward transconductance, and B_w is the effective bandwidth in hertz. In the case of the saturation condition $\gamma = 0.67$, after some computations, we arrive at the sought noise figure

$$F = 1 + \frac{r_{bb} + R_n}{R_s} + \frac{g_m}{R_s} \mid Z_s + r_{bb} + Z_e + Z_{\text{corr}} \mid^2 \qquad 4.60$$

Inspection of Fig. 4.5*c* effectively reveals, for the common emitter circuit arrangement, the same noise figure as above.

4.5.6 1/*f* Noise Up-Conversion

In RF circuits, amplifiers, oscillators, dividers, frequency mixers, and so on, we encounter, close-to-the-carrier phase and amplitude noise, often recalling the DC $1/f$ flicker noise. Some investigations suggest an up-conversion into RF systems. In this connection, we refer to the article by Walls et al. [4.22], where the origin of $1/f$ noise is explained with the inherent modulation of the transconductance, namely, by introducing the current noise fluctuations (4.41 or 4.43) into the conductance relation (4.52) or (4.53), that is, by replacing I_o with $I_o + i_n$, where the noise component encloses both sine and cosine components (cf. Section 1.3.2). The consequence is that after multiplication with the input voltage the output signal would reveal full $1/f$ noise amplitude and phase modulation. Further, by taking into consideration the slowly varying $1/f$ signal, the emitter feedback impedance, Z_e, may reduce the up-conversion process substantially if it presents appreciable impedance in the corresponding frequency range [4.23]. The difficulty is that the overall amplifier gain is also reduced.

From the earlier measurements of the phase noise in oscillators it was found that the phase noise PSD due to the flicker noise was $\sim S_\varphi(f = 1) \times 10^{-11.2}$, which is nearly equal for all transistor types [4.24, 4.25]. Laboratory experiments proved that the flicker noise generated in the corresponding transistor was responsible for the phase modulation of the RF carriers. The improvement (typically > 30 dB) has been achieved by negative feedback with the assistance of the bypassed resistance R_E in the emitter connection (emitter degeneration). This was shown in Fig. 4.8, which introduces a feedback for low frequencies with the effective transconductance

$$g_{m,\text{low}} \approx \frac{1}{r_{ee} + R_E} \qquad 4.61$$

whereas for high frequencies, the transconductance remains to be

$$g_{m,\text{high}} = \frac{\lambda I_{c,oo}}{1+\lambda I_{c,oo} r_{ee}} \approx \frac{1}{r_{ee}} \qquad 4.62$$

and the up-converted, low-frequency phase noise is reduced proportionately:

$$\frac{g_{11}}{g_m} = \frac{r_{ee}}{r_{ee}+R_E} \approx \frac{r_{ee}}{R_E} \qquad 4.63$$

The improvement was earlier experimentally verified by Halford et al. [4.24] and Healey [4.25]. Similarly, there are possibilities of application of the active feedback for reduction of the up-converted noise [4.26].

4.6 MIXERS

Frequency mixers are encountered in nearly all frequency stability devices: precision frequency generators–synthesizers, measurement sys-

Table 4.2 Noise properties of different amplifiers

f_o (MHz)	$S_{\varphi,1/f}(f)$ (dB) $f = 1$	$S_{\varphi,\text{white}}(f)$ (dB)	Reference
DC (at 0.1 Hz)	–170	–176	[4.27]
5	–135	–154 (?)	[4.28]
5/10		–152	[4.29]
10	–140 (?)		[4.30]
40	–130	–163	[4.31]
100	–143		[4.32]
500	–148±136		[4.33]
1,000	–115±5		[4.28]
1,300	–130±5		[4.34]
2,000	–130	–SiGe HBT	[4.35]
2,500	–120±5	–SiGe FET	[4.36]
3,500 (SiGe)	–140	–SiGe HBT	[4.35]
3,500	–125±5	–SiGe FET	[4.37]
4,000	–115±125		[4.38]
10,000	–115±5	SiGe the lowest	[4.39]
10,000 (GaAs)	–105±2.5	GaAs HEMFET	[4.36]
100 (GHz)	–80	–130/–165	[4.39]

tems, communications equipment, an so on. In all of these instruments, low noise and low level of spurious signals are of the foremost importance, particularly since they are based on the nonlinearity between voltage and current of electronic circuits. Both of these properties will be discussed in the following sections.

4.6.1 Multiplicative Mixers

Multiplicative mixers exploit the nonlinearity between voltage and current in many electronic elements (diodes, transistors, etc.). The parabolic dependence is ideal:

$$i = a_0 + a_1 v + a_2 v^2 \tag{4.64}$$

In the case where the voltage v is formed by two sinusoidal signals, then, due to the quadratic term in (4.64), the circuit generates two side bands with sum and difference frequencies

$$\begin{aligned} v_{\text{desired}}(t) = 2a_2 V_i V_o [&\sin[(\omega_i - \omega_o)t + \phi_i(t) - \phi_o(t)] + \\ &+ \sin[(\omega_i + \omega_o)t + \phi_i(t) + \phi_o(t)]] \end{aligned} \tag{4.65}$$

One of them, together with both leaking carriers, is undesired and must be removed by filtering. Note that there are other important properties of mixers that must be taken into account: the reduced level of the desired signal and the noise or spurious contributions of the input signals that appear at the output.

However, this ideal case is not encountered in practice since all mixing devices contain even higher order terms initiated in the generalized (i.e., polynomial) equation (4.64). As an example, consider semiconductor elements that exhibit exponential dependence between current and voltage as explained in Section 4.5.6 [cf. (4.22)]. The consequence is that the mixer output contains higher order terms, several of them quite large [4.40], with frequencies

$$f_{\text{out}} = rf_1 + sf_2 \quad (r, s = 0, \pm 1, \pm 2, \ldots) \tag{4.66}$$

Special attention should be paid to those mixing products whose frequencies lie in the band pass of the output filter as shown in Fig. 4.10. Its application for solving intermodulation problems will be explained with the assistance of the following example:

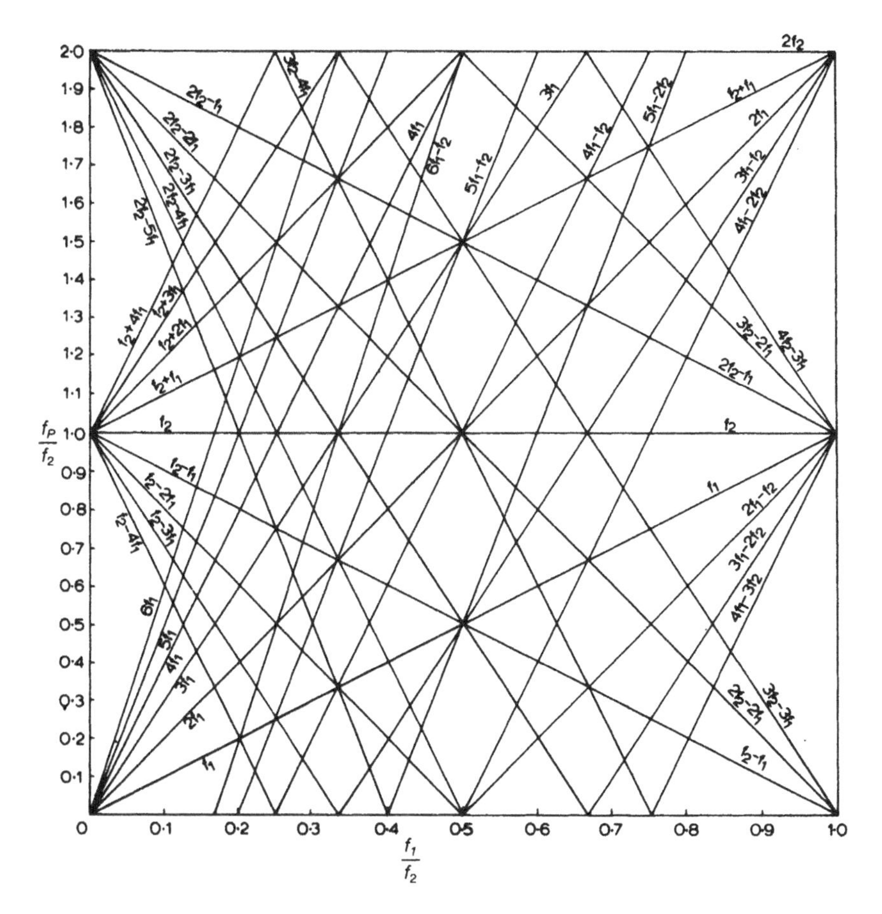

Fig. 4.10 Diagram for investigation of intermodulation signals of the lower order in the output pass band of a mixer (reproduced with permission from [4.44].)

EXAMPLE 4.4

Let us investigate intermodulation properties of the mixer having input signals with frequencies $f_1 = 260$ to 300 MHz and $f_2 = 300$ to 310 MHz, and output low-pass filter with corner frequency at $f_{p,\max} = 50$ MHz. The solution is simple. We mark extremes $(f_1/f_2)_{\min} = 0.84$ and $(f_1/f_2)_{\max} = 1$ on the x-axis and $(f_p/f_2)_{\min} = 0$ and $(f_p/f_2)_{\max} = 0.167$ on the y-axis. We draw through these points horizontal and vertical lines that bound a rectangle in Fig. 4.10. Parameters (r and s) of all straight lines intersecting its area indicate the order of the intermodulation signals that are present in the output:

$$f_{p\min} \leq f_p \leq f_{p\max} \qquad 4.67$$

With the assistance of the diagram in Fig. 4.10, we find $(f_2 - f_1)$, $(2f_2 - 2f_1)$, $(3f_2 - 3f_1)$, and so on. Note that all higher order terms are not indicated. Because of the clarity of reading in the diagram, we have limited the intermodulation order in Fig. 4.10 to

$$|r|+|s|=7 \qquad 4.68$$

The other reason for this restriction is that terms of higher order than those in the above relation are generally below the level of –80 dB and tolerable in most instances (cf. [4.42, 4.43] and Table 4.3). Note that the rectangle that would be drawn in Fig. 4.10 is in reality a polygon. However, we feel that computer applications will provide the desired correct information for investigation of the intermodulation signals.

4.6.2 Switching or Balanced Mixers

A nonlinear electronic element often encountered is a switch, which is schematically depicted in Fig. 4.11. The working mode is such that the input signal either passes to the output or is blocked. The mathematics is very simple:

$$v_2(t) = v_1(t)p_1(t) \qquad 4.69$$

where $p(t)$ is a periodic rectangular function with two levels, either +1 or zero, with the switching frequency Ω_{LO}. Consequently, (4.69) can be rearranged:

$$v_2(t) = V_1 \sin(\omega_s t) \cdot \left[\frac{\Theta}{2\pi} + \sum_{r=1,2,\ldots}^{\infty} \frac{2}{\pi r} \sin \frac{r\Theta}{2} \cos(r\Omega_{\mathrm{LO}} t) \right] \qquad 4.70$$

Discussion of the mixing properties is found elsewhere [e.g., 4.41, 4.42]. But note that each diode in the switching element is a generator of the additive noise as discussed above.

4.6.3 Ring or Double-Balanced Mixers

The ring modulator represented in Fig. 4.12 is equivalent during one half-period of the switching signal to the lattice-resistive network in Fig. 4.12b, and in the following half-period to a similar network with

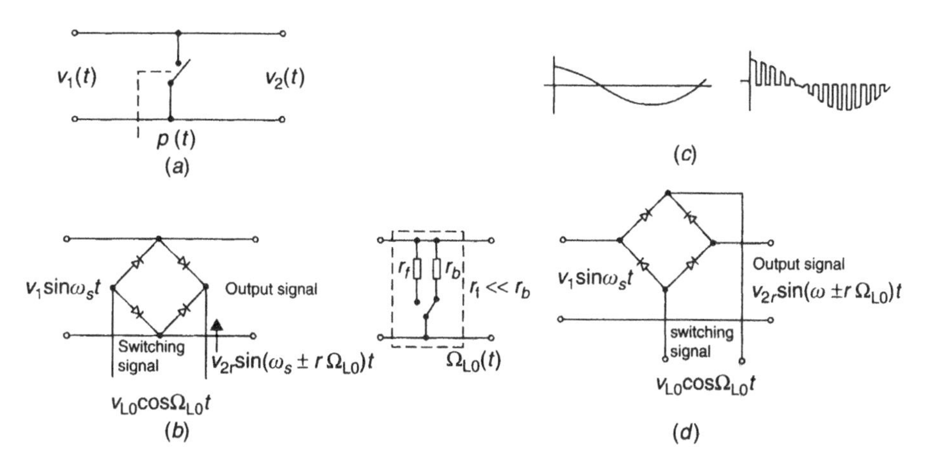

Fig. 4.11 Switching mixer: (*a*) idealized circuit, (*b*) parallel diode switch, (*c*) input and output signal of the parallel switch, (*d*) series diode modulator [4.41].

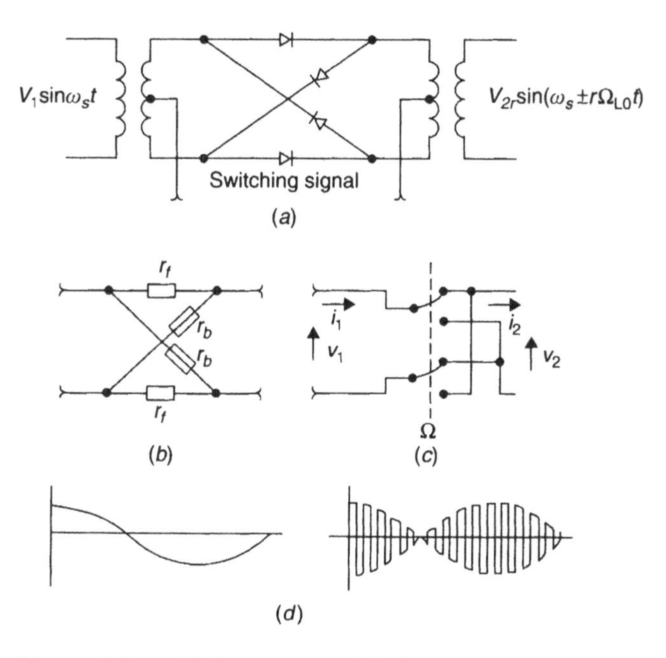

Fig. 4.12 Ring modulator. (*a*) Arrangement of diodes and balanced transformers. (*b*) The equivalent lattice-resistive network during one-half period of the switching signal. In the following half-period, r_b and r_f are interchanged. (*c*) The ideal phase inverter. (*d*) Idealized input and output waveforms. (Reproduced with permission from [4.41].)

r_b and r_f interchanged. However, this is equivalent to a fixed lattice-resistive network followed by the periodic phase inverter (Fig. 4.12c). Being an ideal ring modulator ($r_f = 0$, $r_b = 4$), it reduces to the phase inverter alone and the output voltage is

$$v_2(t) = i_2(t) R_{\text{Loud}} = i_1(t) p_2(t) R_{\text{Loud}} \tag{4.71}$$

where $p_2(t)$ is the switching function

$$p_2(t) = \frac{2}{\pi}\left[\sin(\Omega t) + \frac{1}{3}\sin(3\Omega t) + \frac{1}{5}\sin(5\Omega t) + \ldots\right] \tag{4.72}$$

By investigating properties of this mixer, we arrive at the desired output voltage:

$$v_2(t) = \frac{2}{\pi} I_1 R_{\text{Loud}} \sum_{r=1,3,5,\ldots}^{\infty} \frac{1}{r}\sin[(\omega_s \pm r\Omega_{\text{LO}})t] \tag{4.73}$$

The relation between the amplitude of the input signal and that of the desired side band is called conversion loss, and for the impedance-matched ring modulator (i.e., $V_1 = V_2$) we have

$$L = 20\log\left(\frac{2}{\pi}\right) \approx -4\ \text{dB} \tag{4.74}$$

However, the ring modulator is no ideal switch; consequently, the conversion loss is larger, practically in the range of –5.5 to –7 dB. In addition, due to the balancing properties as evaluated with the assistance of Fig. 4.13 and Table 4.3, a lot of intermodulation signals are reduced or eliminated and no harmonics of the input signals should be present. But the reality is not so bright. Information about suppression of some spurious side bands in the output of the double-balanced mixers in dB is shown in Table 4.4. More information can be found in [4.41] and other references. (Note, that transformers shown in Fig. 4.12*a* are only provided for supplying the switching voltage and are assumed to have a 1:1 ratio.)

4.6.4 Current-Commutating CMOS Mixers

Progress in communications systems, particularly in microwave ranges and the widespread use of IC technology introduced compatible mixers, very often in the doubly balanced form. The so-called Gilbert cell

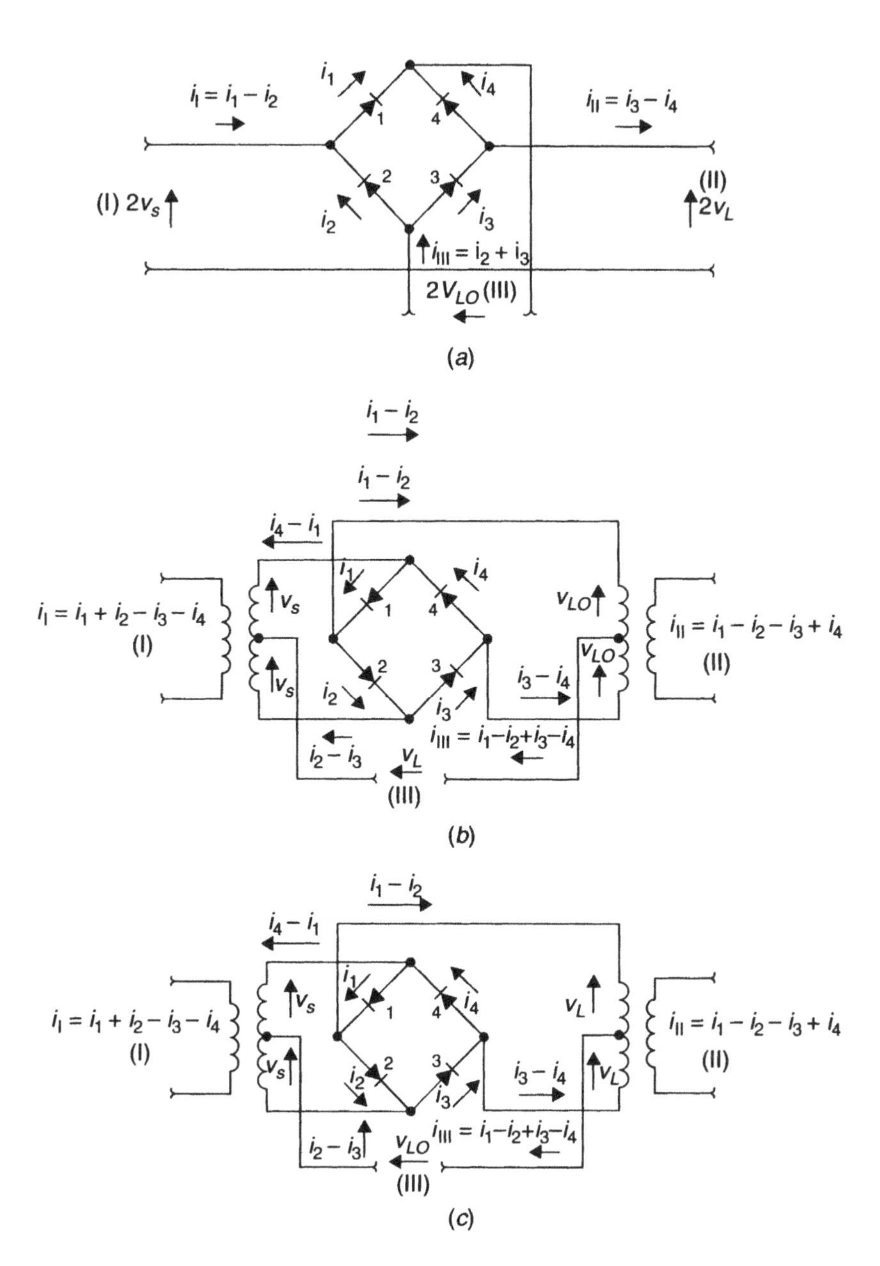

Fig. 4.13 Instantaneous voltages and currents in switching mixers: (*a*) series modulator; (*b*) ring modulator; and (*c*) ring modulator with V_{LO} and V_L interchanged [4.41].

Table 4.3 Balanced properties of the ring modulators

Modulator type	Diode, k	Voltage, v_k	Sign, i_k	Port I, $i_I = i_1 - i_2$	Port II, $i_{II} = i_3 - i_4$	Port III, $i_{II} = i_2 + i_3$
Series modulator	1 2 3 4	$v_{LO} + v_S - v_L$ $v_{LO} - v_S + v_L$ $v_{LO} + v_S - v_L$ $v_{LO} - v_S + v_L$	$(-1)^q$ $(-1)^{s-q}$ $(-1)^q$ $(-1)^{s-q}$	$i_I = i_{II} = K(-1)^q[1 - (-1)^q]$ components with s odd only		$i_{III} = K'(-1)^q$ $\times [1 + (-1)^s]$ components with s even only
				$i_I = i_1 + i_2 - i_3 - i_4$	$i_{II} = i_1 - i_2 - i_3 + i_4$	$i_{III} = i_1 - i_2 + i_3 - i_4$
Ring modulator	1 2 3 4	$-v_{LO} + v_S + v_L$ $+v_{LO} + v_S - v_L$ $+v_{LO} - v_S + v_L$ $-v_{LO} - v_S - v_L$	$(-1)^{s-q}$ $(-1)^q$ $(-1)^{s-q}$ $(-1)^{r+s-q}$	$i_I = K_I(-1)^q$ $\times [1 + (-1)^r]$ $\times [1 - (-1)^s]$ r even s odd only	$i_{II} = K_{II}(-1)^q$ $\times [(-1)^r - 1]$ $\times [1 + (-1)^s]$ r odd s odd only	$i_{III} = K_{III}(-1)^q$ $\times [(-1)^r - 1]$ $\times [1 - (-1)^s]$ r odd s even only
Ring modulator	1 2 3 4	$+v_{LO} + v_S - v_L$ $-v_{LO} + v_S + v_L$ $+v_{LO} - v_S + v_L$ $-v_{LO} - v_S - v_L$	$(-1)^q$ $(-1)^{s-q}$ $(-1)^{s-q}$ $(-1)^{r+s-q}$	$i_I = K'_I(-1)^q$ $\times [1 + (-1)^r]$ $\times [1 - (-1)^s]$ r even s odd only	$i_{II} = K'_{II}(-1)^q$ $\times [1 - (-1)^r]$ $\times [1 - (-1)^s]$ r odd s odd only	$i_{II} = K'_{III}(-1)^q$ $\times [1 - (-1)^r]$ $\times [1 + (-1)^s]$ r odd s even only

Table 4.4 Intermodulation suppression in double-balanced mixers in *dB*

Harmonics s / r	1	2	3
1	0	$\Delta P + 41$	$2\Delta P + 28$
2	35	$\Delta P + 39$	$2\Delta P + 44$
3	10	$\Delta P + 32$	$2\Delta P + 18$
4	32	$\Delta P + 39$	
	14		$2\Delta P + 14$
6	35	$\Delta P + 39$	
7	17		$2\Delta P + 11$

Note: r corresponds to the high-level (LO) input and s corresponds to the low-level (RF) input; ΔP = PLO (dBm) – PRF (dBm).

arrangement (see Fig. 4.14) consists of the driver stage (M3 or M3 and M6) biased at a fixed operating point and two (or four) switching pairs driven by the LO signal. The output current is

$$I_{\text{out}} = I_{o1} - I_{o2} = (I_1 - I_2) - (I_5 - I_4) \qquad 4.75$$

The mixing properties of the switching pairs are the same as in the above discussed ring mixers with conversion loss [cf. (4.74)].The major difference is due to the biasing circuit providing an additional conversion gain [cf. 4.44, 4.45]. Further, direct conversion architecture (zero IF) offers the unique advantage of image rejection. The switching process is induced by a rather large LO signal (V_{LO} input in Fig.4.14). The switching function, $p(t)$, is the same as in Sections 4.6.2 or 4.6.3 and has an approximately rectangular form or a limited sine wave with a nearly trapezoidal form. Its first harmonic is described as above in (4.71) and (4.72) for the double-balanced mixer. The tail current I_B is held constant with the assistance of M3 or M3–M6 transistors. However, when used as the RF input, the AC input voltage V_{in} adds to the output current I_B the component i_s (cf. Fig. 4.14):

$$i_s = V_{\text{in}}\, g_{m3} \qquad 4.76$$

The output signal is generated by switching the bias current in the rhythm of the LO wave and we arrive with the assistance of (4.71) at

$$i_{\text{out}}(t) = [I_B + i_s(t)] \cdot p(t) \qquad 4.77$$

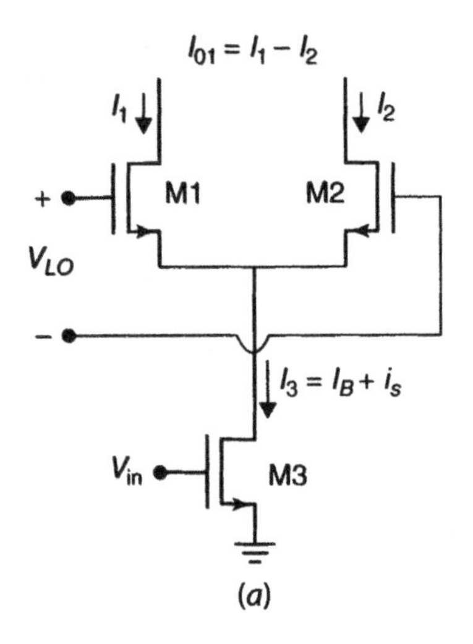

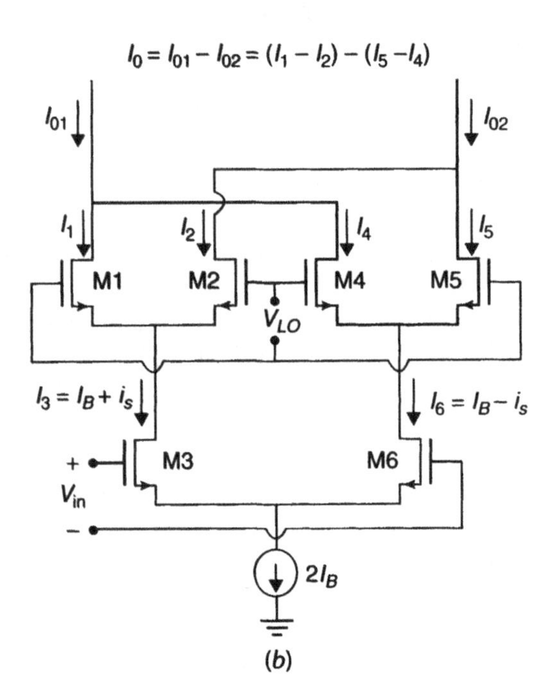

Fig. 4.14 Block arrangement of CMOS (*a*) balanced and (*b*) double-balanced (Gilbert cell) mixers [4.44].

The conversion gain is equal to the conversion loss (4.74) times the input RF gain (4.55), that is,

$$G_M = \frac{P_{out}}{P_{in}}\left(\frac{2}{\pi}\right)^2 = \frac{i_{out}^2 R_L / 2}{V_{in}^2 / 2R_S}\left(\frac{2}{\pi}\right)^2 = \left(\frac{2}{\pi}\cdot g_{m3}\right)^2 R_S \cdot R_L \qquad 4.78$$

or

$$G_M = 20\log\left(\frac{2}{\pi}\cdot g_m 3\right) + 10\log(R_S \cdot R_L) \qquad [\text{dB}] \qquad 4.79$$

4.6.5 Noise Sources in Mixers

The mixer or frequency convector is often a significant noise contributor in a lot of communications and measurement systems. By reverting to the relation (4.65), we see that the output signal also contains phase noise contributions of both mixed signals, of the input and the local oscillator signals as well as of the switching process.

4.6.5.1 Noise in Diode or Switching Mixers

By referring to Fig. 4.11 or 4.12, we see that there are always two diodes in series contributing the noise [cf. (4.25) or (4.42)]. In addition, time shifts generated in switching instances present another contribution to the phase noise accompanying the output signal. The problem is illustrated with the assistance of Fig. 4.15, where the superimposed spurious voltage fluctuations change the reference level. For solution, the first two terms of the Taylor expansion of the effective voltage are sufficient, that is,

$$v(t_i) = v(iT) + \dot{v}(iT)(t_i - iT) \qquad 4.80$$

From (4.80), we compute the time jitter of the ith triggering as:

$$\Delta T_i = \frac{v(t_i) - v(iT)}{\dot{v}(iT)} \qquad 4.81$$

By referring to the limited sine wave (depicted in Fig. 4.16), we get for the tangent slope S, close to the zero level,

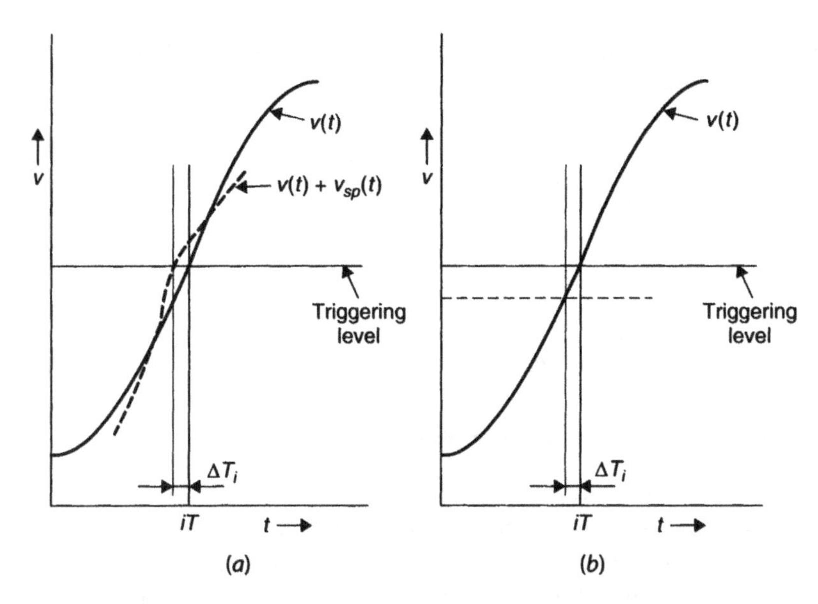

Fig. 4.15 (*a*) Time shift of the triggering level in the presence of the low-frequency interfering signal. (*b*) Time shift of the triggering level due to slow changes of the switching level due to fluctuations in temperature, humidity, and so on [4.41].

$$\dot{v}(t) = \frac{E_o}{d \times t} \approx \frac{E_m \sin(\omega_o \delta t)}{d \times t} \approx E_m \omega_o = \frac{2\pi}{T_o} E_m = S \qquad \left(\text{for } \frac{\delta t}{T_o} < 0.1\right) \tag{4.82}$$

By assuming that noise currents generated in diodes, or in local oscillators, are responsible for the corresponding time fluctuations, we get (4.83)

$$\Delta T = \frac{i_{n,1} + i_{n,2}}{S} = \frac{i_{n,1} + i_{n,2}}{\omega_o I_{\text{peak}}} = \frac{i_{n,1} + i_{n,2}}{I_{\text{peak}}(2\pi / T_o)} \tag{4.83}$$

where we have introduced the slope $S = 2\pi I_{\text{peak}}/T_o$ of the switching signal in the instant of the transition reference level. By assuming the fractional timer jitter as autocorrelation $R(0) - R(\tau) \approx R(0)$, the PSD of the additive phase noise is evaluated as

$$S_{\phi,\text{out}}(f) = \int_0^\infty < \left(\frac{\Delta T}{T_o}\right)^2 > \delta\tau \approx \frac{2S_{i,n}(f)}{S^2} \approx \frac{2S_{i,n}(f)}{(2\pi I_{\text{peak}})^2} \tag{4.84}$$

Note that the autocorrelation is localized around zero, in which case we have simplified the integration with the delta function operation.

EXAMPLE 4.5

Let us investigate the phase noise introduced by the diode ring modulator in instances in which the switching peak current is ≈ 6 mA. For the white noise component, we get

$$S_{\phi}(f) = <\left(\frac{\Delta T}{T_o}\right)^2> \approx \frac{4q(I_{\text{peak}})}{(I_{\text{peak}} \cdot 2\pi)^2} = \frac{q}{I_{\text{peak}}(\pi)^2} \qquad 4.85$$

By introducing $I_{\text{peak}} = 0.006$, $q = 1.6 \times 10^{-19}$, and the corner frequency $f_c = 10{,}000$ (cf. 4.44), we get

$$S_{\phi,D}(f) \approx \frac{10^{-13.5}}{f} + 10^{-17.3} \qquad 4.86$$

which is in good agreement with experimental findings (e.g. [4.46]) for a ring modulator used as a phase detector.

4.6.5.2 Noise in Current-Commutating CMOS Mixers

There are three sources of white noise: (1) the driving stage, (2) the switching pair or pairs, and (3) the output stage. For the single-balanced mixer, we take into account the noise of the source resistance R_s, the gate resistance r_{g3}, possible emitter (source) feedback impedance, and the thermal noise of the transistor $M3$ [cf. Fig.4.14*a* and relation (4.57)]. The corresponding white-noise current PSD is

$$S_{i,n3}(f) = 4kT\left(R_s + r_{g3} + |Z_e| + \frac{\gamma}{g_{m3}}\right)g_{m3}^2 \qquad 4.87$$

In the case of the double-balanced Gilbert cell (Fig. 4.14*b*), we introduce r_{g3} twice.

The situation is similar to that discussed in Section 4.6.5.1. By referring to Fig. 14*a*, one easily finds out that the output current is successively supplied either by $M1$ or $M2$, and in this case is determined by the bias current $I_3 = I_B$. Consequently, the switching pair contributes to the output noise only during the transition interval δ (cf. Fig. 4.16) when both transistors are conducting and generating noise. The corresponding thermal noise may be computed from

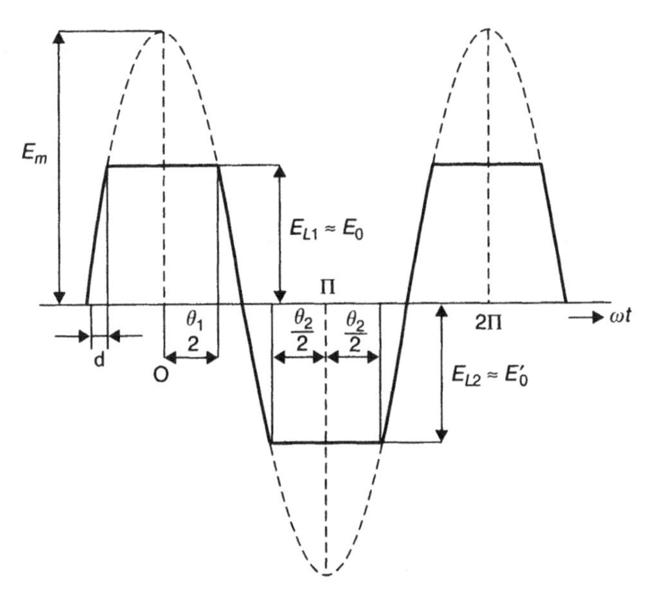

Fig. 4.16 The limited sine wave (adapted from [4.41]).

$$i_{n,\text{out}}^2 \approx 2 < 4kT\gamma g(t) > \tag{4.88}$$

Next, we must estimate the effective transconductance $<g(t)>$. The imperfect switching due to the transistor or noise accompanying the LO signal introduces random time shifts. With the assistance of (4.81), we find for the time interval ΔT, introduced with the effective noise voltage V_{on},

$$\Delta T \approx \frac{V_{\text{on}}}{S} = \frac{V_{\text{on}}}{V_{\text{LO}} 2\pi / T_o} \tag{4.89}$$

and for the corresponding current pulse, averaged over one-half of a period, we get for the noise current $i_{n,\text{out}}$

$$i_{n,\text{out}} \approx \frac{2I_B \Delta T}{T_o / 2} = \frac{2I_B V_{\text{on}}}{V_{\text{LO}} \pi} \approx < G > V_{\text{on}} \tag{4.90}$$

with the effective transconductance

$$G_{\text{eff}} \approx \frac{2I_B}{V_{\text{LO}} \pi} \tag{4.91}$$

After introducing this transconductance into (4.88), we get for the thermal current noise PSD

$$S_{\text{in}}(f) \approx \frac{16kT\gamma}{\pi} \cdot \frac{I_B}{V_{\text{LO}}} \qquad S_{\phi}(f) \approx \frac{16kT\gamma}{\pi I_B} \cdot \frac{1}{V_{\text{LO}}} \tag{4.92}$$

where V_{LO} is the amplitude of the switching oscillator (cf. [4.44, 4.45]) $S_{\varphi}(f) \approx -160$ dB for $V_{\text{LO}} = 0.1$ (V) and 1 mA.

4.6.5.3 Aliasing of the Switching Noise

The above investigation was for noise generated in switching mixers in the T_o or $T_o/2$ period. Consequently, the spectra around the individual harmonics of the sampling frequency overlap (we face the process of aliasing) and the wide-sense stationary process changes into the so-called cyclostationary random process based on periodically time-varying statistics. There are different approaches for solution of the problem. Here, we start with the application of the digital approach [4.41, 4.47]. The Laplace transform of the sampling time function $p(t)$ is

$$\hat{P}(s) = \frac{1}{T} \sum_{n=-\infty}^{\infty} \hat{P}(s - jn\omega_o) \tag{4.93}$$

By replacing variable s with $j\omega$ and applying the transfer function, $H_h(s) = (1 - e^{sT})/sT$, we finally have

$$P(j\omega) = \sum_{n=-\infty}^{\infty} \left| \frac{\sin(\pi\omega / \omega_o)}{(\pi\omega / n\omega_o)} \right| P_n(j\omega - n\omega_o) \tag{4.94}$$

where P_n are amplitudes of individual harmonics of the rectangular wave $p(t)$. Consequently, the PSD at the output of the switches is

$$S_{o,sw}(f) = 2S_{\text{in}}(f) \cdot \left(\frac{2}{\pi}\right)^2 P^2(f) = 2S_{\text{in}}(f) \tag{4.95}$$

$$\cdot \left(\frac{2}{\pi}\right)^2 \sum_{n=-\infty}^{\infty} \left| \frac{\sin(\pi\omega / \omega_o)}{(\pi\omega / n\omega_o)} \right|^2 P_n^2(f - nf_o)$$

From this, we find for the white noise (with the assistance of the Dwight formula 48.13 [4.47])

$$S_{o,sw}(f) \approx 2S_{\text{in}}(f) \cdot \left(\frac{2}{\pi}\right)^2 \left(1 + \frac{1}{3^2} + \frac{1}{5^2} + \frac{1}{7^2} + \ldots\right) \qquad 4.96$$

$$= 2S_{\text{in}}(f) \cdot \left(\frac{2}{\pi}\right)^2 \cdot \frac{\pi^2}{8} = S_{\text{in}}(f)$$

Evidently, the noise increase due to aliasing is compensated for by the loss in the switching mixer.

4.6.5.4 *Flicker or 1/f Noise*

All transistors are generators of $1/f$ noise. This finding is also true for the noise generated in the driving stage and the corresponding $1/f$ component appears at the output. The same is also true for the flicker noise accompanying the LO signal.

A bit different situation is found with the flicker noise generated by the uncertainty of the switching instants [cf. (4.81 or 4.82)]. The flicker noise from the switching pair will appear at DC and even harmonics of the LO frequency, f_{LO}, but around the f_{LO} itself and its odd harmonics, it is compensated for due to biasing (see Fig. 4.17). However, this is not the case with harmonics. Nevertheless, the feedback introduced by a resonant circuit tuned to the second harmonic placed in the driving stage helps to reduce, with feedback, this contribution of the flicker noise. On the other hand, the $1/f$ noise power also may be reduced, with the switching square wave having effectively infinite slopes. A secondary source of the flicker noise is provided with parasitic capacitances shunting the driving stage [e.g., 4.45].

4.6.6 Noise Figure of Current-Commutating CMOS Mixers

The noise figure is defined as the ratio of the overall white output noise to the white noise supplied by the source referred either to the input or output (cf. Secton 4.5.5). The output noise is composed of the noise generated in the driving stage (4.87) plus the noise generated in the switching stages (4.92), both modified by aliasing (4.94), along

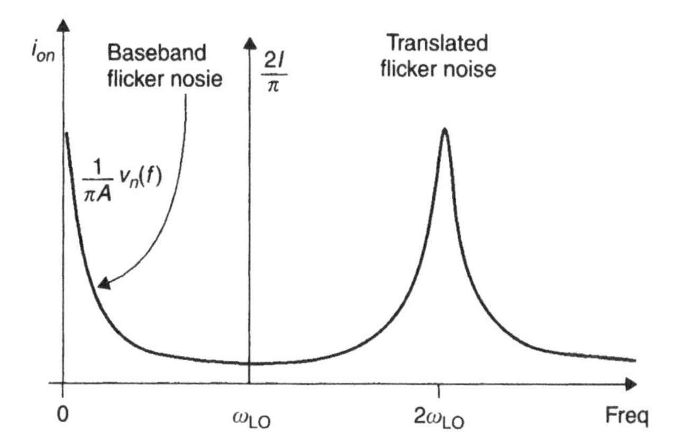

Fig. 4.17 Flicker noise from the switching pair that appears at DC and even harmonics of the LO frequency [4.45]. (Copyright © IEEE. Reproduced with permission.)

with the noise generated in the load resistance itself. For the noise figure, we use the computed output phase noise PSD divided by the input phase noise PSD and the gain (cf. 4.55):

$$FN = \frac{[S_{i,n3}(f)+S_{\text{in}}(f)]R_L^2+8kTR_L}{4kTR_s}\cdot\frac{1}{g_{m3}^2 R_S R_L}=$$
$$\left(1+r_g\frac{3}{R_S}+\frac{|Ze|}{R_S}+\frac{\gamma}{g_{m3}R_S}\right)\frac{R_L}{R_S}+\frac{4}{\pi}\frac{\gamma}{}\cdot\frac{I_B R_L}{V_{LO}R_S^2 g_{m3}^2}+\frac{1}{R_S^2 g_{m3}^2} \qquad 4.97$$

4.6.6 Two-Tone Performance of Current-Commutating CMOS Mixers

In the mixer application, one also encounters the intermodulation between two closely separated input signals f_{R1} and f_{R2}, particularly with approximately the same power. They mix with each other or with the LO f_{LO} to generate third-order intermodulation signals of the type

$$f_{\text{itremod}}=(m+1)f_{R1}-mf_{R2})\pm F_{\text{LO}} \qquad m\geq 1 \qquad 4.98$$

The origin is found in the nonlinearity of the mixing or amplifying characteristic of the systems. Application of the Taylor expansion on the output current reveals

$$i(t)_{\text{out}} = I_o + \frac{di}{dv}(\Delta v) + \frac{1}{2}\frac{di^2}{d^2v}(\Delta v)^2 + \frac{1}{3!}\frac{di^3}{d^3v}(\Delta v)^3 + \cdots = I_o + g_1(\Delta v) + g_2(\Delta v)^2 + g_3(\Delta v)^3 + \cdots \quad 4.99$$

where g_1, g_2, and so on, are the first and higher order transconductances. For the bipolar transistors, we have with the assistance of (4.32)

$$i(t)_{\text{out}} \approx \lambda v(t)_{\text{in}} + \frac{1}{2}\lambda^2 v(t)^2_{\text{in}} + \frac{1}{6}\lambda^3 v(t)^3_{\text{in}} + \cdots \quad 4.100$$

whereas for the field effect transistors with application of (4.37) we arrive at

$$i(t)_{\text{out}} \approx K\left[\frac{V_g(2+\theta V_g)}{(1+V_g\theta)^2} v(t)_{\text{in}} + \frac{2}{(1+V_g\theta)^3} v(t)^2_{\text{in}} + \frac{1}{6}\frac{6\theta}{(1+V_g\theta)^4} v(t)^3_{\text{in}} + \cdots\right] \quad 4.101$$

where we have introduced $V_g = V_{GS} - V_{th}$. In the case of the source or emitter feedback, we introduce an effective input voltage for $v_{in}(t)$, in agreement with (4.53):

$$v_{\text{in,eff}}(t) \approx \frac{v_{\text{in}}(t)}{1+g_1 Z_e} \quad 4.102$$

In the case of two spurious signals,

$$v_1(t) + v_2(t) = V_1 \sin(\omega_1 t) + V_2 \sin(\omega_2 t) \qquad (V_2 \approx V_1) \quad 4.103$$

the fourth-order term in (4.99) reveals the third-order components

$$3V_1^2 \sin^2(\omega_1 t) V_2 \sin(\omega_2 t) + 3V_1 \sin(\omega_1 t) V_2^2 \sin^2(\omega_2 t) \quad 4.104$$

with amplitudes

$$\frac{3}{4}V_1^2 V_2 \approx \frac{3}{4}V_1 V_2^2 \quad 4.105$$

The third-order intercept point, the so-called IIP3, where the power of the spurious signal would be the same as that of the desired signal, provides important information about the linearity of the system. To this end, observing measurements at lower levels and a plot in the log–log diagram, we find that the desired output signal levels would be on a straight line increasing with the slope of 10 dB/dec, whereas the third-order products on the line increase by 30 dB/dec versus the input power. Their intersection is the IIP3 intercept point (see Fig. 4.18 and Example 4.6).

EXAMPLE 4.6

$$\begin{aligned} a_1 \quad & y - y_l = x - x_1 \\ a_3 \quad & y - y_3 = x - x_3 \\ & x_{\mathrm{IIP3}} = \frac{-y_3 + y_1 + 2x_3 - x_1}{2} \end{aligned} \qquad 4.106$$

The straight lines we were looking for are a_1 and a_3.

From Fig. 4.18*b*, we read: $y_3 = -95$ dBm, $y_1 = -25$ dBm, $x_3 = x_1 = -40$ dBm, and $x_{\mathrm{IIP3}} = -5$ dBm.

A general formula for IIP2 and IIP3 (IIPn) magnitudes is given in [4.48]:

$$x_{\mathrm{IIPn}} = \frac{-y_n + y_1 + nx_n - x_1}{n-1} \qquad 4.107$$

4.7 FREQUENCY DIVIDERS

Frequency dividers are encountered in frequency stability measurement systems, however, their major field of application is in communications devices, particularly in modern mobile or satellite microwave systems. In practice, we encounter three major types of dividers, namely, *static digital, injected digital,* and *analogue frequency dividers.* However, digital dividers are preferred for their compatibility with modern IC systems and they are used in *direct digital frequency synthesizers* (DDS) [4.50] and *phase-locked loops* (PLL) with *frequency dividers* (FD) in the feedback path, particularly in the high mega-

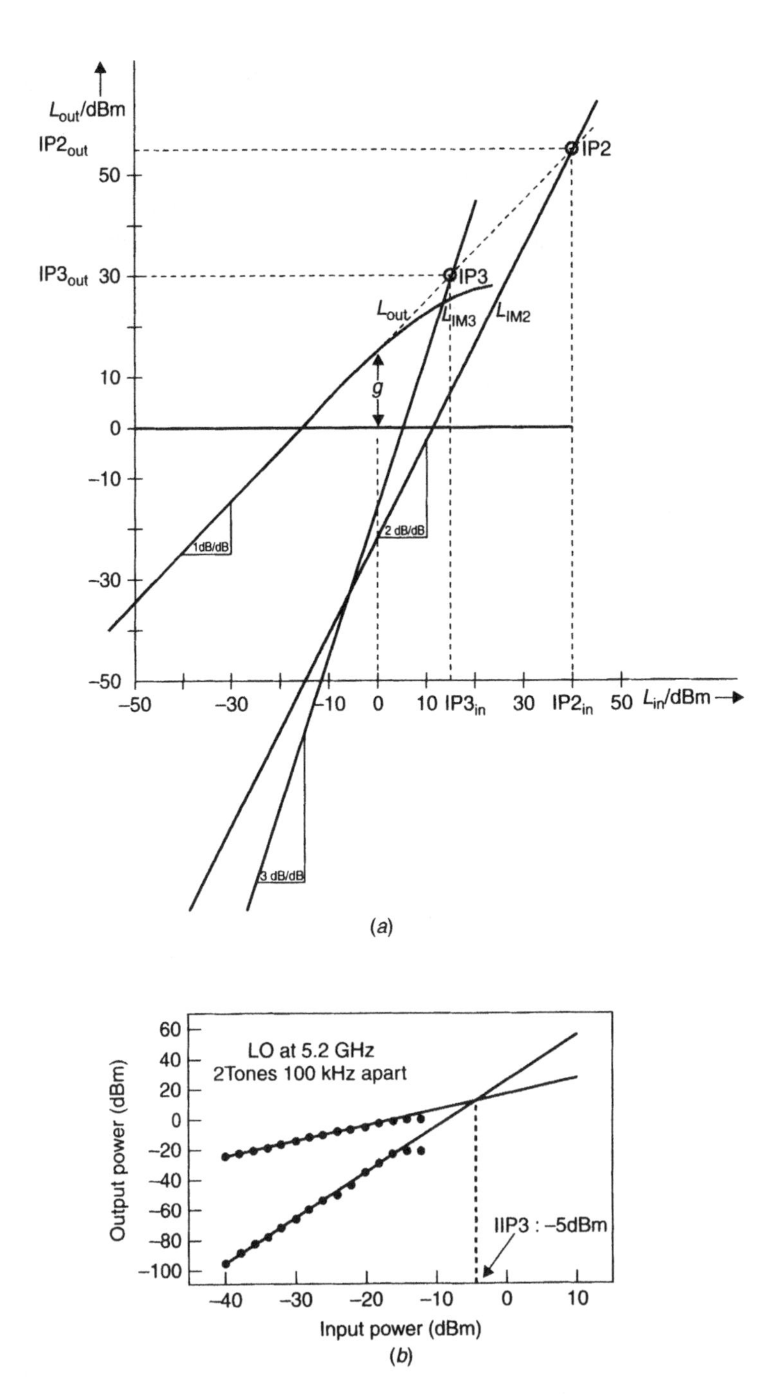

Fig. 4.18 Third-order intermodulation. (*a*) The corresponding straight lines of the the second- and third-order intermodulation plot [4.48]. (Copyright © Rohde & Schwarz, reproduced with permission.) (*b*) A practical example of the third-order intermodulation [4.49]. (Copyright © IEEE. Reproduced with permission.)

hertz and low gigahertz communications bands. In addition, the low-noise and small spurious signals around the carrier, with negligible leakage from channel to channel, are of highest importance for reliable connections. Note that *analogue frequency dividers* are nowadays used in measurement systems at the highest frequencies and in applications where the low additive noise is of importance.

4.7.1 Digital Frequency Dividers

Static digital frequency dividers are based on application of the IC gates forming memory systems. In practical applications, we encounter several basic IC families (see Fig. 4. 19):

TTL (transistor–transistor logic; not recommended for new applications)
CMOS (Complementary metal–oxide surface)
ECL (emitter coupled logic)
SCL (source-coupled logic), CML (current mode logic), and so on

The basic divide-by-two circuits are built from bistable flip-flop circuits, formed generally by two cross-coupled latches. for example, the D-type master–slave configuration (cf. Fig. 4.19*d*). A schematic arrangement is illustrated in Fig. 4.20*a*. Figure 4.20*b* has a chain of n divide-by-two D-latches with the division factor $N = 2^n$. The maximum input operation frequency was recently reported to be in the range of 100 GHz [4.52].

In most practical applications, we encounter the need for an arbitrary or variable division ratio in digital frequency dividers. In instances where small division factors are required, a few D or JK type cells, generally four, are arranged in one package. The output pins are so arranged that division by a small number (2, 3, 4, 5, etc.) is possible [4.53]. Recently, the arrangement for alternative division by 2 and 3 has become more important [4.54].

4.7.1.1 Dual Modulus Dividers

The principal arrangement of dual modulus dividers is shown in Fig. 4.21. The operation is as follows. First the input divider operates with the division factor, $P + Q$. Its output is fed simultaneously into both auxil-

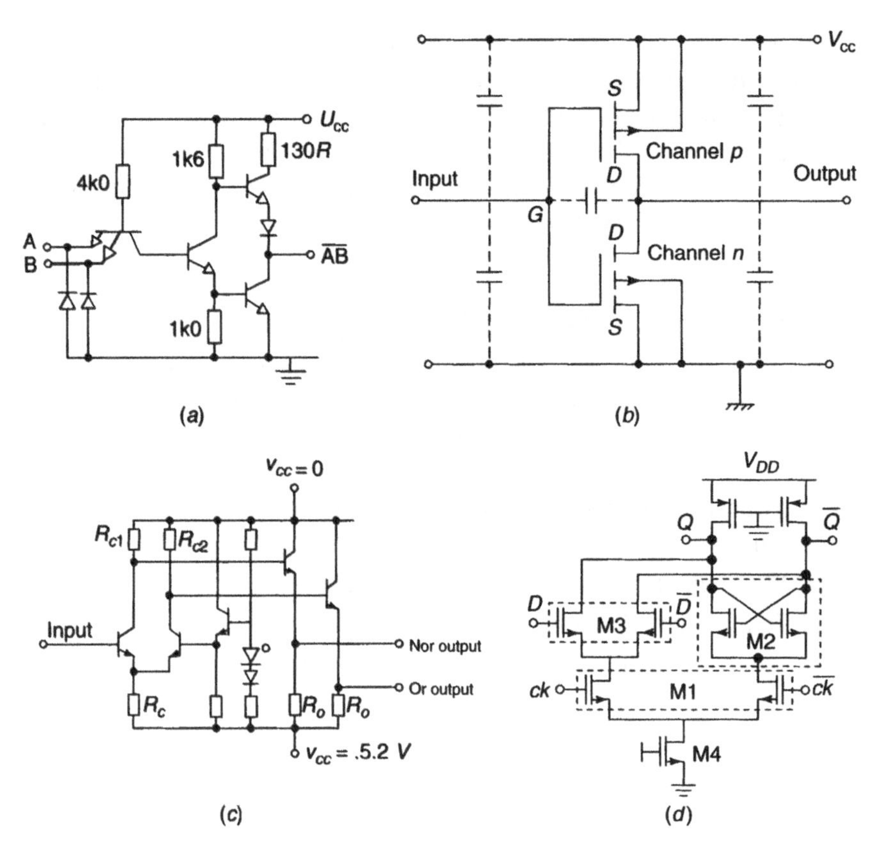

Fig. 4.19 (*a*) The TTL logic technology gate, (*b*) CMOS logic technology gates, (*c*) emitter-coupled logic, and (*d*) current-mode logic [4.51]. (Copyright © IEEE. Reproduced with permission.)

iary counters with the division factors A and M. As soon as the divider A overflows, the output signal changes the division factor, $P+Q$, to P and blocks the input to itself. This state remains unchanged until the divider M is full. Its output signal resets the main divider to the $P+Q$ state again and the auxiliary divider A to zero. Then the cycle starts to repeat. The time available for the change is ideally P periods. The final division ratio from the input to the output is computed from the following relation:

$$N=(P+Q)A+P(M-A)=PM+AQ \quad \text{(output 1)}$$

$$N=P+\frac{AQ}{M} \quad \text{(output 2)} \qquad 4.108$$

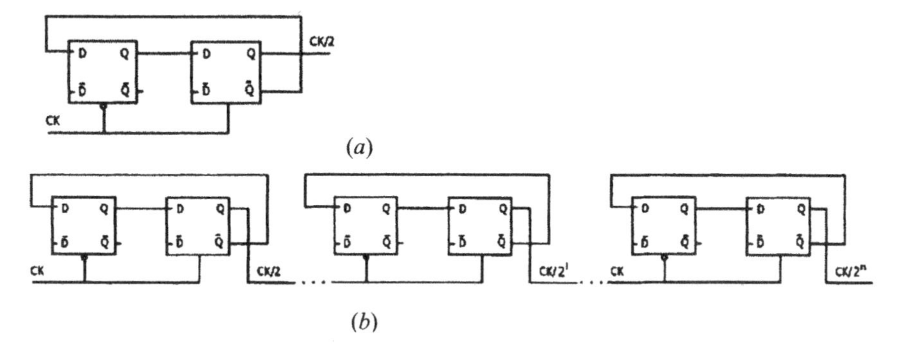

Fig. 4.20 (*a*) A schematic arrangement of divide-by-two circuits formed by two D flip-flops. (*b*) A chain of *n* divide-by-two D-latches with the division factor $N = 2^n$.

The minimum division factor N_{min} will be for $M_{min} = 1$ and $A_{min} = 0$

$$N_{min} = PM_{min} + A_{min} = P \tag{4.109}$$

which leads to the smallest possible dual-modulus divider of $P/P + 1 = 2/3$.

A 2/3 divider is comprised of two blocks, as depicted in Fig. 4.22. In accordance with information delivered by the latter divider block or

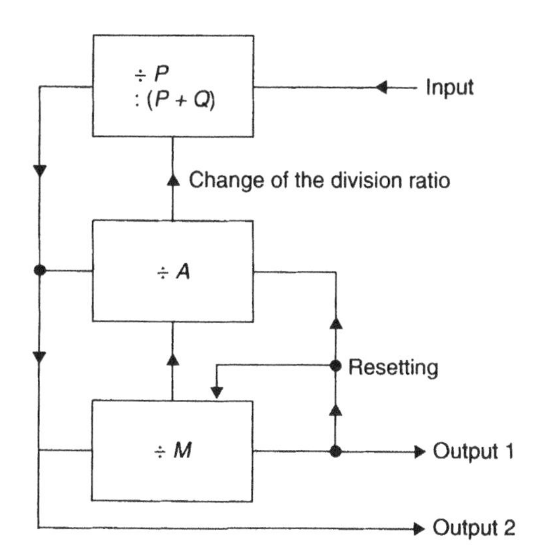

Fig. 4.21 Principle of dual-modulus digital frequency dividers [4.41].

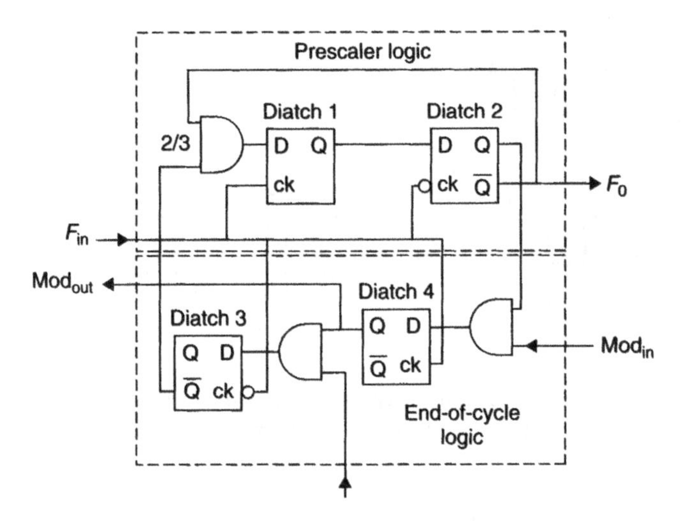

Fig. 4.22 A 2/3 divider cell [4.54]. (Copyright © IEEE. Reproduced with permission.)

supplied by the input p, the circuit performs division by 2 or by 3, as discussed in Example 4.7. A chain of n 2/3 cells provides an output signal with a period of

$$\begin{aligned} T_{out} &= T_{in} \cdot (2+p_o) \cdot (2+p_1) \cdots (2+p_n) = \\ & T_{in} \cdot (2^n p_n + 2^{n-1} p_{n-1} + \cdots + 2p_1 + p_o) \end{aligned} \quad 4.110$$

This equation shows that division by all integers from 2^n to $(2^{n+1} - 1)$ can be realized [4.54].

EXAMPLE 4.7

In this example, we examine a divider by 32 or 33. The principal arrangement is shown in Fig. 4.23. The design uses a 2/3 divider

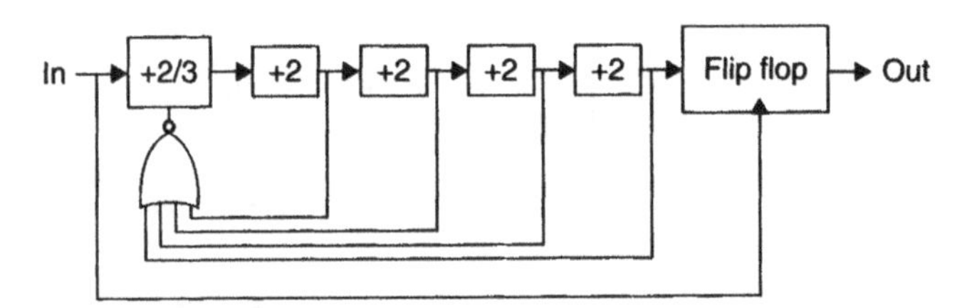

Fig. 4.23 (*a*) Block diagram of the divide-by-eight injection locked frequency divider. (*b*) Corresponding D-latch cell [4.55].

as the input stage. In instances where division by 32 is the desired operation, the 2/3 divider divides by two continuously so that the output is one thirty-second; if the input frequency is divided by 33, the NOR gate switches from the input modulus 0001 to 0000 and G_4 generates a signal that changes the modulus of the 2/3 to three for one cycle out of four cycles, resulting in the total division factor by 33.

4.7.1.2 Fractional-*N* Dividers

The fractional-*N* frequency divider is similar to the divide-by-*N* divider; however, with the assistance of an auxiliary divider, one input pulse is periodically removed (suppressed). The idea and operation of the fractional-*N* divider is explained with the assistance of Fig. 4.24.

For a better understanding of the problem, we assume that the auxiliary DDS frequency synthesizer is a simple divider by an integer F, which divides the reference frequency f_r. By assuming the circuit without pulse swallowing contains the period T_{rN}, the output signal is equal to the N periods T_o of the input frequency:

$$T_{rN} = NT_o \tag{4.111}$$

However, after each overflowing of the divider F, one pulse is missing in the input train T_o and the respective period T_{rN} is prolonged by one T_o. Consequently, the fundamental repetition period of the output signal is exactly

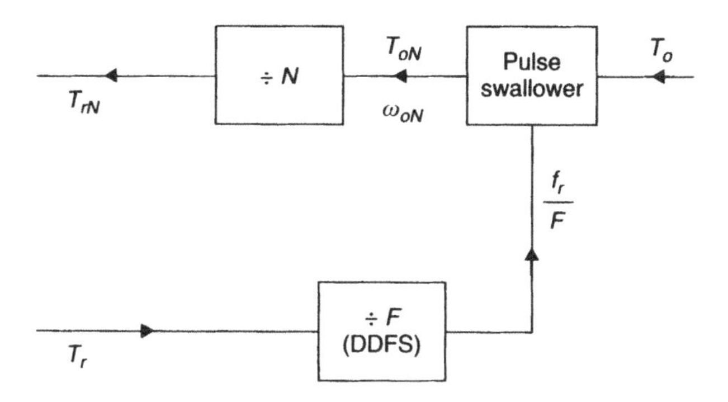

Fig. 4.24 Factional-*N* frequency divider.

$$T_o NF + T_o = T_r F \tag{4.112}$$

and the effective division ratio is easily computed as

$$N_{\text{eff}} = \frac{T_r}{T_o} = N + \frac{1}{F} \tag{4.113}$$

4.7.2 Spurious Signals in Variable-Ratio Digital Frequency Dividers

Inspection of the output period of the variable-ratio digital frequency dividers reveals irregularities introduced by the process of the pulse swallowing. In instances where the output period is only prolonged by one input pulse in the division period, we face a phase modulation by a sawtooth wave with an amplitude of 2π (cf. Fig. 4.25).

The corresponding phase modulation is given by the sawtooth wave with spectral lines [cf. 4.50]:

$$\varphi_{\text{div}} = 2\sum_{m=1}^{\infty} \frac{1}{m} \sin\left(m \frac{2\pi}{T_o NF + T_o} t \right) = 2\sum_{m=1}^{\infty} \frac{1}{m} \sin\left(m \frac{2\pi}{T_r F} t \right) \tag{4.114}$$

with the power of the spurious signals at the output of the divider

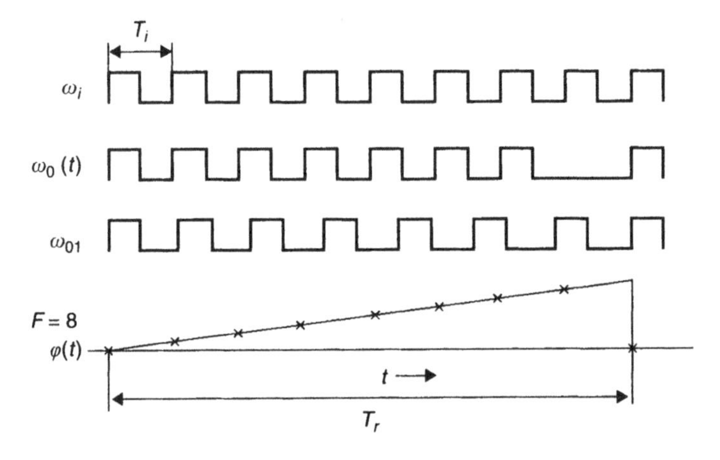

Fig. 4.25 Spurious phase modulation generated by quasiperiodic omission of each eighth pulse [4.50]. (Copyright © IEEE. Reproduced with permission.)

$$\varphi_{sp,1}^2 \approx \frac{1}{2}\frac{4}{m^2}\frac{1}{N^2} \qquad 4.115$$

In instances where F in relation (4.112) is a fraction $F = Y/X$, the effective division factor is

$$N_{\text{eff}} = N + \frac{X}{Y} \qquad 4.116$$

and the output phase is time dependent:

$$\varphi_r(t_k) = \omega_r T_r = \omega_r T_o\left(N + \frac{x}{Y}\right) = \frac{2\pi}{N_{\text{eff}}} s_r(t_k) \qquad 4.117$$

with the time modulation function $s_r(t_k)$ that is similar to the DDS frequency synthesizers:

$$s_r(t_k) = 2\pi\left[k\frac{X}{Y} - \text{integer}\left(k\frac{X}{Y}\right)\right] \qquad 4.118$$

The shape and periods of the spurious signals are investigated with the assistance of the modified continued fraction expansion applied to the factor in brackets in (4.118). The result is a superposition of sawtooth waves. Details were discussed earlier in [4.41] and [4.50].

4.7.3 Noise in Digital Frequency Dividers

Study of the divider noise properties started over 30 years ago [4.56], and the preliminary experimental findings revealed that the power spectral density PSD, $S_{\varphi,D}(f)$, of the output frequency $f_{\text{out}} = f_o$ for the TTL logic family dividers is approximately

$$S_{\phi,D}(f) \approx \frac{10^{-14.7}}{f} + 10^{-16.5} \qquad 4.119$$

which resembles the noise PSD found for the switching mixers [cf. (4.86)]. However, in addition, in each bistable stage there is some delay between the arrival of the input pulse and the instant when the stage settles into its new state. Further, due to inherent semiconductor noises, ambient temperature variations, supplying voltage variations, and so on, there is a small time jitter of the leading and trailing edges of the output rectangular wave form. The remedy provides, at least partially, application of synchronous clocking (cf. Fig. 4.26).

However, the real situation is a bit more complicated since the output noise of frequency dividers is composed of the input signal noise, the PSD of which is reduced by the square of the division factor, N^2, and the contributions added by the division circuits themselves, particularly the last one:

$$S_{\phi,\text{out}}(f) = \frac{S_{\phi,\text{in}}}{N^2} + S_{\phi,\text{div}} \qquad 4.120$$

Very often, especially for large division factors, N, the first term on the rhs of (4.120) disappears. In addition, as early as the 1970s, authors of [4.56] suggested, for larger output frequencies, f_o, that the divider output PSD contained frequency-dependent terms proportional to the in-

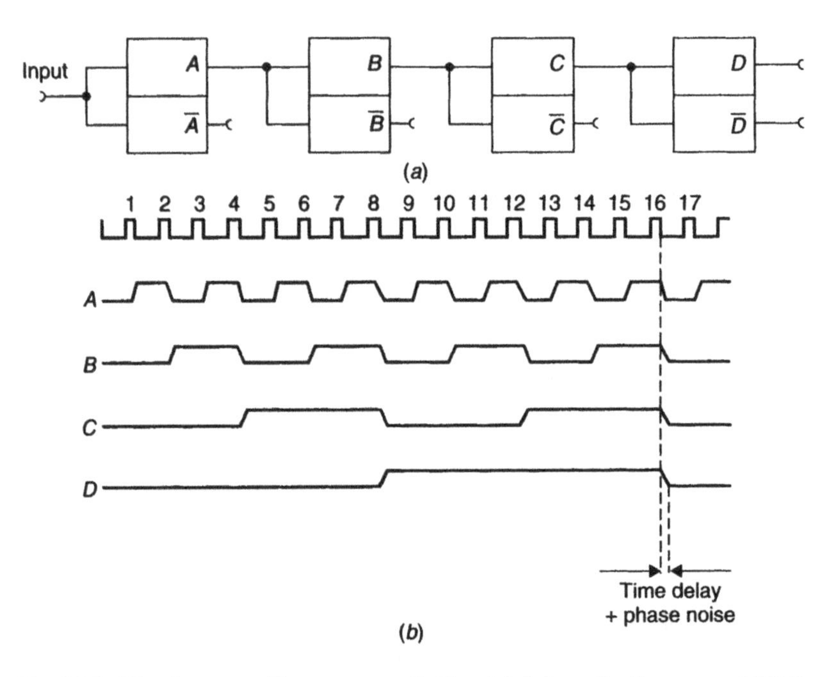

Fig. 4.26 The four-stage binary counterdivider. (*a*) Schematic diagram and (*b*) the respective waveforms with the time jitter [4.41].

creasing f_o. This was verified by later investigations based on much larger experimental sets [4.57, 4.58]. In a deeper discussion of the problem, performed by the present author, after plotting experimental measurements and applying asymptotic approximations (lines), depicted in Fig. 4.27, he arrived at the following equations for the PSD of the output noise of digital frequency dividers:

$$S_{\phi,D}(f) \approx \frac{10^{-14\pm1} + 10^{-27\pm1} f_o^2}{f} + 10^{-16\pm1} + 10^{-22\pm1} f_o \qquad 4.121$$

Finally, for the GaAs divider family (and many other high-frequency dividers) (4.121) requires correction in the first term:

$$S_{\phi,D}(f) \approx \frac{10^{-10\pm1} + 10^{-27\pm1} f_o^2}{f} + 10^{-16\pm1} + 10^{-22\pm1} f_o \qquad 4.122$$

However, until now the origin of the experimental constants, which are functions of the output frequency, $f_{o,\text{out}}$, and hold to the highest input–output frequencies, has not been discussed. By taking into account that the noise contribution of the earlier divider stages slowly disap-

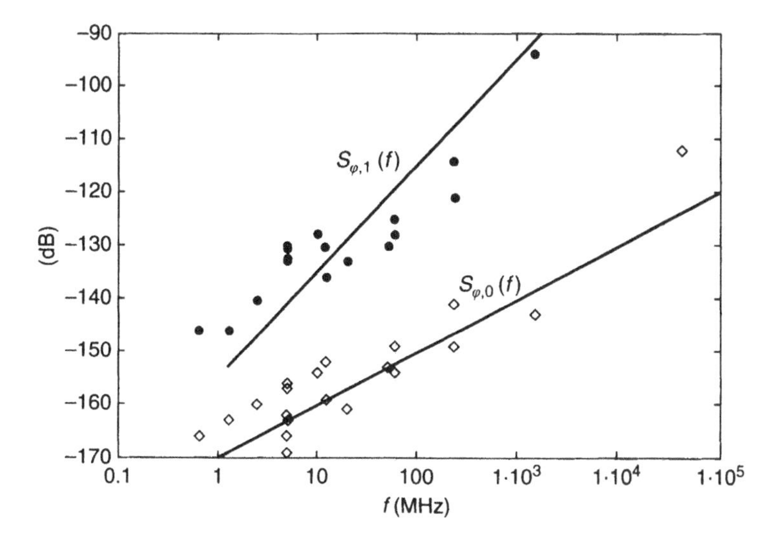

Fig. 4.27 The (PSD of digital frequency dividers: $S_{\varphi,1}(f = 1)$ (dB/Hz) in the $1/f$ noise region (for clarity, we shifted them by +20 dB) and $S_{\varphi,0}(f)$ (dB/Hz) in the white phase noise region (diamonds) of TTL, ELC, and CML digital dividers.

pears, we must assume that the frequency dependent terms in (4.121 or 4.122) are generated in the final stage or stages of dividers. Consequently, evaluation of the output noise PSD requires appraisal of the spurious time jitter. By assuming that the time jitter is generated by the white and flicker phase noise components only, we arrive with the assistance of (5.80) at

$$< \left(\frac{\Delta T}{T_o}\right)^2 > = \frac{f_o^2}{(2\pi)^2} \int_{f_L}^{f_H} \left(\frac{h_{-1}}{f} + h_o\right) df =$$
$$\frac{f_o^2}{(2\pi)^2} h_{-1} \ln\left(\frac{h_H}{f_L}\right) + \frac{f_o^2}{(2\pi)^2} h_o \frac{f_H}{2} = \qquad 4.123$$
$$\frac{f_o^2}{(2\pi)^2} h_{-1} \ln\left(\frac{h_H}{f_L}\right) + a_o \frac{f_H}{2}$$

In Example 4.8, we try to get to the measured numerical approximations.

EXAMPLE 4.8

The white noise constant from (4.123) is

$$\frac{f_o^2}{(2\pi)^2} h_o \frac{f_H}{2} \approx \frac{f_o^2}{(2\pi)^2} \frac{a_o}{f_o^2} \frac{f_{out}}{2} \approx a_o 10^{-2} f_{out} \approx \qquad 4.124$$
$$4kT 10^{-2} f_{out} \approx 10^{-22} f_{out}$$

Assuming that the white noise constant is $a_o = 4kT$, we arrive at the experimental value 10^{-22}. Similarly, we arrive at the sought value of the constant in relation to the flicker noise in (4.121) or (4.122), that is,

$$h_{-1} \approx \frac{10^{-9}}{f_o^2} lg(f_H / f_L) \approx \frac{10^{-8}}{10^{-19}} \approx 10^{-27} \qquad 4.125$$

4.7.4 Injection-Locked Frequency Dividers

Until now, we have discussed frequency dividers composed of latches that can store the logic state for an infinite time and with no lower limit to the input frequency, which are designated as *static digital frequency dividers*. Major difficulties with these static digital frequency dividers are finite upper-bound frequencies in the gigahertz ranges [cf. 4.52, 4.59] and rather large operation currents. The latter, the effective large power consumption, is in contradistinction to the modern digital communications in microwave ranges, such as GSM (Global System for Mobile Communications) and other emerging communications systems in the gigahertz ranges. The remedy is provided by injection-locked frequency dividers usually consisting of oscillating flip-flops (cf. Fig. 4.28) or oscillators phase-locked on their harmonic or subharmonic equal to the input frequency.

The situation is discussed in more detail in Section 6.10.3.2. The major advantage is low power consumption (in contradistinction to the static digital frequency dividers) without impairing the expected output phase noise. The difficulty is the reduced sensitivity of the input voltage versus input frequency, as depicted in Fig. 4.29, together with operation on low harmonics only. The latter difficulty may be alleviated by using ring oscillators for synchronization [e.g., 4.60, 4.61].

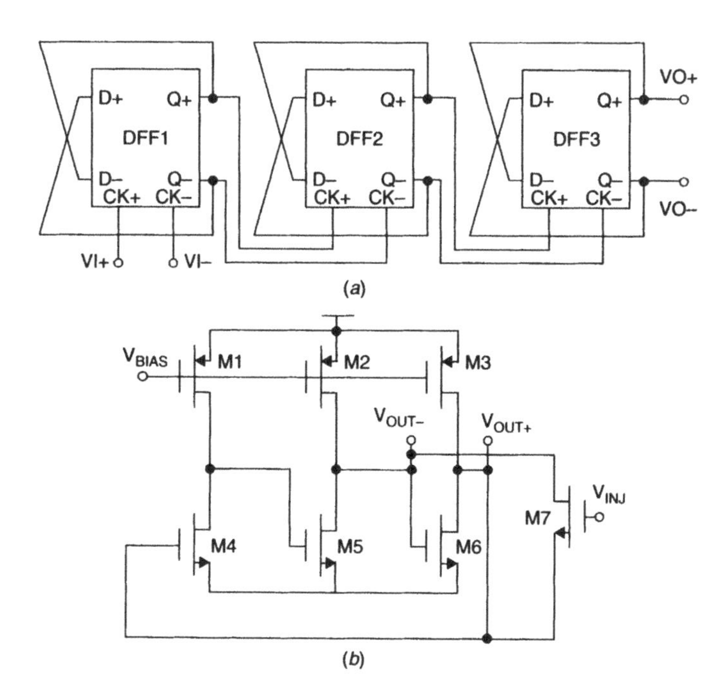

Fig. 4.28 (*a*) Block diagram of the divide-by-eight injection locked-frequency divider. (*b*) Corresponding *D*-latch cell [4.55]. (Copyright © IEEE. Reproduced with permission.)

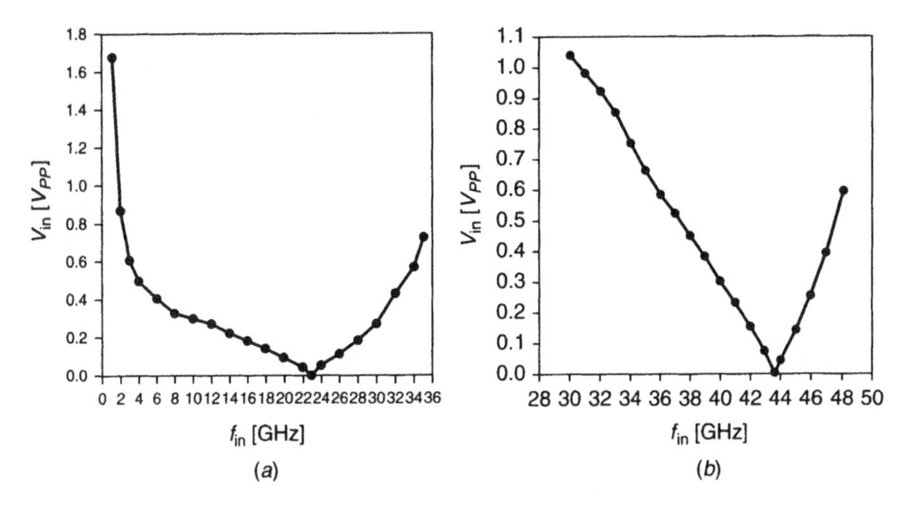

Fig. 4.29 Sensitivity of input voltage versus input frequency of (*a*) the static frequency dividers and (*b*) the dynamic dividers [4.55]. (Copyright © IEEE. Reproduced with permission.)

4.7.5 Regenerative Frequency Dividers

Regenerative (or Miller [4.62]) dividers use positive feedback to generate an output signal with a fractional frequency. A block diagram is shown in Fig. 4.30 and the operation was discussed in [4.41]. First, without the input signal the regenerative action cannot take place. When the signal is switched on, a small signal, $a_n(t)$, is generated through the transient phenomenon at the second input of the mixer (double balanced) and, due to the feedback frequency division, is accomplished if the overall gain exceeds one and saturation effects limit the amplitude, as in a conventional oscillator. In addition, the oscillation requires zero-phase shift around the loop, that is,

$$\omega t + \phi_i - \left(\frac{n-1}{n} \omega t + \phi_{n-1} \right) + \phi_a = \frac{\omega t}{n} + \phi_a \qquad 4.126$$

and

$$\left(\frac{\omega t}{n} + \phi_o \right)(n-1) + \phi_b = \phi_{n-1} + \frac{n-1}{n} \omega t \qquad 4.127$$

After eliminating φ_{n-1} from (4.126) and (4.127), we get the relation between the output and input phases ϕ_o and ϕ_i, respectively:

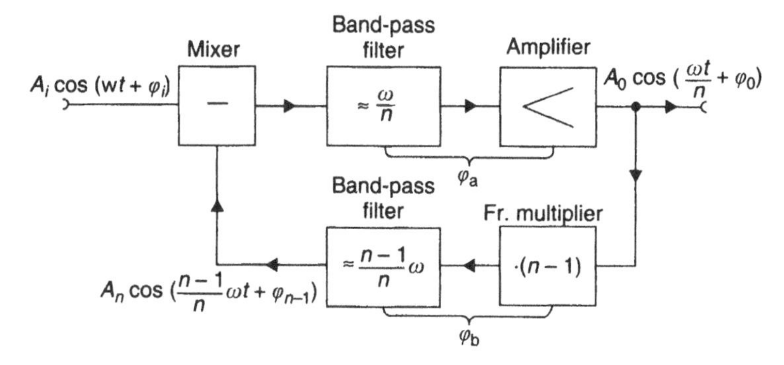

Fig. 4.30 Block diagram of a regenerative frequency divider [4.41].

$$\phi_o = \frac{\phi_i + \phi_a - \phi_b}{n} \tag{4.128}$$

The corresponding noise PSD is

$$S_\phi(f) \approx \frac{10^{-13.5}}{f} + 10^{-16.5} \tag{4.129}$$

4.7.5.1 Spurious Signals in the Regenerative Frequency Dividers

If no multiplier is included in the loop frequency, division is by two. For larger division factors, the chain of such dividers is often used [4.63]. Given appropriate loop filters, other division factors can be realized, such as one-third and two-thirds, generally $1/n$ and $(n - 1)/n$, n being an integer, However, the operational bandwidth of the loop decreases with an increasing division ratio [4.64]. The latter depends predominantly on the selectivity of the band-pass filters, (ω/n) and $[\omega(n - 1)/n]$, and, in addition, on the loop gain for the next possible working modes, that is, $\omega(n \pm 1)$ [and naturally for all others $\omega(n \pm k)$ as well], which must be safely below *one*. A block diagram of such a conjugate frequency divider is reproduced in Fig. 4.31. With higher division ratios, we face a comb of frequencies in Fig. 4.32 and reproduce an output spectrum of the divide-by-five divider. Note that loop filters eventually comprise additional time delays that may reduce the tuning range or even cause an asynchronous operation mode, particularly if the effective Q of the band-pass filters is too high (cf. [4.64] and Fig. 4.33).

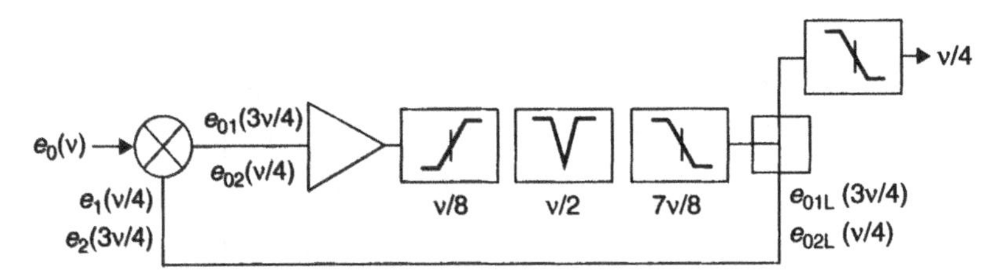

Fig. 4.31 Schematic diagram of divide-by-four regenerative divider [4.50]. (Copyright © IEEE. Reproduced with permission.)

4.7.5.2 Noise Sources in Regenerative Frequency Dividers

Regenerative dividers are potentially very-low-noise circuits. Note the similarity between (4.129) and (4.120). Again, the input noise is reduced by the squared division factor. However, due to feedback, the same reduction is applied to all component noises as well. This property groups analogue frequency dividers into a divider set with the lowest output noise. This might be one reason for their application at very high frequencies [4.65 or 4.66], at which digital systems are noisy (cf. 4.123 or 4.124) and cumbersome.

Experimental investigation of the output noise of the regenerative dividers performed by several authors reveals that (4.86) is valid for

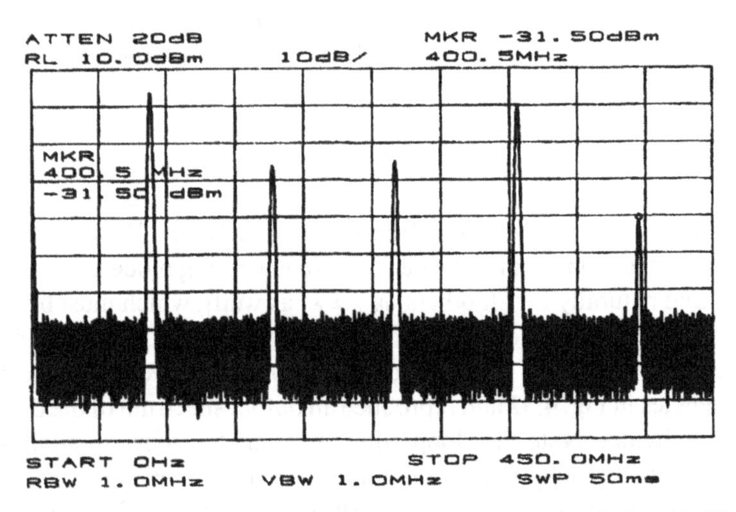

Fig. 4.32 Schematic diagram of a divide-by-four regenerative divider [4.63].

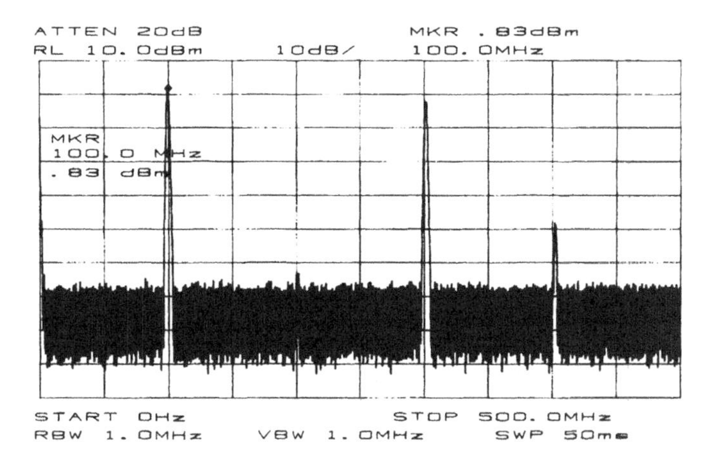

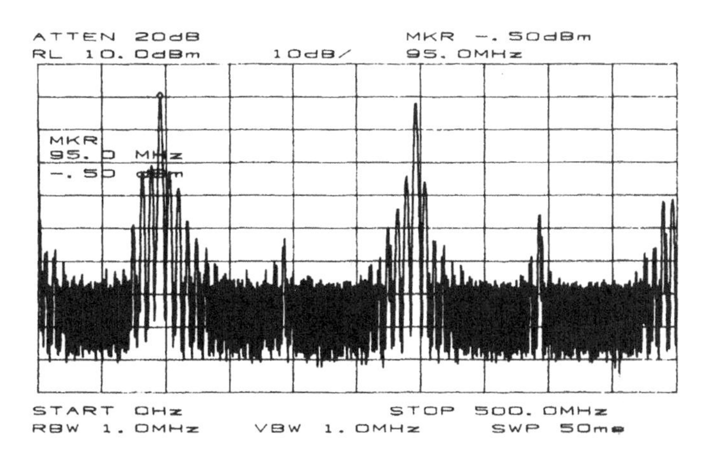

Fig. 4.33 Output of the regenerative divide-by-four divider: (*a*) synchronous output; (*b*) asynchronous operation mode [4.63].

mixers and digital dividers (4.119) with low output frequency and also could be applied to analogue frequency dividers since the components generating noise are mainly mixers and amplifiers. Actual background noise was analyzed by several authors in the ranges from hundreds of megahertz to tens of gigahertz. It is interesting that the phase noise in the investigated regenerative frequency dividers is approximately

$$S_\phi(f) \approx \frac{10^{-13.5}}{f} + 10^{-16.5} \tag{4.130}$$

4.8 FREQUENCY MULTIPLIERS

Since the start of practical applications of electromagnetic waves in communications or, more properly, in radio communications, we encounter frequency multipliers used to generate more channels with higher harmonic frequencies, but retaining the frequency stability of the original oscillators, particularly after introduction of crystal resonators. These early frequency multipliers used electronic circuits with nonlinear voltage–current characteristic that distorted output wave forms and filtered out the desired harmonic signal. At present, nearly all frequency multiplications are performed with the assistance of PLLs, which will be discussed in detail in Chapter 6.

Reverting to the frequency multiplication via generation and filtering of harmonics, still encountered in standard frequency laboratories and institutes, we recall that an ideal frequency multiplier would change the input signal $A_{\text{in}}\cos \omega_{\text{in}}t$ directly into the desired harmonic multiple $A_{\text{out}}\cos n\omega_{\text{in}}t$ with the assistance of the nonlinear elements and Chebyshev polynomials[1] [4.67]:

$$Y = \cos(n \cdot \cos^{-1}X) = T_n(X) \qquad 4.131$$

Graphical solution (e.g., on CRT) leads to the well-known Lissajous figures. Simple realization of condition (4.131) is only encountered for the generation of the second harmonic that requires the parabolic nonlinearity easily realized through a push–push connection of diodes often encountered. Very effective for frequency multiplication are voltage or current pulses generated by Schottky barrier diodes producing combs of higher harmonics still used for generation of microwave signals in the 100 GHz and higher frequency ranges [cf. 4.68, 4.69].

REFERENCES

4.1. M.S. Gupta, *Electrical Noise: Fundamentals and Sources,* New York: IEEE Press, 1977.

4.2. J.B. Johnson, Thermal Agitation of Electricity in Conductors. *Phys. Rev., 32* (July 1928), pp. 97–109.

[1] $T_1(X) = X$, $T_2(X) = 2X^2 - 1$, and $T_3(X) = 4X^3 - 3X$.$\cdots$ $T_{n+1}(X) = 4XT_n(X) - T_{n-1}(X)$ [cf 4.70].

4.3. J. Bernamont, *Ann. Phys.* [Paris], *7,* 71 (1937).
4.4. A. Van der Ziel, *Noise in Solid State Devices and Circuits,* New York: Wiley, 1986.
4.5. J. Pierce, Physical Sources of Noise, *Proc. IRE* (May 1956), 601–608 (reprinted by Gupta).
4.6. H.D Ascarrunz, A. Zhang, E.S. Ferre-Pikal, and F.L. Walls, PM Noise Generated by Noisy Components, in *1998 IEEE Int. Frequency Control Symp.* (May 1998), pp. 210–217.
4.7. S.M. Sze, *Physics of Semiconductor Devices* (1969), Wiley, p. 43.
4.8. J.J. Ebbers and T.L. Moll, Large Signal Behavior of Junction Transistors, *Proc. IRE, 42* (1954), pp. 1761–1772.
4.9. BSSIM4.6.1 MOSFET Model, User's Manual, Available at www-device.eecs.berkeley.edu/bsim4.
4.10. J-D Jin and S.S.H. Hsu, A 0.18-μm CMOS Balanced Amplifier for 24-GHz Applications, *IEEE J. Solid-State Circuits, 43* (Feb. 2008), pp. 440–445.
4.11. S-S Song, D-G. Im, H-T Kim, and K. Lee, Highly Linear Wideband CMOS Low-Noise Amplifier Based on Current Amplification for Digital TV Tuner Applications, *IEEE Microwave Wireless Components Letts., 18* (Feb. 2008), p. 118–120.
4.12. A. Van der Ziel, Noise in Solid-State Devices and Lasers, *Proc. IEEE, 58* (Aug. 1970), pp. 1178–1206 (Reprinted by Gupta).
4.13. O. Mueller, On 1/*f* noise in Diodes and Transistors, in *Proceedings of the 28th Annual Frequency Control Symposium,* (1974), pp. 166–176.
4.14. M.T. Sebestian, A-K Axelsson, and N. Mc *N* Alford, List of Microwave Dielectric Resonator Materials and their Properties, South Bank University UK. (Available at http://www.lsbu.ac.uk/dielectric-materials).
4.15. P.L. Olsen, Evaluate 1/*f* Effects on Oscillator Phase Noise, *Microwave & RF* (August 1997), p. 8.1/*f* Noise Characteristic Influencing Phase Noise, CEL—California Eastern Laboratories, Application Note 1026. Available at htpp:/www. CEL.COM.
4.16. M. Sanden, F. Jonsson, M. Oestling, O. Marinov, and M.J. Deen, Low-frequency Noise in Polysilicon Emitter Bipolar Transistors and Up-conversion to Phase Noise in Oscillators, *IEEE Trans. Electron Devices, 49* (March 2002), pp. 514–20.
4.17. N.H. Hamid, A.F. Murray, and S. Roy, Time-Domain Modeling of low Frequency Noise in Deep-Submicrometer MOSFET, *IEEE Tr. On Circuit and Systems—I, 55* (Feb.2008), pp. 233–245.
4.18. L. Forbes, Ch. Zhang B. Zhang, and Y. Chandra. Comparison of Phase Noise Simulation Techniques on BJT LC Oscillators, *IEEE Tr. UFFC* (June 2000), p. 716.
4.19. P.J. Topham, Ga As Bipolar Trasistors for Microwave And Digital Circuits, *GEC J. Res., 9*(2) (1991), p. 74.
4.20. R.A. Anderson, S-parameters techniques for Faster, More Accurate Network Design, *Hewlett-Packard J.* (Feb. 1967) (Application Note 95-1).
4.21. A. Van der Ziel, Unified Presentation of 1/*f* Noise in Electronic Devices: Fundamental 1/*f* Noise Sources, *Proc. IEEE, 76* (1988), pp. 233–58.

4.22. Walls, E.S. Ferre-Pikal, F.L. and S.R. Jefferts, Origin of $1/f$ AM and PM Noise in bipolar Junction Transistor Amplifiers, *IEEE Tr. UFFC, 44* (March, 1997), pp. 326–34.
4.23. E. Rubiola and V. Giordano, On $1/f$ Frequency Noise in Ultra-sable Quartz Oscillators, *IEEE Tr. UFFC, 54* (Jan. 2007), pp. 15–22.
4.24. D. Halford, A.E. Wainwrite, and J.A. Barnes, Flicker Phase Noise in RF Amplifiers and Frequency Multipliers: Characterization, Cause, and Cure, in *Proceedings of the 22nd Annual Frequency Control Symposium* (1968), pp. 340–341.
4.25. SD.J. Healey III, Flicker Frequency and Phase and White Frequency Phase Fluctuations in Frequency Sources, in *Proceedings of the 26th Annual Frequency Control Symposium* (1972) pp. 29–42.
4.26. E.S. Ferre-Pikal, Reduction of Phase Noise in Linear HBT Amplifiers Using Low-Frequency Active Feedback, *IEEE Tr. On Circuit and Systems—I*, 51 (Aug. 2004), pp. 1417–23.
4.27. E. Rubiola, C. Francese, and A. De Marchi, Long-term Behavior of Operational Amplifiers, *IEEE Tr. Intrum. Measur* (2001), pp. 89–94.
4.28. E.S. Ferre-Pikal, Reduction of phase noise in linear HBT amplifiers using low-frequency active feedback, IEEE Tr., 52 (2004), 1417–1421.
4.29. C.W. Nelson, F.L. Walls, M. Sicarrdi, and A. De Marchi, A New 5 and 10 MHz High Isolation Distribution Amplifier, *Frequency Control Symp.* (1994), pp. 567–571.
4.30. A. Hati, D.A. Howe, F.L. Walls, and D.K. Walker, Merits of PM Noise Measurement over Noise Figure: A Study at Microwave Frequencies, *IEEE Tr. UFFC, 51* (2006), pp. 1889–1894.
4.31. T.D. Tomlin, K. Fynn, and A. Cantoni, A Model for Phase Noise Generation in Amplifiers, *IEEE Tr. UFFC, 48* (2001), pp. 1574–1554.
4.32. A. De Marchi, F. Mussino, and M. Siccardi, A High Isolation Low Noise Amplifier with Near Unit Gain up to 100 MHz, in *Proceedings of the Annual Frequency Control Symposium* (1993), pp. 216–219.
4.33. G.K. Montress and T.E. Parker, Design Techniques for Achieving State-of-the-Art Oscillator Performance, in *Proceedings of the 44th Annual Symposium on Frequency Control*, pp. 522–535.
4.34. W.J. Tanski, Develoment of a Low Noise L-band Dielectric Resonator Oscillator, *Frequency Control Symp.* (1994), pp. 472–477.
4.35. G. Cibiel, L. Escotte, and O. Llopis, A Study of the Correlation Between High-Frequency Noise and Phase Noise in Low-Noise Silicon-Based Transistors, *IEEE Tr., Microwave Theory Tech., 52* (2004), pp. 183–190.
4.36. A. Hati, D.A. Howe, F.L. Walls, and D.K. Walker, Merits of PM Noise Measurement over Noise Figure: A Study at Microwave Frequencies, *IEEE Tr. UFFC, 53* (2006), pp. 1889–1894.
4.37. R. Boudot, S. Gribaldo, V. Giordano, O. Llopis, C. Rocher, N. Bazin, and G. Cibiel, Sapphire Resonators + SiGe Transistors Based Ultra Low Phase Noise Microwave Oscillators, in *Frequency Control Symp. Expos.* (2005), pp. 865–871.
4.38. C. Chambon, L. Escotte, S. Gribaldo, and O. Llopis, Band Noise-Parameter Measurement of Microwave Amplifiers Under Nonlinear Condition, *IEEE Tr. Microwave Theory Tech. 55* (2007), pp. 795–800.

4.39. D.A. Howe and A. Hati, Low-noise x-band Oscillator and Amplifier Technologies: Comparison and Status, in *Frequency Control Symp. Expo., 2005* (Aug. 2005), pp. 481–487.
4.40. V.F. Kroupa, Amplitude of the General Intermodulation Product, *IEEE Proc., 58* (1970), p. 851–852. Or *Frequency Synthesis: Theory, Design et Applications,* C. Griffin, London, 1973.
4.41. V.F. Kroupa, *Frequency Synthesis: Theory, Design and Applications,* C. Griffin, London (1973). *Phase Lock Loops and Frequency Synthesis* (2003), New York: Wiley.
4.42. B.C. Henderson, Predicting Intermodulation Suppression in Double-Balanced Mixers, 2001 Watkins-Johnson Communications.
4.43. Mixer Application Information, Tech-note, Watkins-Johnson Co. 1976. Reprinted 2001.
4.44. M.T. Terrovitis and R.G. Meyer, Noise in Current Commutating CMOS <ixers *IEEE J. Solid-State Circuits, 34* (June, 1999), pp. 772–783.
4.45. H.D. Darabi and A.A. Abidi, Noise in RF CMOS Mixers: A Simple Physical Model, *IEEE J. Solid-State Circuits, 35* (Jan, 2000), pp. 15–25.
4.46. F.L. Walls and C.M. Felton: Low Noise Frequency Synthesis, in *Proceedings of the 41st Annual Frequency Control Symposium* (May 1987), pp. 512–518.
4.47. H.B. Dwight, *Tables of integrals and other mathematical data*, 4th ed., (1961), New York: The MacMillan Company.
4.48. Ch. Rauscher, *Fundamentals of Spectrum Analysis,* Rohde & Schwarz, Muenchen, 2002.
4.49. J. Park, Ch-Ho Lee, B-S Kim, and J. Laskar, Design and Analysis of Low Flicker-Noise CMOS Mixer for Direct-Conversion Receivers, *IEEE Tr., MTT-54* (Dec., 2006), pp. 4372–4380.
4.50. V.F. Kroupa, ed. *Direct Digital Frequency Synthesizers*. IEEE Press, 1999.
4.51. S. Levantino, L. Romano, S. Pellerano, C. Samori, and A. Lacaita, Phase Noise in Digital Dividers, *IEEE J. of Solid-State Circuits, 39* (May 2004), pp. 775–783.
4.52. A.Rylyakov and T. Zwick, 96 GHz Frequency Divider in SiGe Bipolar Technology, *IEEE J. Solid-State Circuits* (Oct. 2004), pp. 1712–1715.
4.53. U.L. Rohde, *Microwave and Wireless Synthesizers,* (1997), New York: Wiley.
4.54. C.S. Vaucher, I. Ferecic, M. Locher, S. Sedvallson, Urs Voegeli, and Z. Wang, A family of Low-power Truly Modular Programmable Dividers in Standard 0.35-μm CMOS Technology, *IEEE J. Solid-State Circuits, 35* (July 2000), pp. 1039–1045.
4.55. S. Cheng, H. Tang, J. Silva-Martinez, and A.I. Kasrsilayin, A Fully Differential Low-power Divide-by-8 Injection Locked Frequency Divider up to 18 GHz, *IEEE J. of Solid-State Circuits* (March 2007), pp. 583–591.
4.56. V.F. Kroupa, J. Pavlovec, and L. Sojdr, Noise in Standard Frequency Sources, in *Digest of the Conference on Precision Electromagnetic Measurements,* Braunschweig (1980), pp. 147–151.
4.57. W.F. Egan: Phase Noise Modeling in Frequency Dividers, in *IEEE Proc. of the 45th Frequency Control Symposium* (May 1991), pp. 629–635.
4.58. V.F. Kroupa, Jitter and Phase Noise in Frequency Dividers, *IEEE Tr. Insrt. Meas.* (Oct. 2001), pp. 1241–1243.

4.59. T. Suzuki, T. Takahashi, T. Hirose, and M. Takikawa, A 80-Gbit/s D-type Flip-Flop Circuit using HEMT Technology, IEEE J. Solid-State Circuits (Oct. 2004), pp. 1706–1711.

4.60. Z. Lao, W. Bronner, A. Thiedem, M. Schlechtweg, A. Huelsmann, M. Rieger-Motzer, G.. Kaufel, B. Raynor, and M. Sedler, 35-GHz Static and 48-GHz Dynamic Frequency Divider IC's using 0.2-μm AlGaAs/GaAs-HEMT's, *IEEE J. of Solid-State Circuits* (Oct. 1997), pp. 1556–1562.

4.61. S. Sim, D-W. Kim, and S. Hong, A CMOS Direct Injection-locked Frequency divider with high Division Ratios, *IEEE Microwave and Wireless Components Letts., 19* (May 2009), p. 314–316.

4.62. R.L. Miller, Fractional Frequency Generators Utilizing Regenerative Modulation, *Proc. IRE* (1939), pp. 446–457.

4.63. M. Mossammaparast, McNeilage, P. Stockwell, J.H. Serls, and E. Suddaby, Low Phase Noise Divisions from X-band to 640 MHz, in *Proceedings of the 2000 International Frequency Control Symposium,* pp. 685–688.

4.64. A.S. Gupta, J.F. Garcia Nava, and F.L. Walls, Conjugate Regenerative Dividers, *IEEE Tr. on UFFC, 51* (March 2004), pp. 271–276.

4.65. E.S.7-Pikal and F.L. Walls, Low PM Noise Regenerative Dividers, in *Proceedings of the 2000 International Frequency Control Symposium* (1997), pp. 426–432.

4.66. M.M. Driscoll, Spetral Performance of Frequency Multipliers and Dividers, in *Proceedings of the International Frequency Control Symposium* (1992), pp. 193–199.

4.67. F.S. Banes and G.F. Eiber, An Ideal Harmonic Generator, *Proc. IEEE, 53,* pp. 693–695.

4.68. Ch. Mao, Ch.S. Nallani, S. Sankan E. Seak, and K.O. Kenneth, 125 GHz Frequency Doubler in 0.13 μm CMOS, *IEEE J. Solid-State Circuits, 44* (May 2009), pp. 1531–1538.

4.69. E. Monaco, M. Pozzoni, F. Svelto, and A. Mazzanti, Injection-Locked CMOS Frequency Doublers for μ-Wave and mm-Wave Applications, *IEEE J. of Solid-State Circuits, 45* (August 2010), pp 1565–1574.

4.70. G.A. Korn and T.M. Korn, *Mathematical Handbook* (1958), New York: McGraw-Hill.

CHAPTER 5

Time Domain Measurements

There is a general agreement that the phase noise power spectral density, PSD, in the range of the Fourier frequencies close to the carrier, f_o, is the basic measure of the frequency stability of oscillators, frequency synthesizers, and other electronic systems or components. However, the possibility of direct measurements of the PSD, $S_\varphi(f)$, with the assistance of common frequency analyzers, usually stops somewhere between 1 and 10 Hz. This difficulty is solved by using the time domain measurements [5.1, 5.2]. The disadvantage of the time domain measurement is a rather poor resolution caused by a low Q of the equivalent filter function, $H(s)$, and its spurious responses at harmonics of the equivalent fundamental frequency (Fig. 5.1).

The commonly accepted measure is the Allan or two-sample variance, $\sigma^2(\tau)$, with the assumption that the phase noise PSD $S_\varphi(f)$ is piecewise linear. Barnes [5.3] computed a conversion chart between $\sigma^2(\tau)$ and $S_\varphi(f)$.

It was Baugh [5.4] who, several years later, suggested the remedy by using a more sample variance; the corresponding procedure is generally designated as the *Hadamard transform*. The major achievement was a larger effective Q of the measurement system, but the problem with spurious responses remained and was not satisfactorily solved even with a more elaborate system of more sample variances [5.5] (Fig. 5.2).

In the following sections, we summarize properties of the important two-sample variance, generally designated as the *Allan variance*, and we make reference to connections between these time domain variances and their PSD counterparts.

Frequency Stability. By Venceslav F. Kroupa

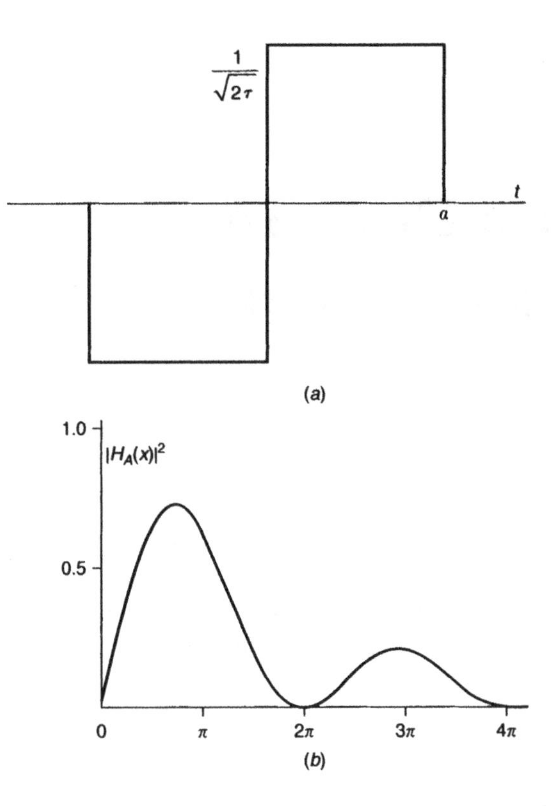

Fig. 5.1. Two-sample variance, (*a*) The effective time domain function and (*b*) the corresponding transfer function.

In Section 5.3, we discuss the problem of time domain measurements, namely, of the *time jitter* so important in modern communications in microwave ranges, and other applications of digital circuits.

5.1 BASIC PROPERTIES OF SAMPLE VARIANCES

Because of the divergence difficulties caused by the flicker phase noise or the random walk in oscillators and standard frequency generators with time domain variances, Allan [5.6] suggested the following two-sample variance as a measure of the frequency stability in the time domain:

$$\sigma_y^2(\tau) = \frac{1}{2} < (\bar{y}_{k+1} - \bar{y}_k)^2 > \qquad 5.1$$

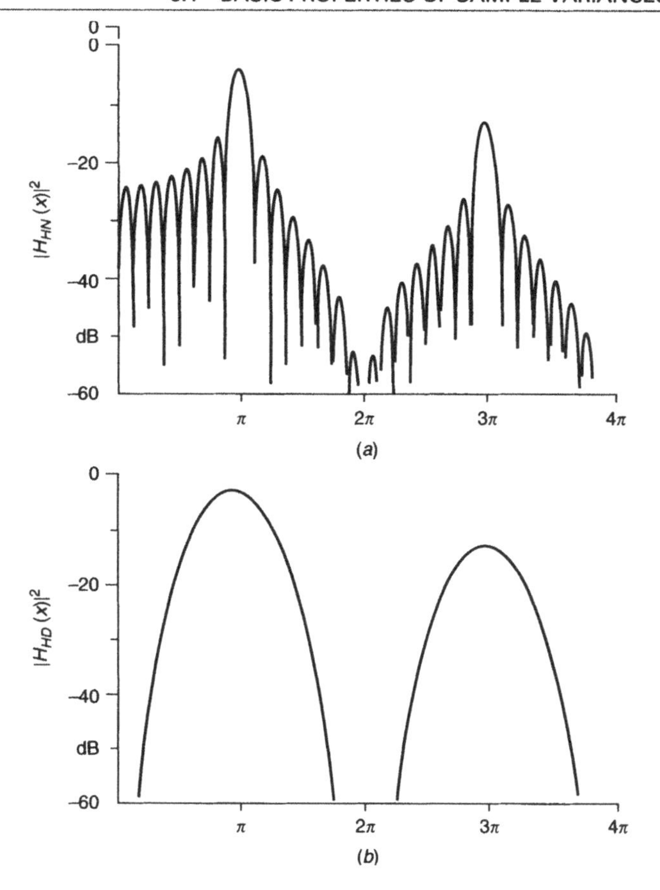

Fig. 5.2. The transfer function of the Hadamard variance formed by eight rectangles in the time domain: (*a*) without additional weighting (*b*) with binomial weighting.

where $\bar{y}$ is the mean fractional frequency in the time interval τ:

$$\bar{y}_k = \frac{1}{\tau} \int_{t_{k-1}}^{t_k = t_{k-1} + \tau} \frac{\Delta\omega(t)}{\omega_o} dt = \frac{1}{\tau} \int_{t_{k-1}}^{t_k = t_{k-1} + \tau} y(t) dt \approx \frac{1}{\tau} \frac{\Delta\phi}{\omega_o} \qquad 5.2$$

With the assistance of the convolution theorem, we may rearrange the above equation as

$$\bar{y}_k = \frac{1}{\tau} y(t) \otimes p(t) \qquad 5.3$$

where $p(t)$ is the unit rectangular pulse starting at time t and ending at $t + \tau$. In the case of the variance of more samples, we get

$$\sigma_y^2(\tau)=\frac{1}{\tau^2}<[a_0y(t)\otimes p(t)+a_1y(t+\tau)\otimes p(t+\tau)+\cdots+ a_{n-1}y(t)\otimes p(t+(n-1)\tau)]^2> \quad 5.4$$

where $a_0, \ldots, a_{n-1}$ are weighting coefficients that will be dealt with later in more detail. After reverting to (5.1), its simple rearrangement leads to

$$\sigma_y^2(\tau)=<[y(t)\otimes h(t)]^2>=<z^2(t)> \quad 5.5$$

With the assistance of Parceval's theorem, stating that total power over all time must equal power over all frequencies [a conservation principle, applied to (5.5)], we have

$$<z^2(t)>=\int_{-\infty}^{\infty} Z(\omega)Z^*(\omega)df \quad 5.6$$

By using the theorem that the Fourier transform of a convolution is equal to the product of the corresponding Fourier transforms, we arrive at the relationship between the time and frequency domain stability measures, that is,

$$\sigma_y^2(\tau)=\int_0^{\infty} S_y(\omega)[H(\omega)H^*(\omega)]df=\int_0^{\infty} S_y(\omega)\,|H(\omega)|^2df \quad 5.7$$

Here $S_y(\omega)$ is the one-sided power spectral density of the instantaneous fractional frequency fluctuations $y(t)$ and $H(\omega)$, which is the equivalent transfer function of the time domain measurement system.

5.2 TRANSFER FUNCTIONS OF SEVERAL TIME DOMAIN FREQUENCY STABILITY MEASURES

For the founding fathers of the frequency stability theory [5.1], the evaluation of the integral in (5.7) was difficult to compute; therefore, they used the piecewise properties of the oscillator phase noise (cf. Fig. 2.13) for solving the Allan variance. The integration is not a problem with modern computers. However, the piecewise approach makes it possible to understand better the stability problem at very low Fourier frequencies.

5.2.1 Two-Sample (Allan) Variance

The two-sample variance was first introduced into frequency stability measurement by Allan [5.6] in the form of (5.1). The Fourier transform of the rectangular unit pulse $p(t)$ of width τ is

$$P(\omega)=\frac{1}{\tau}\int_{-(\tau/2)}^{+(\tau/2)} e^{-j\omega t}dt=\frac{1}{\tau}\cdot\frac{1}{-j\omega}(e^{-j\omega\tau/2}-e^{j\omega\tau/2})=\frac{\sin(\omega\tau/2)}{\omega\tau/2} \quad 5.8$$

Since the Allan variance is composed of one negative and one positive pulse, the height of the rectangular pulses $h_A(t)$ in the two sample (Allan) variance is $1/\sqrt{2}$ because of the normalization, $\frac{1}{2}$, in (5.1) (see Fig. 5.1*a*). Thus, by applying the shifting theorem we find the transfer function $H_A(\omega)$ to be

$$H_A(\omega)=P(\omega)(e^{j\omega\tau/2}-e^{-j\omega\tau/2})\frac{1}{\sqrt{2}}=\sqrt{2}\frac{\sin^2(\omega\tau/2)}{\omega\tau/2} \quad 5.9$$

By introducing the normalized frequency, x, which simplifies computation, we get

$$|H_A(x)|=\sqrt{2}\frac{\sin^2 x}{x} \qquad x=\omega\tau/2=\pi\tau f \quad 5.10$$

The plot of the square of the transfer function $|H_A(x)|^2$ is shown in Fig. 5.1*b*. After introduction of the above relation into (5.7), we get for the Allan variance

$$\sigma_y^2(\tau)=\int_0^\infty S_y(x)\frac{2\sin^4 x}{x^2}dx \quad 5.11$$

where $S_y(x)$ is the PSD of the fractional frequency noise.

5.2.2 Evaluation of the Transfer Functions

Another approach for evaluation of the above transfer functions in the closed form may proceed with the assistance of autocorrelation. After reverting to (5.1) and (5.2), we obtain

$$\sigma_y^2(\tau)=\frac{1}{2}<(\bar{y}_{k+1}-\bar{y}_k)^2>=\frac{1}{(2\pi\tau f_o)^2}<[\phi(2\tau)-\phi(\tau)-\phi(\tau)+\phi(0)]^2>=$$

$$\frac{1}{(2\pi\tau f_o)^2}<[\phi^2(2\tau)+4\phi^2(\tau)+\phi^2(0)-4\phi(2\tau)\phi(\tau)-4\phi(\tau)\phi(0)+2\phi(2\tau)\phi(0)]>=$$

$$\frac{1}{2(2\pi\tau f_o)^2}[6R_\phi(0)-8R_\phi(\tau)+2R_\phi(2\tau)]= \qquad 5.12$$

$$\frac{1}{(2\pi\tau f_o)^2}\int_0^\infty S_\phi(f)[3-4\cos(2\pi f\tau)+\cos(4\pi f\tau)]df$$

and finally (with the Dwight formula 404.14 [5.7]),

$$\sigma_y^2(\tau)=\frac{2}{(\pi\tau f_o)^2}\int_0^\infty S_\phi(f)\sin^4(\pi f\tau)df=\frac{2}{(\pi\tau)^2}\int_0^\infty \frac{S_y(f)}{f^2}\sin^4(\pi f\tau)df \qquad 5.13$$

Since noise characteristics of oscillators have a piecewise characteristic (cf. Section 2.3.3),

$$S_y(f)=\sum_{i=-2}^{2}\frac{h_i}{f^{2-i}} \qquad 5.14$$

we easily find that Allan variances also have a piecewise property:

$$\sigma_{y,i}^2(\tau)=\frac{2}{(\pi\tau)^2}\int_0^\infty \frac{h_i}{f^{2-i}}\sin^4(\pi f\tau)df=\frac{2h_i}{(\pi\tau)^{i+1}}\int_0^\infty \frac{\sin^4(x)}{x^{2-i}}dx \qquad 5.15$$

which is applied for evaluation of individual contributions [5.8].

5.2.2.1 *White Phase Noise*

The PSD of the phase and fractional frequency noise in this range is

$$S_y(f)=S_\phi(f)\left(\frac{f}{f_o}\right)^2=h_2f^2 \qquad 5.16$$

With the assistance of (5.12) or the Korn formula No. 189 [5.9], we evaluate the partial Allan variance as

$$\sigma_{h_2}^2(\tau)\approx\frac{3h_2f_H}{(2\pi\tau)^2} \qquad 5.17$$

5.2.2.2 Flicker Phase Noise

The PSD of the Flicker phase noise (FPN) is

$$S_\phi(f) = \frac{h_1 f_o^2}{f} \qquad \text{and} \qquad S_y(f) = h_1 f \tag{5.18}$$

and with the assistance of (5.12) the partial Allan variance is

$$\sigma_{h_1}^2(\tau) = \frac{3}{(2\pi\tau)^2}\int_0^{f_H} \frac{h_1}{f} df - \frac{4}{(2\pi\tau)^2}\int_0^{f_H} \frac{h_1}{f}\cos(\omega\tau) df \tag{5.19}$$

$$+ \frac{1}{(2\pi\tau)^2}\int_0^{f_H} \frac{h_1}{f}\cos(2\omega\tau) df =$$

$$\frac{h_1}{(2\pi\tau)^2}\int_0^{x_H}\left[4\frac{1-\cos(x)}{x} - \frac{1-\cos(2x)}{x}\right]dx =$$

$$\frac{h_1}{(2\pi\tau)^2}[4S_1(x_H) - S_1(2x_H)]$$

where $S_1(x)$ is estimated using the *cosine integral* [5.9]; for small arguments of x, its value is approximately

$$S_1(x) \approx \frac{x^2}{4} \tag{5.20}$$

However, for large arguments of x the value is

$$S_1(x) \approx lg(x) + \gamma \qquad (\gamma \approx 0.5772) \tag{5.21}$$

After introduction into (5.19), we finally arrive at

$$\sigma_{h_1}^2(\tau) = \frac{h_1}{(2\pi\tau)^2}\ 3\left(lg(\pi\tau f_H) + \gamma\right) - lg(2) \tag{5.22}$$

$$= \frac{h_1}{(2\pi\tau)^2}[1.038 + 3lg(\pi\tau f_H)] \approx \frac{3h_1}{(2\pi\tau)^2} lg(\pi\tau f_H)$$

Note that both partial variances, for white and flicker phase noises, are inversely proportional to τ^2, however, for stable crystal oscillators the first one generally prevails.

5.2.2.3 White Frequency Noise

The PSD of the white noise fractional frequency fluctuations is

$$S_y(f) = S_\phi(f)\left(\frac{f}{f_o}\right)^2 = h_o \qquad 5.23$$

For the closed-form solution of the Allan variance, we apply (5.15) for $i = 0$ and use tables of the trigonometric relations [5.7] and of definite integrals

$$\sigma_{h_o}^2(\tau) = 2\frac{h_o}{\pi\tau}\int_0^{f_H}\frac{\sin^4 x}{x^2}dx = 2\frac{h_o}{\pi\tau}\int_0^{f_H}\frac{\cos(4x) - 4\cos(2x) + 3}{8x^2}dx =$$
$$2\frac{h_o}{\pi\tau}\int_0^{f_H}\frac{-2\sin(2x)^2 + 8\sin(x)^2}{8x^2}dx = \frac{h_o}{2\tau} \qquad 5.24$$

5.2.2.4 Flicker Frequency Noise

This time, we start again with the PSD of the fractional frequency noise

$$S_y(f) = \frac{h_{-1}}{f} \qquad 5.25$$

and after its introduction into (5.11 or 5.15), we arrive at

$$\sigma_{h_{-1}}^2(\tau) = h_{-1}\int_0^{f_H}\frac{2\sin^4 x}{x^3}dx = 2h_{-1}\int_0^{f_H}\frac{\cos(4x) - 4\cos(2x) + 3}{8x^3}dx \qquad 5.26$$

Solution of the integral is performed with the assistance of trigonometry and the Korn formulas (2.97, 2.87, and 2.96) in several steps [5.9] until we eventually arrive at

$$\sigma_{h_{-1}}^2(\tau) = h_{-1}2lg(2) \qquad 5.27$$

5.2.2.5 Random Walk Frequency Noise

The PSD of the corresponding fractional frequency is

$$S_y(f) = \frac{h_{-2}}{f^2} \qquad 5.28$$

and after its introduction into (5.15), we obtain

$$\sigma^2_{h_{-2}}(\tau) = h_{-2}\pi\tau \int_0^{f_H} \frac{2\sin^4 x}{x^4}\, dx \qquad 5.29$$

Solution of the integral is again performed with the *by parts theorem* (or with the assistance of the Dwight formula 431.9 [5.7]), applied in several steps, with the result

$$\sigma^2_{h_{-2}}(\tau) = h_{-2}\frac{(2\pi)^2\tau}{6} \approx 6.58\tau h_{-2} \qquad 5.30$$

5.2.2.6 Aging

In all secondary frequency standard generators, we encounter aging, which generally is controlled by an exponential law. However, in tens or hundreds of seconds we may consider a linear change of frequency with time:

$$f_o(t) = f_o + \frac{df_o}{dt}\Delta t \qquad 5.31$$

from which the fractional frequency is

$$y_{n+1} = y_n + \frac{1}{f_o}\int_0^{\tau} \frac{\Delta f_o}{\Delta t}\, dt = y_n + D_y\tau \qquad 5.32$$

With the assistance of (5.1) the Allan variance is

$$\sigma^2_{\text{aging}}(\tau) = \frac{(\tau D_y)^2}{2} \qquad 5.33$$

where the secular frequency change, D_y, is in fractional frequency per second or per day.

5.2.2.7 Sinusoidal Frequency Modulation

By considering a sinusoidal frequency modulation (FM) with frequency f_m, we obtain for its PSD

$$S_y(f)=\frac{1}{2}\left(\frac{\Delta\omega}{\omega_o}\right)^2\delta(f_m) \qquad 5.34$$

where $\delta(f_m)$ is the power of the spurious modulation designated as a delta function, and after introduction into (5.11), we arrive at the corresponding Allan variance:

$$\sigma^2_{FM}(\tau)=\left(\frac{\Delta\omega}{\omega_o}\right)^2\frac{\sin^4(\pi f_m\tau)}{(\pi f_m\tau)^2} \qquad 5.35$$

Examination of (5.35) reveals zeros for $f_m\tau$ = integer. From this condition, we easily find the carrier of the modulation frequency (Fig. 5.3*a*) with τ plotted linearly and τ plotted logarithmically (Fig. 5.3*b*). A practical example is illustrated in Fig. 5.3*c*, where the FM is generated by the oven switching on and off.

5.2.3 Piecewise Property of the Allan Variance

The mutual relationship between the time domain frequency stability measurement, with the assistance of the Allan variance, and the corresponding PSD of the fractional frequency noises, expressed as $S_y(f)$, is summarized in Table 5.1

In the previous sections, we evaluated the Allan variance as a function of the different types of noises encountered in practice. Since both PSDs $S_y(f)$ and $S_\varphi(f)$ can be piecewise approximated (cf. Chapter 1), we conclude the same property for the Allan variances as well (Fig. 5.4*a*):

$$\sigma^2(\tau)=\frac{b_{-2}}{\tau^2}+\frac{b_{-1}}{\tau}+b_o+b_1\tau+b_2\tau^2 \qquad 5.36$$

Note that b_i constants, read from the time domain measurements and summarized in Table 5.2, are proportional to the fractional frequency noise constants h_i. A short investigation reveals that b_{-2} is generated by white and flicker phase noises and is proportional to the upper bound

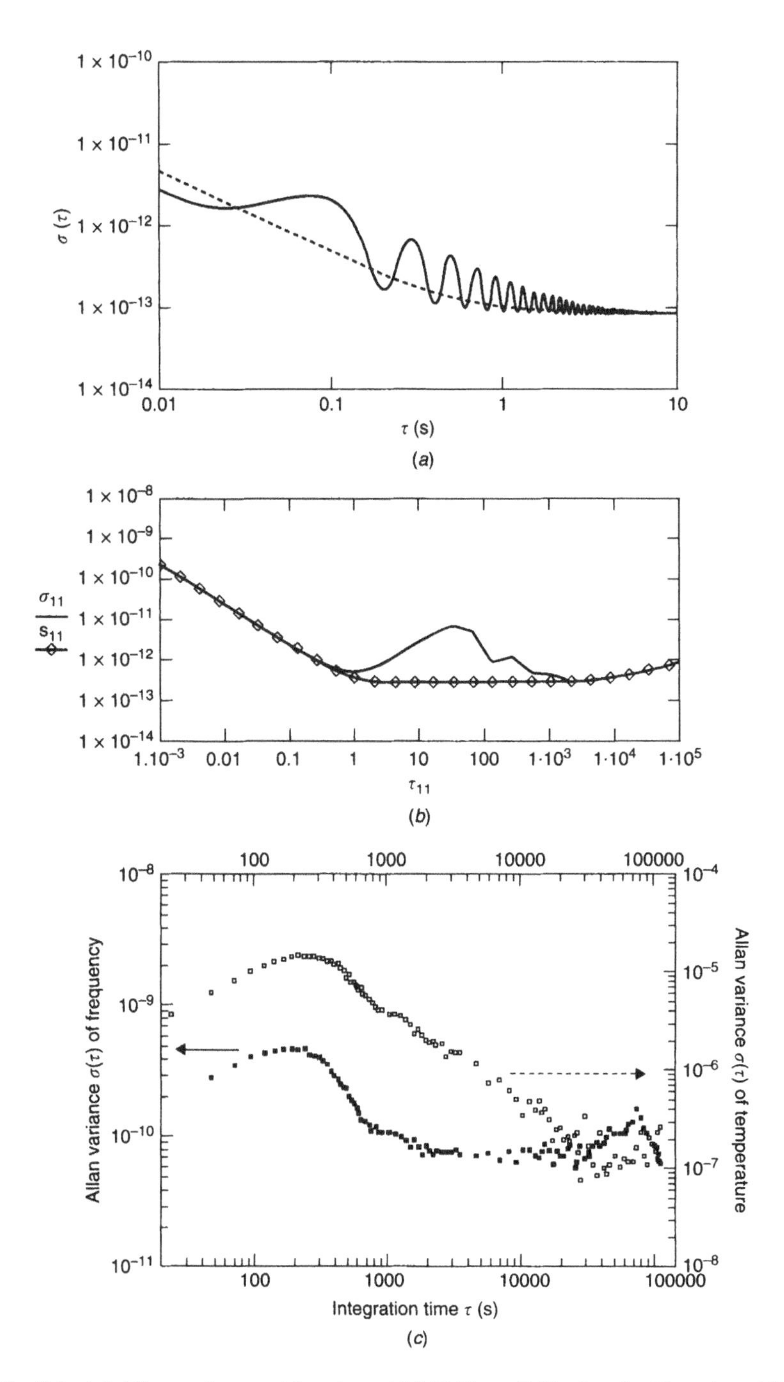

Fig. 5.3. (*a*) Allan variance with a sinusoidal FM by a 5-Hz signal with τ plotted linearly. (*b*) Allan variance with a sinusoidal FM by a 100-Hz signal with τ plotted logarithmically. Finally, a practical example is illustrated in (*c*), where the FM modulation may be generated by the oven switching on and off [5.10]. (Copyright © IEEE. Reproduced with permission.)

Table 5.1 Conversion between spectral densities, $S_y(f)$, two-sample variances, $\sigma_y^2(\tau)$, and modified Allan variances, Mod $\sigma_y^2(\tau)$, for a power law noise spectral density model

$S_y(f)$	$\sigma_y^2(\tau)$	Slope of $\sigma_y^2(\tau)$ vs. τ (dB)	Mod $\sigma_y^2(\tau)$	Slope of $\sigma_y^2(\tau)$ vs. τ (dB)
$h_2 f^2$	$3h_2 f_H/(2\pi\tau)^2$ $(2\pi\tau f_H \gg 1)$	–20	$6\pi\, \tau_o h_2 f_H/(2\pi N\tau_0)^3$	–30
$h_1 f$	$h_1[1.38 + 3\ln(\pi\tau f_H)]/(2\pi\tau)^2$	–20	$h_1[1.38 + \ln(2Nf_H\tau_{oo})]$ $1(2\pi N\tau_o)^2$	–20 $(N \gg 1)$
h_0	$h_0/2\tau$	–10	$h_0/2N\tau_0$	–10
h_{-1}/f	$2h_{-1}\ln 2 \approx 1.39\, h_{-1}$	Zero change	$h_{-1}2\ln 2 \approx 1.39\, h_{-1}$	Zero change
$h_2 f^2$	$(2\pi)^2\tau h_{-2}/6 \approx 6.58\tau\, h_{-2}$	+10	$(2\pi)^2 N h_{-2}\tau/6 \approx 6.58N\tau\, h_{-2}$	+10
Sinusoidal PM modulation	$(\Delta\omega/n\omega)^2\, \delta(f - f_m)$ times $[\sin^2(\pi f_m\tau)/2(\pi f_m\tau)]^2$	Periodic change	$(\Delta\omega/n\omega_o)^2$ times $[2\sin^3(N\pi f_m\tau_o)/2(N\pi f_m\tau_0)]^2$	Periodic change
Aging $dy/dt = D_y$	$(\tau D_y)^2/2$	+20	$(N\tau_o D_y)^2/2$	+20

of the low-pass measurement filter f_h. For large corner frequency f_h the white phase noise (WPN) dominates, whereas for small f_h the FPN prevails. The value of b_{-1} is generated by white frequency noise (WFN) and, for example, for crystal oscillators in the 5- or 10-MHz ranges its contribution is small. The Allan variance plateau provided by the b_o constant corresponds to the flicker frequency noise (FFN). Often this is the most important information about frequency stability at low Fourier frequencies of the investigated generators. However, care is necessary since its magnitude might often be smeared by the neighboring terms [cf. (5.36)]. The constant b_1 originating in the random walk of frequency is generally related to environmental fluctuations and infrequently with additional nonlinearity in the oscillating network. The last right-hand side (rhs) term in (5.36) is generated by aging and provides very important information about the quality of the

Table 5.2 Allan variance and slopes of the reconstructing asymptotes

b_{-2}	$3fh_H h_2/(2\pi\tau\tau)^2 +$ $h_1[1.38 + 3\ln(\omega_H\tau)]/(2\pi)^2$	–20 dB/ decade	$h_{-2} = 0.076\, b_{-2}/h_H$ $h_{-1} = 7.15\, b_{-2}$
b_{-1}	$h_0/2\tau$	–10 dB/ decade	$h_o = 2\, b_{-1}$
b_o	$2h_{-1}\ln 2$	Zero change	$h_{-1} = b_o$
b_1	$(2\pi)^2\tau h_{-2}/6$	+10 dB/decade	$h_{-2} = 0.15\, b_1$
b_2	$(\tau D_y)^2/2$	+20 dB/decade	$D_y = 2\, b_2$

investigated precision frequency standards. Summarizing, we are able to read from the Allan variance stability plot at least a few frequency domain characteristic constants, particularly for very low Fourier frequencies where the type of the originating noise is often uncertain. The situation is illustrated in Example 5.1.

EXAMPLE 5.1

As an illustration, we investigate the phase noise and time domain characteristic for a 5-MHz ultrastable crystal oscillator discussed by Candelier et al. [5.11]. The mean phase noise readings are summarized in Tables 5.3 and 5.4, the respective time domain measurements. With their assistance, we plot the piecewise approximation of the corresponding Allan variance (see Fig. 5.4*a*); from intersections with the vertical line, $\tau = 1$, we find out the individual b_i constants summarized in Table 5.5(*a*):

$$\sigma^2(\tau) \approx \frac{10^{-27.5}}{\tau^2} + \frac{10^{-27.2}}{\tau} + 10^{-26.4} + 10^{-28.3}\tau + 10^{-32.3}\tau^2 \qquad 5.37$$

Note that (5.37) enables us to estimate the aging to be ~ $10^{-16.15}$/s, (i.e., ~ 10^{-11}/day). Further, we may compute the random walk of frequency (RWF) and FFN noise coefficients h_{-2} and h_{-1}, and, surprisingly, also the WFN h_o coefficient (Table 5.5*b*). However, the usefulness of b_{-2} is doubtful since it combines the FPN and WPN noises with the measurement filter passband f_H. The phase noise characteristic plotted with the assistance of Table 5.3 together with the information about the RWF from the time domain measurements reveals (Fig. 5.4*b*)

Table 5.3 PDS of an ultrastable 5-MHz crystal oscillator[a]

Frequency offset f in (Hz)	1	10	100	1,000	10,000
PSD $S_\varphi(f)$ (dB/Hz)	−131	−147	−156	−156	−150

[a]Adapted from [5.11].

Table 5.4 Allan variances of an ultrastable 5-MHz crystal oscillator[a]

τ (s)	1	2	4	8	16	100	400	1000	5000
$\sigma(\tau) \times 10^{-13}$	0.67	0.6	0.58	0.65	0.75	1.1	1.5	2	6

[a]Mean values from [5.11].

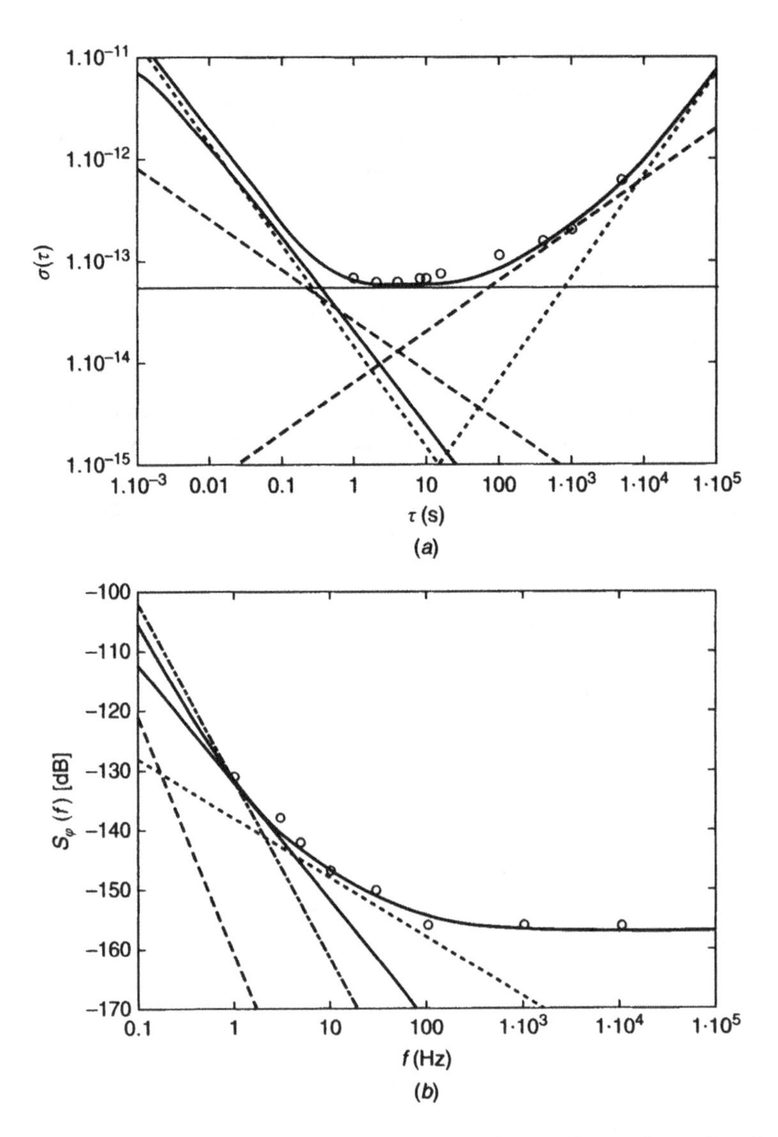

Fig. 5.4. (*a*) Plot of the Allan variance. Circles mark the Allan variance data from Table 5.4; points indicate contribution of the WPN; thin full line asymptote contribution from FPN; dashed line asymptote from WFN; the horizontal thin full line, contribution from FFN; the dashed line, asymptote with the positive slope RWN contribution; points with positive slope, aging (*b*). The corresponding phase noise characteristic based on the data from Table 5.4 with the RWN asymptote estimated from the Allan variance plot. Circles mark the phase noise data from Table 5.3; full line the estimated PSD; asymptotes; points are the FPN; dash–dot FFN; dash RWN.

Table 5.5 (*a*) Constants b_i of a 5-MHz oscillator read from Fig. 5.4a

i	−2	−1	0	1	2
b_i	$10^{-27.6/-27.4}$ (f_H = 316 Hz)	$10^{-27.2}$	$10^{-26.4}$	$10^{-28.3}$	$10^{-32.3}$; Aging $10^{-16.15}$ (Hz/s); 10^{-11} (Hz/day)

Table 5.5 (*b*) The frequency and phase noise coefficients of a 5-MHz oscillator computed from the Allan variances read from Fig. 5.4*b*

i	−2	−1	0	1	2
h_i	$10^{-29.1}$	$10^{-26.7}$	$10^{-26.9}$	$10^{-27.2}$	10^{-29}
$S_v(f=1)$	$10^{-15.7}$	$10^{-13.3}$	$10^{-13.4}$	$.10^{(13.8}$	$10^{-15.6}$

$$S_\phi(f)=\frac{a_4}{f^4}+\frac{a_3}{f^3}+\frac{a_2}{f^2}+\frac{a_1}{f^1}+a_o\approx \frac{10^{-15.7}}{f^4}+\frac{10^{-13.3}}{f^3}+\frac{10^{-13.3}}{f^2}+\frac{10^{-13.8}}{f}+10^{-15.6} \qquad 5.38$$

The value of the coefficient $a_4 = h_{-2}f_o^2$ is found from the Allan variance plot for large $\tau = s$. We expect that the term a_4/f^4 is due to the environmental variations, probably temperature. The authors indicate the thermal sensitivity of the frequency to be ~3 × 10^{-13}/°C, which corresponds to the temperature fluctuations of ~ 0.01°C.

5.2.4 Confidence Interval

The true Allan variance is defined as the mean from an infinite number of samples. This is not the case in real life, particularly, for larger evaluation times. Consequently, introduction of the confidence intervals is of importance, especially in cases where only a few measurements are available. The general approach is based on χ^2 statistics (e.g., [5.12 or 5.13]):

$$\frac{\chi^2}{d.f.}=\frac{s^2}{\sigma^2}; \qquad E[\chi^2]=d.f.; \qquad \mathrm{Var}[\chi^2]=2(d.f.) \qquad 5.39$$

where s^2 is the sample variance, σ^2 is the true variance, $E[\chi^2]$ is the average or the expectation value, and $d.f.$ is the number of degrees of freedom.

In short, here we recapitulate the investigation performed by several authors [5.14–5.16]. The effort was to find the value of *d.f.* for different types of noises. The results for the normalized variances are summarized in Table 5.6. Generally,

$$\frac{\mathrm{Var}[s^2]}{\sigma^4} = \frac{2}{d.f.} = F(M) \qquad 5.40$$

Note that for the white and flicker phase noises, degrees of freedom are defined as

$$d.f. \approx \frac{M}{2} = \frac{m-1}{2} \qquad 5.41$$

where m is the number of measurements of individual mean frequencies ($\bar{y}$). However, for flicker and random walk frequency noises, individual samples are nearly independent of each other and the processes retain the full number of degrees of freedom, that is, $d.f. = m - 1$ (see Table 5.6 and Example 5.2).

The relative uncertainty on the estimated Allan variance provides another approach for estimation of the confidence limits [5.14]:

$$\Delta = \frac{\sigma^2(\tau) - \frac{1}{m-1}\sum_{i=1}^{m}(y_i - y_{i-1})^2}{\sigma^2(\tau,m)} = \frac{\sigma^2(\tau) - s^2(\tau,m)}{\sigma^2(\tau,m)} \qquad 5.42$$

Table 5.6 Normalized variances for different types of noises, evaluated by several authors [5.14–5.16]

Noise type	Lesage and Audoin [5.14]	$M \gg 1$	Yoshimura [5.15]	$M \gg 1$	Howe, Allan, and Barnes [5.16]	$M \gg 1$
WPN	$(35M - 18)/9M^2$	$3.89/M$	$((35\text{-}18)/9M^2$	$3.89/M$	$4(M+1)/M(M+3)$	$4/M$
FPN	$35M - 18)/9M^2$	$3.89/M$	$M = 2$ $2.83/M$	$M = 100$ $3.73/M$		
WFN	$(3M - 7)/M^2$	$3/M$	$(3M - 1)/M^2$	$3/M$	$9(M+2)/(3M^2 + 7M + 6)$	$3/M$
FFN	$(2.3M - .3)/M^2$	$2.3/M$	$M = 2$ $2.1/M$	$M = 100$ $2.27/M$	$(2.3M - 0.3)/M^2$	$2.3/M$
RWN	$(9M - 1)/4M^2$	$2.25/M$	$2/M$	$2/M$	$2(M-1)/$ $[M(M+1)^2 - 3(M+1) + 4]$	$2/M$

With the assistance of (5.41), we get for the expected true variance

$$\sigma(\tau) \approx s(\tau)\left[1 \pm \sqrt{\frac{2}{d.f.}}\right] \tag{5.43}$$

Another simple method for computing the confidence interval, applicable for *d.f.* > 10, which assumes a symmetric normal (Gaussian) distribution, uses the relation

$$I_\alpha \approx s(\tau)\frac{K_\alpha}{\sqrt{m}} \tag{5.44}$$

where I_α is the confidence interval (for 1σ or 68% confidence interval) and K_α the experimental constant evaluated by Lesage and Audoin [5.14, 5.18] are shown in Table 5.8.

EXAMPLE 5.2

Evaluate the confidence limits for $\sigma(\tau)$ of the corresponding $s(\tau) = 5 \times 10^{-12}$ computed from 10 measurements, that is, $m - 1 = 9$ degrees of freedom. From Table 5.7, we find for the range between 5 and 95% for the 9 degrees of freedom:

$$\chi^2(0.05) = 3.33 \text{ and } \chi^2(0.95) = 16.92$$

where the corresponding confidence interval of $\sigma(\tau)$ is

$$3.6 \times 10^{-12} < \sigma(\tau) < 8.2 \times 10^{-12}$$

Table 5.7 χ^2 Distribution for the 1-α confidence Interval for σ^2 for 1–10 d.f. (i.e., for the 90, 80, and 60% confidence intervals) [5.4, 5.14]

d.f.	1	2	3	4	5	6	7	8	9	10
$\alpha = 0.95$	0.0039	0.103	0.352	0.711	1.145	1.635	2.167	2.733	3.325	3.940
$\alpha = 0.05$	3.841	5.991	7.8.15	9.488	11.070	12.592	14.067	15.507	16.919	18.307
$\alpha = 0.90$	0.016	0.211	0.584	1.064	1.610	2.204	2.833	3.490	4.168	4.865
$\alpha = 0.10$	2.706	4.605	6.251	7.779	9.236	10.645	12.017	13.362	14.684	15.987
$\alpha = 0.80$	0.064	0.446	1.005	1.649	2.343	3.070	3.822	4.594	5.380	6.179
$\alpha = 0.20$	1.642	3.219	4.642	5.989	7.289	8.558	9.803	11.030	12.242	13.442

Table 5.8 Constants K_α for approximate evaluation of the confidence interval in accordance with relation (5.44)[a]

Type of noise (α)	WPN (2)	FPN (1)	WFN (0)	FFN (−1)	RWF (−2)
K_α	0.99	0.99	0.87	0.77	0.75

[a]See [5.18].

Evaluation of the true variance in accordance with (5.43) reveals the confidence limit

$$2.64 \times 10^{-12} < \sigma(\tau) < 7.36 \times 10^{-12}$$

whereas with (5.44) for $m = 9$ and FFN

$$3.72 \times 10^{-12} < \sigma(\tau) < 6.28 \times 10^{-12}$$

5.2.5 More Sample Variances

Investigation of Fig. 5.1*b* reveals that the effective Q of the Allan variance is rather low (actually ~ 0.75). Much activity was dedicated to mending this drawback. One remedy is to use more sample variances, especially the so-called *Hadamard transform* [5.4]. It is based on (5.4) with equal weighting coefficients a_i. Its transfer function is shown in Fig. 5.2*a*. The difficulty with spurious responses could be solved by applying a binomial weighting on coefficients a_i (see Fig. 5.2*b*). Nowadays, these special time variances [5.5] are rarely encountered. However, this is not the case with the *modified Allan variance* or the *discrete Fourier transform*.

5.2.6 Modified Allan Variance

The modified Allan variance is based on the overlapping of time samples, as depicted in Fig. 5.5*a* [5.19, 5.20].

The original definition of (5.1) is enlarged to

$$\text{Mod}\,\sigma_y^2 = \frac{1}{2\tau^2} < \left[\frac{1}{N}\sum_{i=1}^{N}(x_{i+2N} - 2x_{i+N} + x_i)\right]^2 \geqq \frac{1}{2N^2} < \left[\sum_{i=1}^{N}(\bar{y}_{i+N} - \bar{y}_i)\right]^2 > \qquad 5.45$$

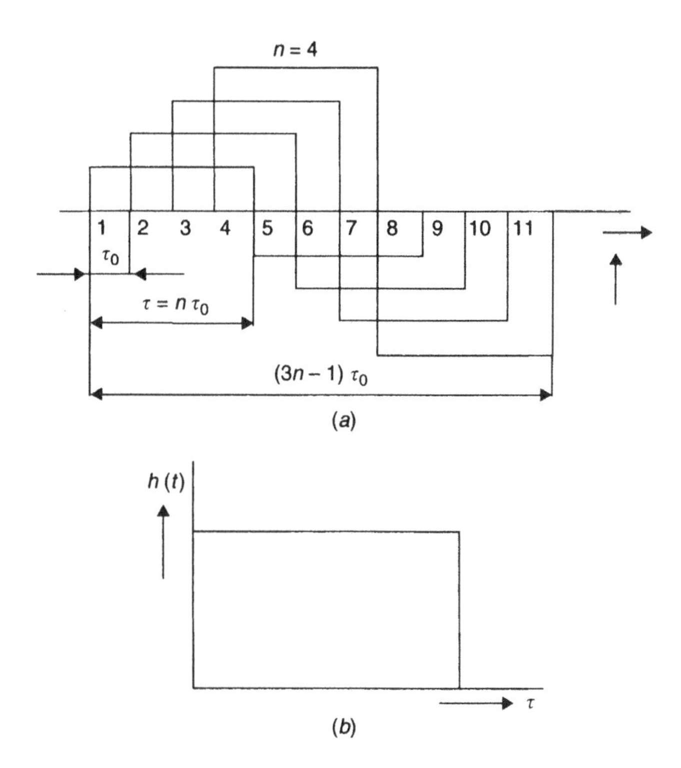

Fig. 5.5. (*a*) Time domain function of the modified Allan variance with overlapping samples. (*b*) Time response *h*(*t*) of one pulse.

where

$$\bar{y}_{i+N} = \frac{x_{i+2N} - x_{i+N}}{\tau} = \frac{\phi(t + i\tau + 2N\tau) - \phi(t + i\tau + N\tau)}{2\pi\nu_o\tau} \qquad 5.46$$

Evidently, the time response $h(t)$ is formed by a superposition of N pulses of the duration (depicted in Fig. 5.5*b*)

$$\tau = N\tau_o \qquad 5.47$$

and delayed successively by τ_o (cf. Fig. 5.5*a*). The corresponding Fourier transform of the delaying process is

$$P_M(\omega) = \frac{1}{N}[1 + e^{j\omega\tau_o} + e^{2j\omega\tau_o}\ldots + e^{(N-1)j\omega\tau_o}] = \frac{1}{N} \cdot \frac{e^{jN\omega\tau_o} - 1}{e^{j\omega\tau_o} - 1} \qquad 5.48$$

After introducing the negative parts of the pulses, correspondingly delayed, we arrive at the transfer function of the modified Allan variance:

$$|H_M(\omega)|^2 = 2\frac{\sin^4(N\omega\tau_o/2)}{(N\omega\tau_o/2)^2}\frac{1}{N^2}\left|\frac{e^{jN\omega\tau_o/2}-1}{e^{j\omega\tau_o/2}-1}\right|^2 =$$

$$2\frac{\sin^4(N\pi f\tau_o)}{(N\pi f\tau_o)^2}\cdot\frac{1}{N^2}\cdot\frac{\sin^2(N\pi f\tau_o)}{\sin^2(\pi f\tau_o)} = \frac{1}{N^4}\cdot\frac{2}{(\pi f\tau_o)^2}\cdot\frac{\sin^6(N\pi f\tau_o)}{\sin^2(\pi f\tau_o)} = \quad 5.49$$

$$\frac{1}{N^4}\cdot\frac{2}{x^2}\cdot\frac{\sin^6(Nx)}{\sin^2(x)} \qquad (x = \pi f\tau_o)$$

For small values of $N\pi\tau_o f$, the last factor, $\sin^2(Nf\pi\tau_o)/\sin^2(f\pi\tau_o)$, simplifies to N^2. Thus, for small upper bounds, the *modified Allan variances* are equal to simple *Allan variance* if we introduce in (5.15) the variable Nx. However, for larger values of N, the contribution of the mentioned factor is negligible (cf. Fig. 5.6). Consequently, the transferfunction of the investigated type of the Allan variance changes into (5.15) and we arrive at

$$\sigma_y^2(N\tau_o) = \int_0^{f_H} S_y(x)\frac{1}{N^4}\cdot\frac{2}{x^2}\cdot\frac{\sin^6(Nx)}{\sin^2(x)}d(Nx)$$

$$= \frac{1}{N^3}\frac{2h_i}{(\pi\tau_o)^{i+1}}\int_0^{f_H}\frac{x^i}{x^2}\frac{\sin^6(Nx)}{\sin^2(x)}dx \quad 5.50$$

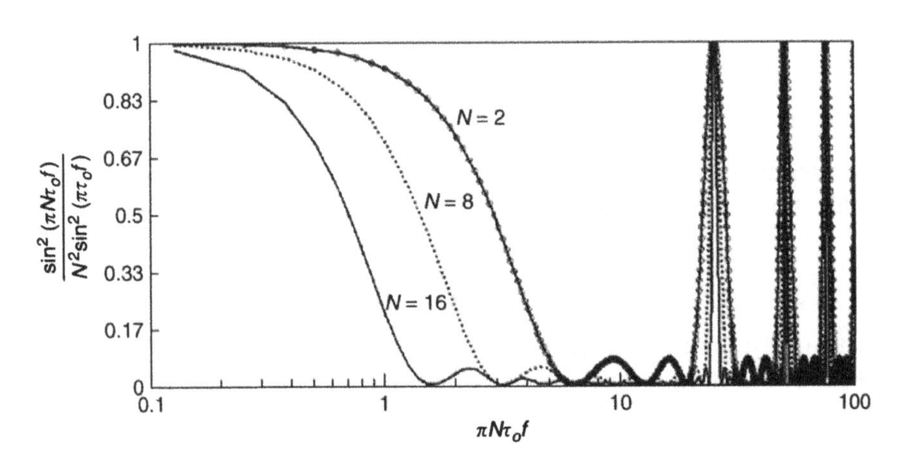

Fig. 5.6. Contributions to the modified Allan variance [cf. (5.49)].

In the next step, we evaluate the ratio $\sin^6(Nf\pi\tau_o)/\sin^2(f\pi\tau_o)$ and get

$$\frac{\sin^6(Nx)}{\sin^2(x)}=\frac{1}{8}[\cos(4Nx)-4\cos(2Nx)+3]= \frac{1}{8}[-2\sin^2(2Nx)+8\sin^2(Nx)] \tag{5.51}$$

Finally, we arrive at the value of the modified Allan variance as in (5.15):

$$\sigma_y^2(N\tau_o)=\frac{1}{N}\frac{2h_i}{(\pi\tau_o)^{i+1}}\frac{1}{8}\int_0^{f_H}\frac{x^i}{(Nx)^2}[\cos(4Nx)-4\cos(2Nx)+3]d(x) \tag{5.52}$$

5.2.6.1 White Phase Noise

With the assumption that the contribution of trigonometric terms in (5.52) is small, we evaluate the white phase noise (WPN) component of the modified Allan variance as

$$\sigma_M^2(N\tau_o)\approx\frac{6\pi\tau_o f_H h_2}{(2\pi N\tau_o)^3} \tag{5.53}$$

5.2.6.2 Flicker Phase Noise

The modified Allan variance follows immediately from (5.52) and (5.19):

$$\sigma_M^2(N\tau_o)\approx\frac{2h_1}{(\pi N\tau_o)^2}\frac{1}{8}[-S_i(4Nx_H)+4S_i(2Nx_H)]\approx 0 \tag{5.54}$$

However, for large sampling times $N\pi\tau_o$, we arrive at

$$\sigma_M^2(N\tau_o)\approx\frac{h_1}{(2\pi N\tau_o)^2}3([lg(2Nx_H)+1.038] \tag{5.55}$$

5.2.6.3 White Frequency Noise

With the assistance of (5.51) and (5.24), we find for the modified Allan variance

$$\sigma_M^2(N\tau_o) \approx \frac{2h_o}{\pi N\tau_o} \int_0^{\infty} \frac{-2\sin^2(2Nx) + 8\sin^2(Nx)}{8(Nx)^2} d(Nx) \approx \frac{h_o}{2N\tau_o} \quad 5.56$$

5.2.6.4 Flicker Frequency Noise

Solution of the integral will be performed by applying the same procedure as used in connection with solution of (5.27):

$$\sigma_M^2(N\tau_o) = 2h_{-1} \int_0^{f_H} \frac{\sin^4(Nx)}{(Nx)^3} dx = 2h_{-1} \int_0^{f_H} \frac{\cos(4Nx) - 4\cos(2Nx) + 3}{8(Nx)^3} dx \quad 5.57$$

that is, with the assistance of Korn formulas 2.97, 2.87, and 2.96 in several steps [5.9]) until we eventually arrive at a result that is not dependent on the upper-bound frequency:

$$\sigma_M^2(N\tau_o) \approx h_{-1} 2 lg(2) \quad 5.58$$

5.2.6.5 Random Walk (RW)

We proceed in the same way as above and find that the simple Allan variance and the modified Allan variance are the same with respect to sampling time:

$$\sigma_M^2(N\tau_o) \approx h_{-2} N\tau_o \frac{(2\pi)^2}{6} \quad 5.59$$

5.2.6.6 Relation Between Allan Variance and the Modified Allan Variance

Comparing modified Allan variance relations (5.52)–(5.57) with the time domain measure of frequency stability with the simple Allan variances, discussed in Section 5.2.2, reveals a major difference only for the WPN and in instances where the cut-off frequency, f_h, of the measurement device is rather large. In that case, we encounter the change of the slope from 20 to 30 dB/decade (see Table 5.1). In all other cases, the differences are rather small (see Fig. 5.7). However, the confidence interval is narrower because of increased d.f., since $N = M$ in Section 5.2.4.

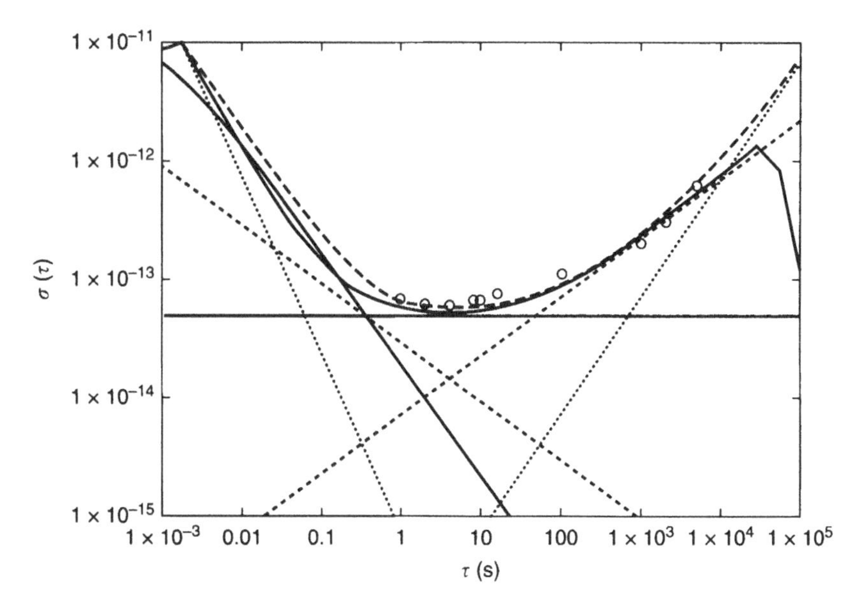

Fig. 5.7. Allan (dashed) and modified Allan variances ($N = 16$ full line). Points indicate the contribution of the WPN, a thin full line contribution is from FPN, a dashed line is that from WFN, the horizontal thin full line contribution is from FFN, and the dashed line is with the positive slope RWN contribution. Circles mark the phase noise data from Table 5.3.

5.2.7 Triangular Variance

In the digital age, the transfer function of the triangular variance is effectively a sampled one. Consequently, one must expect that the triangular variance is equal to the modified Allan variance with a very large N.

5.3 TIME JITTER

Time jitter measurement is another time domain method for evaluation of the frequency stability. Precise timing information is required in many applications, particularly in those dealing with digital signals. The consequence is the increased demand on the frequency stability of clock oscillators, modulation processing, transmission hardware, and prompt information about the quality of the connection. Degradation of the digital information depends on the presence of the phase or frequency noises, particularly by shifting the switching levels between individual 1's and 0's (cf. Fig. 5.8) Consequently, the corresponding *time jitter* of individual bit edges becomes a significant and very important parameter in modern communications, in high-performance computers, and many other digital devices. To this end, new measurement systems also were introduced. Evaluation of the time jitter (generally specified in picoseconds) from the phase noise spectra yields the most important method for its evaluation [5.21, 5.22]. A very important technique is the investigation of the HF signal traces on oscilloscopes, the so-called eye diagrams, and application of the statistics of the timing errors as a tool for decisions about the origin of the time jitter, particularly for evaluation of the expected *bit error resolution* (BER), for histogram appraisal, and for investigation of some problems in digital communications [5.23–5.25].

5.3.1 Types of the Time Jitter and Defining Units

First, we consider the carrier transferring information

$$A\sin\left[\omega_o t + \int_0^{T_o}\Delta\omega(t)dt + \varphi(t)\right] \approx A\sin[\omega_o t + \Delta\omega(t)T_o + \varphi(t)] = A\sin\left[\omega_o(t + \Delta(t) + \frac{\varphi(t)}{\omega_o}\right] \quad 5.60$$

where $\Delta(t)$ and $\varphi(t)/\omega_o$ are slowly varying time functions representing the time jitter. As to their origin, they are composed of the bounded de-

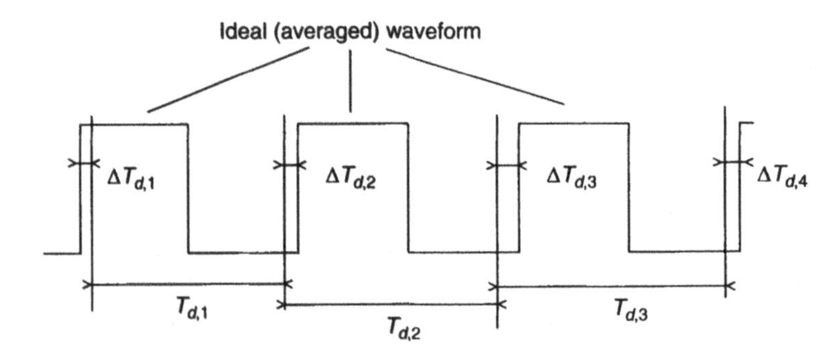

Fig. 5.8. Time jitter definitions.

terministic parts and the unbounded random time changes (thermal or shot-noise phase fluctuations and $1/f$ components):

$$\Delta(t) = J_D + J_R \qquad 5.61$$

The origins of the deterministic jitter include

- Duty cycle distortion (DCD): asymmetric rise/fall times
- Intersymbol interference (ISI): from channel dispersion, filtering, and so on
- Periodic jitter (sinusoidal spurious signals): power supply, leaking reference signal
- Uncorrelated: cross talk coupled to an adjacent signal-carrying conductor
- Electromagnetic interference (EMI): unwanted radiated or leaking emissions
- Reflections due to the signal interfering with itself

5.3.1.1 Period Jitter

Period jitter simply measures the duration of each period of the investigated waveform (i.e., $T_{d,1}$, $T_{d,2}$, $T_{d,3}$) (cf. Fig. 5.8, e.g., in the instants of positive zero crossings) with respect to zero crossings of an ideal clock period:

$$\Delta T_{\text{jitter}} = \frac{1}{n}\sum_{1}^{n}(T_{\text{per}} - T_{\text{clock}}) \qquad 5.62$$

It would be zero for the random behavior of ΔT_{per}, but not for the corresponding variance. The published or sought results are the calculated rms difference.

5.3.1.2 Cycle-to-Cycle Jitter

Cycle-to-cycle jitter is the time difference between successive periods of the investigated signal, that is (cf. Fig. 5.8),

$$\Delta C_n = T_{d,n+1} - T_{d,n} \qquad 5.63$$

and can be found by applying the first-order difference operation on the period jitter.

5.3.1.3 Peak-to-Peak Jitter

Peak-to-peak jitter (the worst case of cycle-to-cycle jitter) is the measure of the random jitter combined with deterministic components. However, estimation is not a clear-cut one because of the observation time spent on unbounded random components (oscilloscope persistence). It is obtained from a histogram of successive edge measurements:

$$\Delta t_{pp} \approx J_{D,pp} + J_{R,pp} \qquad 5.64$$

The knowledge of the peak-to-peak jitter is needed to appreciate the expected *time interval error.* To this end, the total time jitter is defined as

$$\Delta t_{pp,\text{total}} \approx J_{D,pp} + 14\sigma \qquad 5.65$$

5.3.1.4 Time Interval Error

The time interval error is the time difference between the actual and expected zero crossing (see ΔT_{di} in Fig. 5.8). Its importance is in the cumulative effect; once the time interval error reaches ± 0.5 of unit intervals, the eye on the sought oscilloscope is closed and the system experiences bit errors (more on this in Section 5.3.3).

5.3.1.5 Jitter in the Unit Interval

The unit interval is the ratio of the spurious phase noise in radians, $\Delta\varphi$, divided by 2π (cf. Fig. 5.9):

$$J(\text{UI}) = \left(\frac{\Delta\varphi}{2\pi}\right) \qquad 5.66$$

5.3.1.6 Jitter in Units of Time

Jitter in units of time results by dividing the unit interval by the pulse frequency:

$$J(t) = \frac{\Delta\varphi}{2f_d} \quad [f_d = \text{bit period in } (\text{s}^{-1})] \qquad 5.67$$

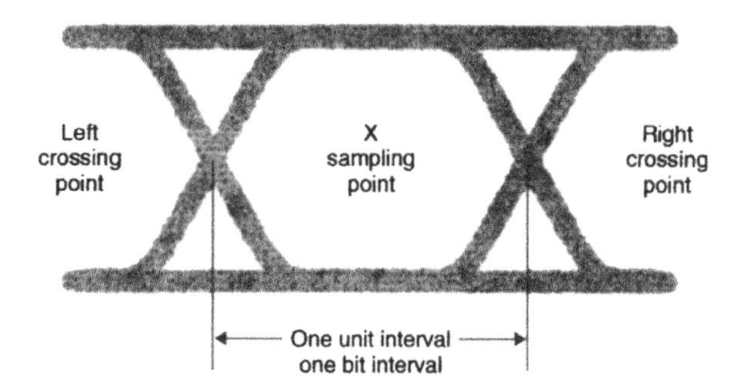

Fig. 5.9. Simplified eye diagram with definition of the unit interval (UI) [5.23].

5.3.2 Probability Density of the Time Error

By considering both types of timing errors, that is, the deterministic jitter φ_{jD} and the unbounded random jitter φ_{jR} (thermal noise, shot noise, $1/f$ noise), we have

$$< \Delta t > = \frac{< \varphi_j(t) >}{2\pi f_d} = \frac{< \varphi_{jD}(t) + \varphi_{jR}(t) >}{2\pi f_d} \qquad 5.68$$

with the variance

$$< \Delta t^2 > = \frac{< \varphi_{jD}(t)^2 > + 2 < \varphi_{jD}(t) >< \varphi_{jR}(t) > + < \varphi_{jR}(t)^2 >}{(2\pi f_d)^2} \approx \frac{< \varphi_{jD}(t)^2 > + < \varphi_{jR}(t)^2 >}{(2\pi f_d)^2} \qquad 5.69$$

The above approximation takes into account that the mean value of the random jitter is zero. However, the *deterministic jitter* has different sources (cross-talk, spurious magnetic fields, power-supply switching, etc.) (cf. Section 5.3.1). It is of a bounded nature usually specified in terms of the peak-to-peak value. In statistics, it may be approached with the assistance of the uniform distribution (see Fig. 5.10 and Section 1.4.2.1):

$$p(x) \approx \frac{1}{\Delta T_n - (-\Delta T_{n-1})} \approx \frac{1}{2\Delta T_x} \qquad 5.70$$

with mean value $(\Delta T_n + \Delta t_{n-1})/2$ and variance $(\Delta T)^2/3$, (1.54).

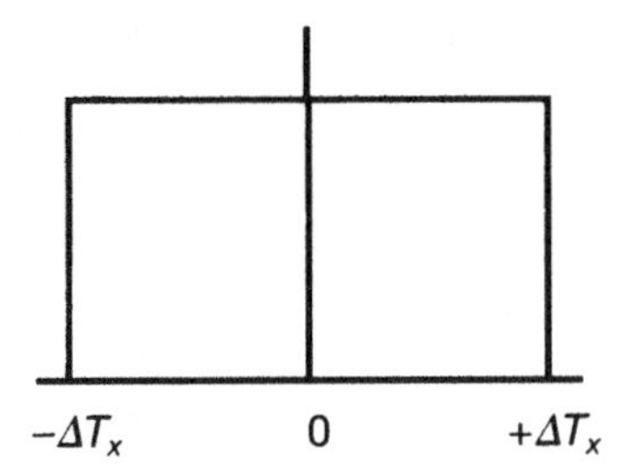

Fig. 5.10. Uniform distribution of the discrete noise jitter.

In the case of the *random time jitter,* the probability density presents a Gaussian distribution (Section 1.4.2.4) with mean value μ and dispersion σ (see Fig. 5. 11*a*):

$$p_R(x) = \frac{1}{\sigma\sqrt{2\pi}} e^{-0.5[(x-\mu)/\sigma]^2} \tag{5.71}$$

with the probability of the investigated event being between points *a* and *b* (cf. Table 5.9):

$$P_R(x) = \frac{1}{\sigma\sqrt{2\pi}} \int_{x_a}^{x_b} e^{-0.5[(x-\mu)/\sigma]^2} dx = \frac{1}{\sqrt{2\pi}} \int_{t_a}^{t_b} e^{-0.5t^2} dt \quad \left(t = \frac{x-\mu}{\sigma}\right) \tag{5.72}$$

and the corresponding variance σ^2.

To estimate the dispersion σ, in instances where we only know the mean value μ and the plot of the probability density $p_R(x)$, we find on the plot the point

$$p_R(x) = p_{R,\max} e^{-0.5} = 0.607 p_{R,\max} \tag{5.73}$$

and the distance on the x-axis between $p_{R\max}$ and $p_{/R=0.607}$ is just equal to σ (see Fig. 5. 11*a*). Nevertheless, there are other points on the x-axis that are multiples of small fractions of σ. This information is summarized in Table 5.10.

Often, estimation of the dispersion is evaluated from the peak-to-peak jitter, that is, from the difference between the end points x_{max} and x_{min}. However, resolution depends on the use of a DAC (digital-to-ana-

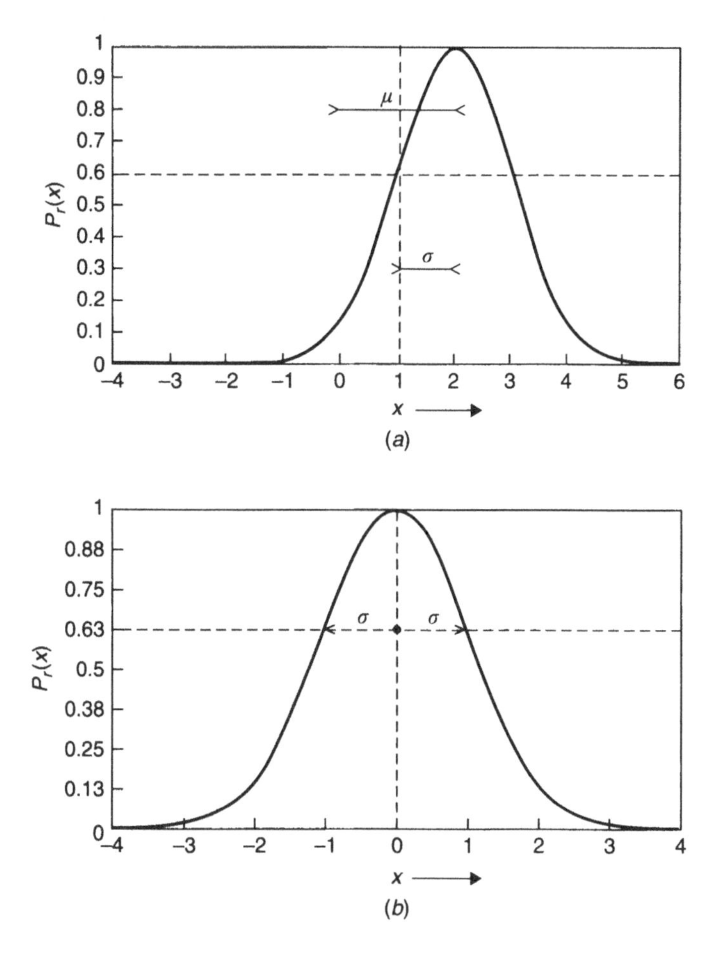

Fig. 5.11. Plot of the Gaussian probability density: (*a*) with the mean value and variance σ, and (*b*) normalized with $\mu = 0$ and variance $\sigma = 1$.

logue convertor) for the $p_R(x)$ reading. From the last two rows in Table 5.10, we may conclude that

$$\sigma \approx \frac{x_{\max} - x_{\min}}{7.75 \pm 0.25} \qquad 5.74$$

The difficulty with the latter appraisal is that the reading of the end points $x_{\max}$ and $x_{\min}$ is rather vague and often blurred with deterministic jitter components.

Table 5.9 Gaussian distribution: probability for different ranges from 1 to 3σ

$\pm 1\sigma$	68.27%	$1-\sigma$	31.7%
$\pm 2\ \sigma$	95.45%	$1-2\sigma$	4.6%
$\pm 3\ \sigma$	99.%	$1-3\sigma$	0.27%

5.3.3 Bit Error Ratio (BER)

The BER defines the probability of one bit error, that is,. the probability for success or failure of the bit transition. Investigation must take into account both discrete and random noises. By considering the sampling point just in the center of the unit interval (cf. Fig. 5.9) and the random time jitter only, we get for one bit error probability

$$P_{\text{error}} = P(T_{\text{obs}}) = \frac{1}{2}\left(1 - \frac{2}{\sqrt{2\pi}} \int_0^{T_{\text{obs}}} e^{-0.5t^2} dt\right) \qquad 5.75$$

where T_{obs} is the observation time and the factor $\frac{1}{2}$ is for the one branch only. Evaluation of the BER for observation times extending over several sigmas is summarized in Table 5.11 and the respective plot is reproduced in Fig. 5.12*a*. Note that even for BER = 10^{-18}, both branches do not close even for a 10-Gbps system. It would take over 3×10^8 s (~ 10 years) to arrive at one BER. However, in real life the situation is not as bright because of the discrete time jitter components. In instances where the sampling point is just at the edge of the unit interval of the

Table 5.10 Relations between the normalized *y*-values, y/y_o, the Corresponding *x*-values, and the matched σs

x/σ	y/y_o
0.5	0.882
1.0	0.607
1.5	0.325
2.0	0.135
2.5	0.044
3.0	0.011
3.7	0.001
4.0	0.00034

discrete jitter, uniform distribution prevails in the range from 0 to ΔT_x (cf. Fig. 5.10) and one bit error probability is just $\frac{1}{2}$. The corresponding plot is shown in Fig.5.12*b* and is labeled as *the bathtub plot* since it resembles the cross section of a bathtub (numerical values are recalled in Table 5.11).

5.3.4 Eye Diagrams

Eye diagrams provide important information about time jitter. Effectively, they are based on the Lissajous figures and present a composite view of superimposed waveforms, with a deeper knowledge about the bit periods. This knowledge enables a quick appreciation of the total jitter. Advantages of the process are speed and easy measurement arrangement (even with live data). Generally, no more than one or two unit intervals are displayed (see Fig. 5.13), whereas investigation of Fig. 5.13*a* reveals four different trajectories on the bottom in contradistinction to only two in the upper part (one may expect four zeros, but only two ones in the analyzed waveform part). Further, we notice two different slopes for rising and falling edges caused by the deterministic jitter. In addition, intersections below the threshold level indicate DCD distortion. The upper part of the diagram, recalled in Fig. 5.13*b*, allows more detailed analysis of the random and deterministic jitter. Note that peak-to-peak deterministic jitter components are identified and labeled $J^{\mathrm{D}}_{\mathrm{App}}$ and $J^{\mathrm{D}}_{\mathrm{Dpp}}$ (for advanced and delayed peak–peak deterministic jitter). Finally, Fig. 5.13*c* is used for explaining the total time jitter. We close by stating

Table 5.11 Bit error ratios of Gaussian distribution: probability for a different number of σs.

BER	$2N\sigma$
10^{-4}	7,438
10^{-6}	9,507
10^{-7}	10,399
10^{-9}	11,996
10^{-11}	13,412
10^{-12}	14,069
10^{-14}	15,304
10^{-18}	17,512

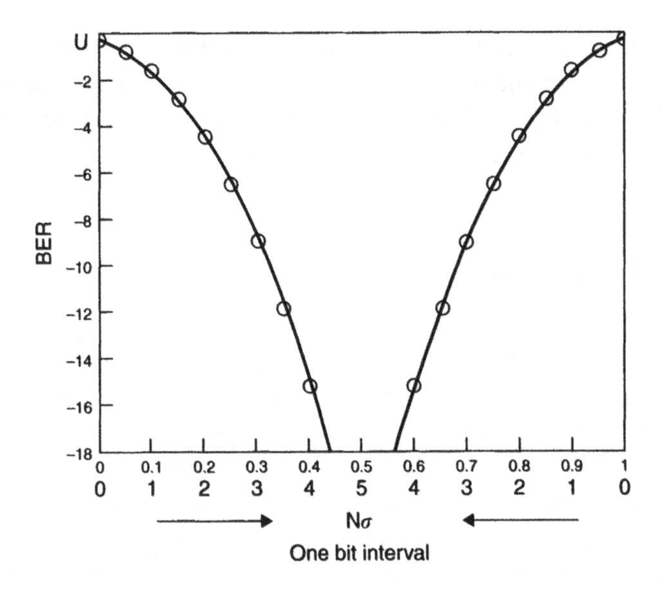

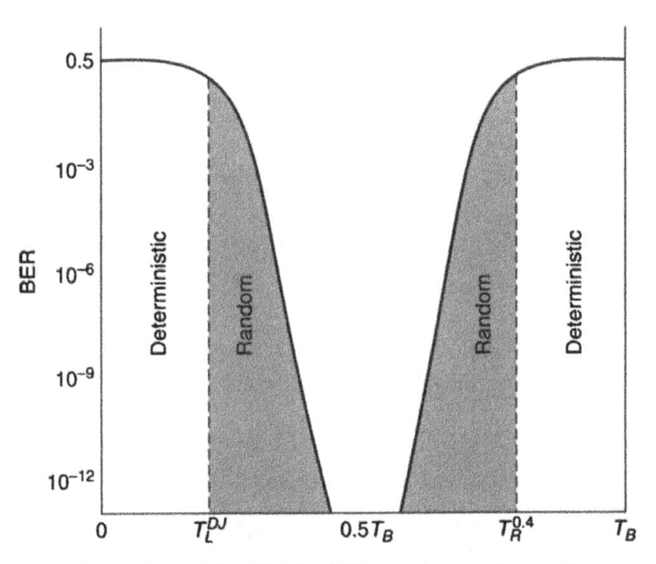

Fig. 5.12. (*a*) Evaluation of the BER for different sigma values in the presence of random noise only. (*b*) The bathtub plot. (Adapted from Agilent Technologies publications [5.23].

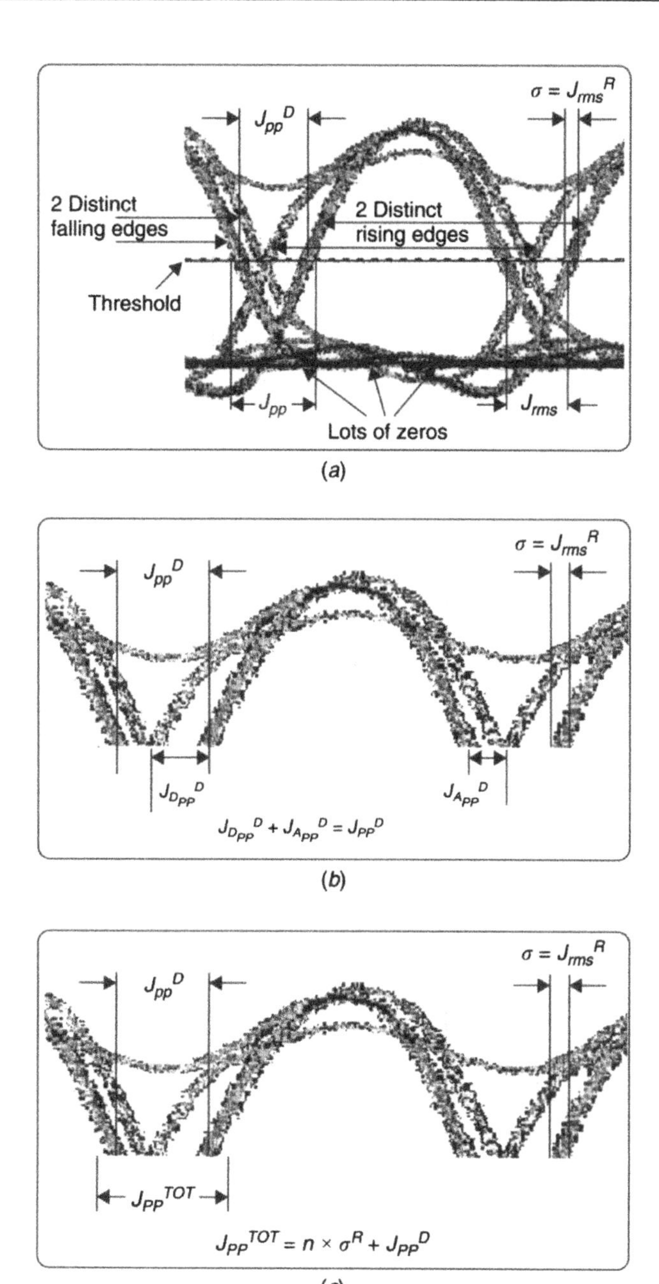

Fig. 5.13. (*a*) An eye diagram with an irregular shape; (*b*) the upper part only, allowing analysis of its random and deterministic jitter; and (*c*) used for explaining the total time jitter ([5.23], Agilent Technologies publications).

that an oscilloscope with a long-persistent display mode is the easiest way to create eye diagrams, that the closing eye points to the presence of random time jitter (not to be mistaken for the time-dependent deterministic jitter, periodic spurious modulation, etc.), and that multiple rising and falling edges result from data-dependent jitter (DDJ).

5.3.5 Histograms

A histogram is a plot of the measured set of data (events) on the y-axis with respect to the number of bins of the parameter (digitized time jitter and others) on the x-axis. Examples are depicted in Fig. 5.14. The envelope provides a good estimate of the probability density function (PDF) of the process. For a set of the measured data, M, the recommended number of bins (for the Gaussian distribution) on the x-axis is

$$n_b \approx 2 \cdot M^{1/3} \tag{5.76}$$

The histogram in the event of a Gaussian distribution of the disturbing noise due to triggering on one trajectory only is recalled in Fig. 5.14*a*. In the case of time jitter investigations, such a situation would be unique (cf. Fig. 5.9 with singular crossing points without any discrete components). However, there is generally a set of triggering points (instants) brought about by deterministic jitter components (cf. Fig. 5.13). Each triggering point is accompanied by random noise fluctuations: The corresponding histogram of the two neighboring triggerings is depicted in Fig. 5.14*b*.

The resulting dissimilar height of the peaks might be caused by different slopes of trajectories since the dispersion, σ, of both noises is expected to be the same. Further, the distance between the peaks presents the corresponding deterministic jitter. Reverting to the two peaks here may require a different origin, namely, the presence of a harmonic or harmonics in the carrier signal. (The block diagram of a histogram principal arrangement is reproduced in Fig. 5.15.)

5.3.6 Separation of the Random and Deterministic Jitter

Separation of random and deterministic jitter enables a deeper understanding of spurious processes in the system and in many instances

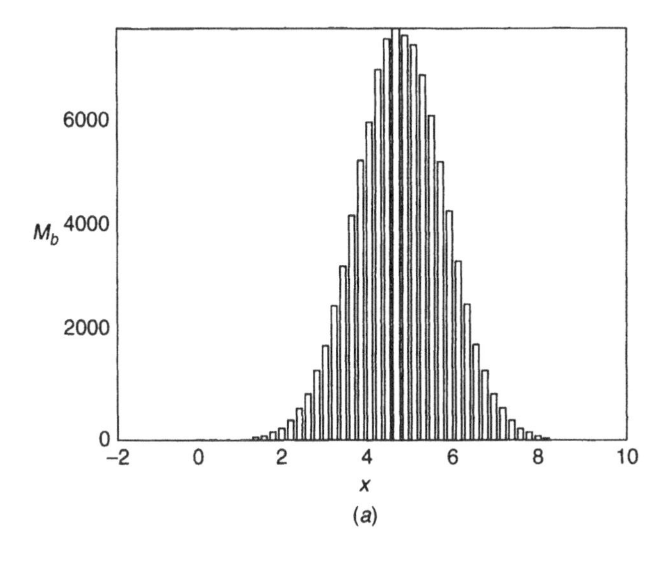

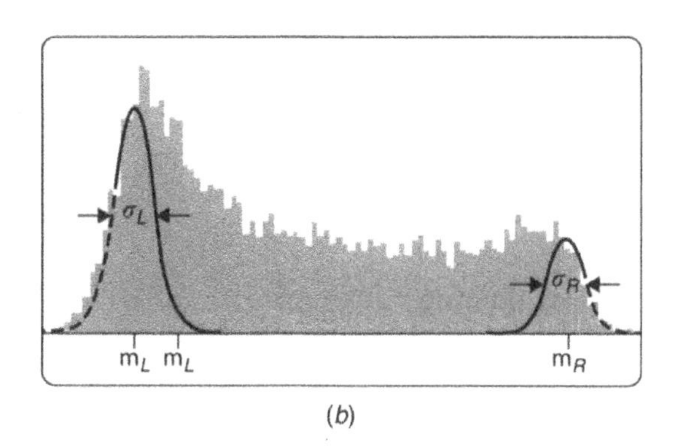

Fig. 5.14. (*a*) The histogram in the event of a Gaussian distribution of the disturbing noise due to triggering on one trajectory only and (*b*) a histogram of the two neighboring triggerings.

the decomposition may spare difficult or time-consuming measurements. The jitter model most commonly used is based on the hierarchy shown in Fig. 5.16. The total jitter of the investigated system is first separated into bounded and unbounded families. Furthermore, the deterministic group is subdivided in to several categories: for example, *periodic jitter (PJ), DDJ,* and *DCD* (see Section 5.3.1). Statistical theory teaches that if two or more random processes are inde-

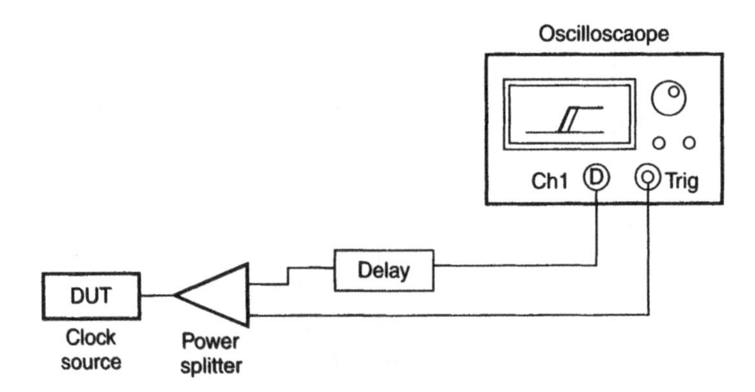

Fig. 5.15. Block diagram of the principal arrangement of a histogram measurement.

pendent, the resulting distribution is equal to the convolution of individual distributions.

The Gaussian distribution was already discussed. The periodic jitter originating from the phase modulation by a sinusoidal signal has the following distribution (cf. Fig. 5.17 and [5.13]):

$$f_y(y) = \frac{2}{2\pi\sqrt{1-y^2}} \qquad \text{for} \qquad |y| < 1 \qquad 5.77$$

On the other hand, the triggering processes (DDJ, DCD, etc.) result in simple vertical lines, as indicated in Fig. 5.17.

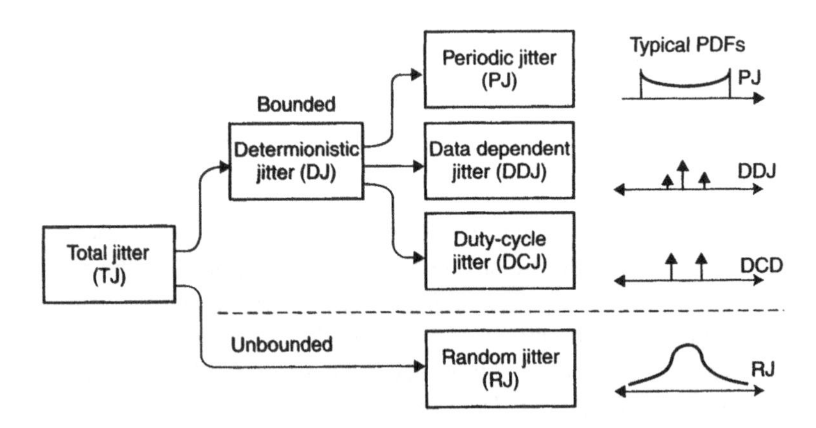

Fig. 5.16. The hierarchy of the time jitter model (Adapted from [5.24]).

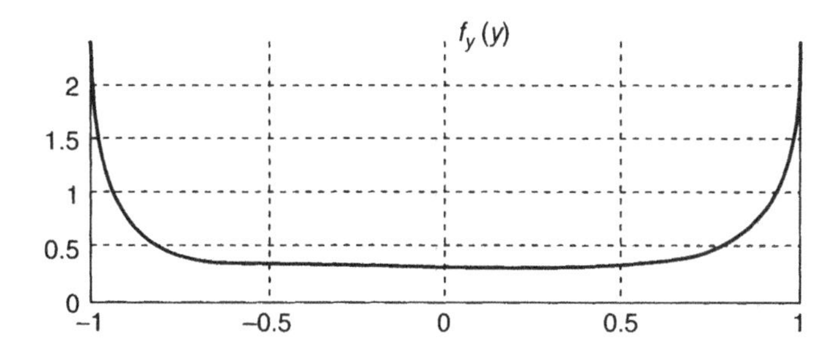

Fig. 5.17. The periodic jitter originating from phase modulation by a sinusoidal signal (cf. Fig. 5.16 and [5.13]).

Separating the random jitter and the deterministic components, depicted in the combined histogram, results in finding the corresponding dispersion, σ, of the random noise. To this end, it is recommended that we evaluate the dispersion sigmas of the far left and right tails of the histogram with the experimental fit to the Gaussian distribution graph. This is difficult because of the scarcity and insufficient precision of the data in the tails due to quantization. The method is described in Example 5.3.

EXAMPLE 5.3

Estimate the random noise distribution, σ, from the histogram in Fig. 5.18. In the first approximation, we compute the peak–peak difference p–p = 57 ps and with the assistance of (5.74) we estimate the dispersion: $\sigma = 7.4$ ps. For more precise results, we read several normalized y-values, y/y_o, on the left side of the histogram (Fig. 5.18), preferably for quantities suggested in Table 5.10, and the corresponding x-values minus the expected mean value $x_{y/yo}$. Then we map them on the normalized Gaussian distribution plot (Fig. 5.19*a*) and find the optimum compliance by dividing the difference $(x - x_{y/yo})$ with the anticipated σ_L (left-hand term). Note that changing the estimated $x_{y/yo}$ shifts the curve to the right or left, but compliance with the tail values is being set by modifying the sought σ. We apply the same procedure to the right-hand tail of the histogram and evaluate σ_R. The mean value of both sigmas is taken for the actual sigma:

$$\sigma = \frac{1}{2}(8.0 + 5.5) = 6.75 \text{ (ps)}$$

In addition, we estimate the difference between both mean values as the deterministic jitter

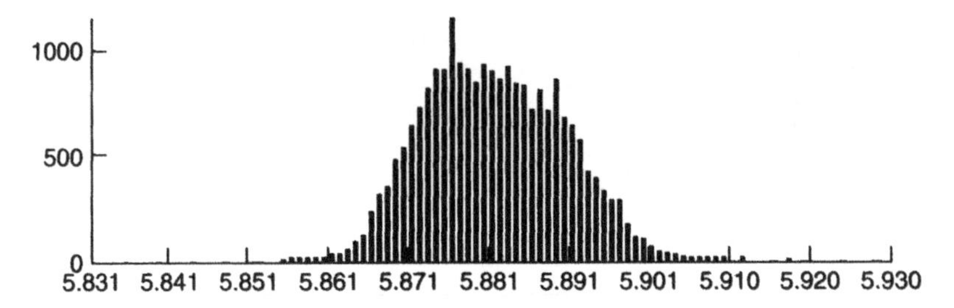

Fig. 5.18. A histogram to assist computation of σs (Adapted from [5.25]).

$$\mu_R - \mu_L = 5886 - 5875 = 11 \text{ ps}$$

With respect to these difficulties, rareness, and granularity of the data, the final solution via a log–log system is recommended (cf. Fig. 5.19*b*, bottom).

Another possibility for resolving random and deterministic jitter provides application of the fast Fourier transform (FFT) on the TIE. In this way, deterministic spectral lines (cf. Fig. 5.16) are removed from the corresponding output and the inverse FFT reveals only the random noise.

5.3.7 Time Jitter Evaluation from PSD Noise Characteristics

Evaluation of oscillator jitter of phase-locked loops (PLL's) and digital communications devices from the measured-phase noise characteristic provides the most accurate values for the corresponding time jitter. Starting with (5.69) and after limitation to the random process only, we arrive at

$$< \Delta t^2 > \omega_o^2 = < \varphi(t_{i+1})^2 - 2\varphi(t_{i+1})\varphi(t_i) + \varphi(t_i)^2 > = \tag{5.78}$$

$$2[R_\varphi(0) - R_\varphi(\tau)] \qquad (\tau = t_{i+1} - t_i)$$

As long as the time readings φ_{i+1} and φ_i are uncorrelated, the autocorrelation $R_\varphi(\tau)$ approaches zero and the root-mean-square (rms) time error is

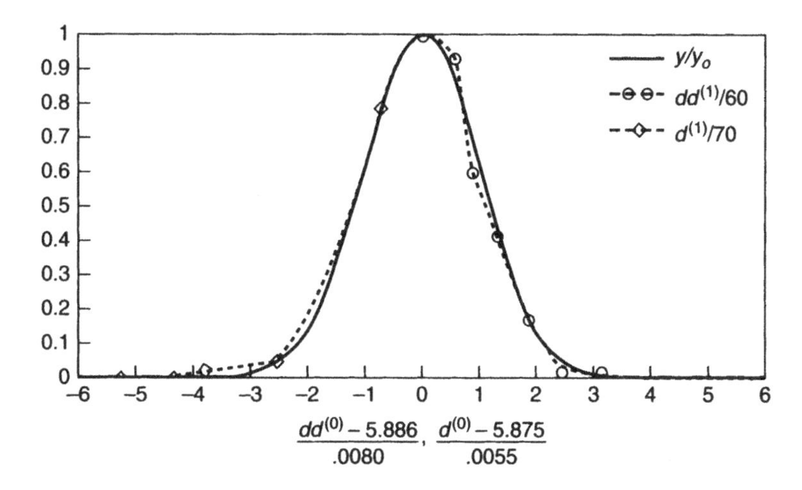

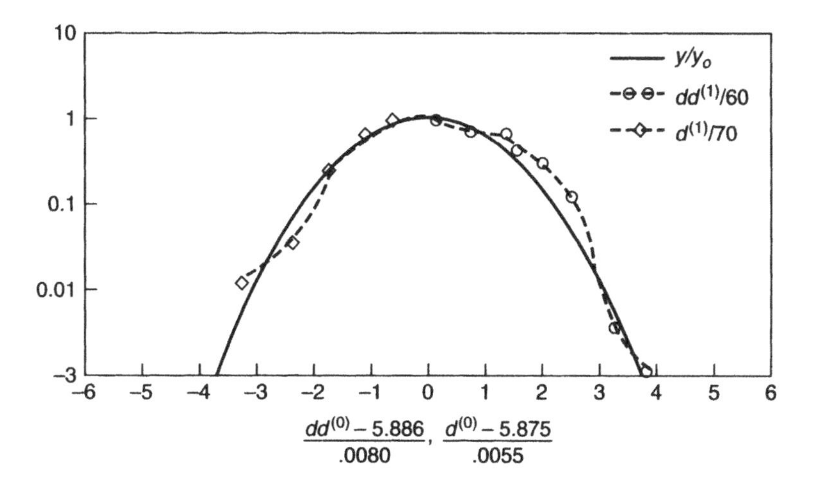

Fig. 5.19. Top: Mapping values x_{y/y_o} on the normalized Gaussian distribution plot. Bottom: Mapping in the log–log presentation.

$$\Delta t_{rms}^2 = \frac{1}{\omega_o^2} \int_{f_L}^{f_H} S_\varphi(f) df \qquad 5.79$$

In instances where the PSD's of the noises of the investigated system are known (oscillators, phase-locked loops, PLL, etc.) the estimation of the mean value is replaced by integration, that is,

$$\Delta t_{\text{rms}}^2 = \frac{1}{\omega_o^2}\int_{f_L}^{f_H} S_\varphi(f)df =$$

$$\frac{1}{(2\pi)^2}\int_{f_L}^{f_H}\frac{S_y(f)}{f^2}df = \frac{1}{(2\pi)^2}\left[\frac{h_{-1}}{2f_L^2}+\frac{h_o}{f_L}+h_1\log\left(\frac{f_H}{f_L}\right)+h_2 f_H\right] \quad 5.80$$

Note that introduction of the fraction frequency PSD to the time jitter, being a function of the fractional frequency constants h_i only, is not changed by frequency multiplication or division. In addition, when we know the phase noise characteristic, that is, the power spectral density $S_\varphi(f)$ in the entire frequency range of the investigated device, the time jitter evaluation by the relation (5.78) or (5.79) provides dependable results. The effective time jitter is generally slightly influenced by the lower integration bound. Consequently, it should be defined (e.g., for SONET the bandwidth is 12 kHz to 20 MHz), whereas the upper integration bound is determined by the cut-off frequency, f_H, of the measurement pass band. In instances where the approximation of the PSD characteristic is known or can be estimated (e.g., the measured phase noise characteristic of the PLL, as reproduced in Fig. 5.20) the random time jitter can be evaluated with either the assistance of *h-factors* (5.80) or integrated with the assistance of the measured values of $S_\varphi(f)$:

$$\Delta t = \frac{1}{\omega_o}\sqrt{\sum_{f_i=f_L}^{f_H} 10^{S_\phi(f_i)/10}\left[\frac{f_{i+1}-f_{i-1}}{2}\right]} \quad 5.81$$

Note that $S_\varphi(f) = 2\mathcal{L}(f)$ is equal to the two noise side bands. Finally, the dispersion of the phase noise in degrees evaluated from the time jitter is

$$\Delta\phi_{\text{degree}} = \Delta t_{rms}\cdot f_o\cdot 360° \quad 5.82$$

EXAMPLE 5.4

Evaluate the period jitter of the precision 5-MHz oscillator discussed in Example 5.1. By introducing the numerical noise factors h_i from Table 5.5(*b*) into (5.80), we find that the only important contribution to the jitter in the entire frequency range from 1 Hz to 1 MHz is from white phase noise, that is,

$$(\Delta t)_{\text{rms}} \approx \frac{1}{2\pi}\sqrt{10^{-29.1}+10^{-29.2+6}} \approx 0.4\times 10^{-12}\ (\text{s}) \quad \Delta\phi \approx 0.72° \quad 5.83$$

EXAMPLE 5.5
Evaluate the period jitter of the PLL having output frequency in the range of 3.5 GHz with the 50-MHz crystal oscillator reference. With the assistance of the phase-noise characteristic, shown in Fig. 5.20, and (5.81) in the range from 10 kHz to 100 MHz, we get for the rms of the time jitter and degrees in the unit interval

$$\Delta t_{\text{rms}} = 1.56 \times 10^{-12}\ \text{s} \qquad \Delta\varphi_{\text{deg}} = 2°$$

5.3.7.1 Time Jitter with Spurious Signals

Spurious side bands generated by phase or frequency modulation by small sinusoidal signals are uncorrelated with the random components. Consequently, evaluation of the time jitter proceeds in accordance with (5.80), that is,

$$\Delta t_{rms}^2 \approx \frac{1}{(2\pi)^2} \int_{f_L}^{f_H} \left[\frac{S_y(f)}{f^2} + \sum \frac{1}{2} \left(\frac{\Delta\omega_m}{\omega_m} \right)^2 \delta(f_m) \right] df \qquad 5.84$$

5.3.8 Time Jitter Evaluation From the Time Domain Measurements

The time jitter evaluation in the time domain is commonly measured between zero crossing of two or more periods. In that case, the difference between the consecutive readings is

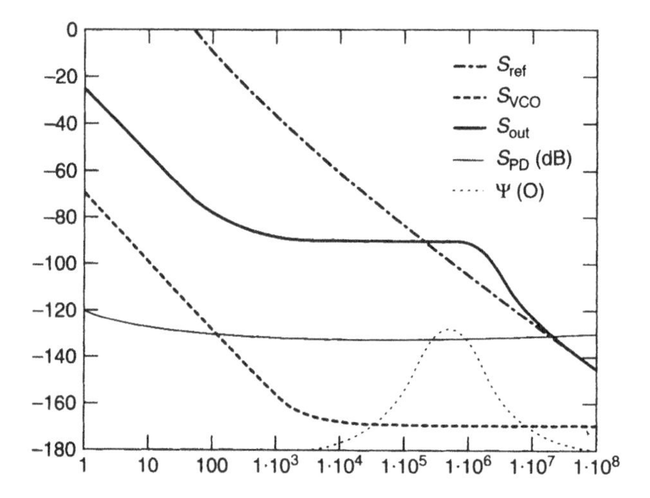

Fig. 5.20. Phase-noise characteristic of the PLL.

$$2\pi f_o t_1 + \varphi(t_1) = 0 \quad \text{and} \quad 2\pi f_o t_2 + \varphi(t_2) = 2\pi N \tag{5.85}$$

(cf. Fig. 5.8) from which

$$2\pi f_o(t_2 - t_1) + \varphi(t_2) - \varphi(t_1) = 2\pi N \tag{5.86}$$

The time jitter accumulated during the N periods [cf. 5.1, 5.26] is

$$\Delta t = (t_2 - t_1) - NT_o \tag{5.87}$$

and with the assistance of (5.78) we arrive at

$$<\Delta t^2> = \frac{2}{\omega_o^2}[R_\varphi(0) - R_\varphi(\tau)] = \frac{1}{\omega_o^2}\int_0^\infty S_\phi(f)[1 - \cos(2\pi f \tau)]df = \frac{2}{\omega_o^2}\int_0^\infty S_\phi(f)\sin^2(\pi f \tau)df \qquad (\tau = t_{i+1} - t_i = NT_o) \tag{5.88}$$

Note again the similarity with (5.12), but the filtering function is simpler, given by (5.8) (cf. Fig. 5.5*b*). However, in real life actual integration bounds are reduced by the measurement system to f_H and by the measurement time to f_L. By considering that the overall phase is equal to the measurement time τ, we may assume that

$$f_L \approx \frac{1}{2\pi\tau} \qquad x_L = \pi\tau f_L = 0.5 \tag{5.89}$$

After introduction of the PSD of fractional frequency fluctuations, we change relation (5.88) into

$$<\Delta t^2> = \frac{1}{2\pi^2}\int_{f_L}^{f_H} \frac{S_y(f)}{f^2}\sin^2(\pi f \tau)df \tag{5.90}$$

and with the assistance of (5.14) and (5.11) we arrive at

$$\Delta t_{rms}^2 = \frac{h_i}{2\pi^2}\int_{f_L}^{f_H} \frac{1}{f^{2-i}}\sin^2(\pi f \tau)df = \frac{h_i \tau^2}{2(\pi\tau)^{i+1}}\int_{x_L}^{x_H} \frac{\sin^2(x)}{x^{2-i}}dx \tag{5.91}$$

Next, we investigate the time jitter brought about by individual noise segments.

5.3.8.1 White Phase Noise (WPN)

In this range, the phase noise PSD $S_\varphi(f)$ is independent of the Fourier frequency. However, the upper bound f_H depends on the measurement system. Consequently, its contribution is easily evaluated with the assistance of (5.88):

$$\Delta t_{\mathrm{rms}}^2 \equiv \frac{h_i \tau^2}{2(\pi\tau)^3} \int_{x_L}^{x_H} \sin^2(x)dx = \frac{h_2}{(2\pi)^2} f_H \equiv \frac{1}{\omega_o^2} S_\phi(f) f_H \qquad 5.92$$

However, in the steady state, the corresponding time jitter variance is proportional to f_H (i.e., to the upper bound of the integration range).

5.3.8.2 Flicker Phase Noise

We start with the assistance of (5.88) and (5.91), and with application of the *cosine integrals* (e.g., [5.9] p. 866) we arrive at ($i = 1$)

$$\begin{aligned} < \Delta t^2 > &= \frac{h_1}{(2\pi)^2} \int_{2x_L}^{2x_H} \frac{1-\cos(2x)}{2x} d(2x) = \frac{h_1}{(2\pi)^2} [S_1(2x_H) - S_1(2x_L)] = \\ &\frac{h_1}{(2\pi)^2} [lg(2x_H) + \gamma - Ci(2x_H) - lg(2x_L) - \gamma + Ci(2x_L)] \approx \\ &\frac{h_1}{(2\pi)^2} lg(f_H / f_L) \quad (x_L = 0.5; \gamma = 0.5772) \end{aligned} \qquad 5.93$$

5.3.8.3 White Frequency Noise

By introducing from (5.14) the corresponding PSD of fractional frequency fluctuations into (5.91), we get

$$\begin{aligned} < \Delta t^2 > &= \frac{h_o \tau}{2\pi} \int_{x_L}^{x_H} \frac{\sin^2(x)}{x^2} dx \approx \frac{h_o \tau}{2\pi} \left(\frac{\pi}{2} - \int_0^{x_L} \frac{\sin^2(x)}{x^2} dx \right) \approx \\ &\frac{h_o \tau}{2\pi} \left(\frac{\pi}{2} - x_L \right) \approx \frac{h_o \tau}{2\pi} \approx \frac{h_o}{(2\pi)^2 f_L} \quad [x_L \approx 0.5] \end{aligned} \qquad 5.94$$

In instances where the integration limits are from zero to infinity, the time jitter would be

$$<\Delta t^2> \approx \frac{h_o \tau}{4} \qquad 5.95$$

5.3.8.4 Flicker Frequency Noise

After introducing the coefficient h_i $(i = -1)$ into relation (5.91), we arrive at

$$<\Delta t^2> = \frac{h_{-1}\tau^2}{2} \int_{x_L}^{x_H} \frac{1-\cos(2x)}{2x^3} d(x) =$$

$$\frac{h_{-1}\tau^2}{4} \left[\left| -\frac{1}{2x^2} + \frac{\cos(2x)}{2x^2} - \frac{\sin(2x)}{x} \right|_{x_L}^{x_H} + 2\int_{x_L}^{x_H} \frac{\cos(2x)}{x} d(x) \right] \approx \qquad 5.96$$

$$\frac{h_{-1}\tau^2}{4}[1.5 - 2C_i(2x_L)] \approx \frac{h_{-1}\tau^2}{2} \qquad [x_L = 0.5]$$

with the assistance of the Korn formula 287 and 297 [5.9] at $(i = -1)$.

5.3.8.5 Random Walk Frequency (RWF)

Finally, we pay attention to the case of the RWF noise, however, from the artificial rather than the practical point of view. In the same way, we proceed as with the FFN and arrive at

$$<\Delta t^2> = \frac{h_{-2}\tau^2}{2(\pi\tau)^{-1}} \int_{x_L}^{x_H} \frac{\sin^2(x)}{x^4} dx = \frac{h_{-2}\pi\tau^3}{2} \int_{x_L}^{x_H} \frac{1-\cos(2x)}{2x^4} d(x) =$$

$$\frac{h_{-2}\pi\tau^3}{4} \left[\left| -\frac{1}{3(x)^3} + \frac{\cos(2x)}{3(x)^3} \right|_{x_L}^{x_H} + \frac{2}{3}\int_{x_L}^{x_H} \frac{\sin(2x)}{x^3} dx \right] =$$

$$\frac{h_{-2}\pi\tau^3}{4}\frac{2}{3} \left[\left| -\frac{\sin(2x)}{2(x)^2} - \frac{\cos(2x)}{x} \right|_{x_L}^{x_H} - 2\int_{x_L}^{x_H} \frac{\sin(2x)}{x} d(x) \right] \approx \qquad 5.97$$

$$\frac{h_{-2}\pi\tau^3}{3}\left[\frac{1}{x_L} - \left(\frac{\pi}{2} - S_i(2x_L)\right)\right] \approx$$

$$\frac{h_{-2}\pi\tau^3}{2} \ldots [x_L = 0.5]$$

5.3.9 Correspondence between Time Jitter Measurements

Principles of the frequency stability of precision oscillators, frequency synthesizers, and other frequency generators and devices are based on both frequency and time domain systems. The principles were discussed in a special issue of the *IEEE Proceedings,* Feb. 1966 [5.1], and have been finalized since then. The efforts to eliminate the convergence problem in instances of the $1/f^{\alpha}$ or pink noises initiated in the time domain frequency stability measures of the two-sample variance, generally designated as the Allan variance. Here, we have summarized the corresponding properties and provided an analytical solution with the goal of finding the closest relations to the frequency domain measures. We have paid particular attention to the confidence intervals and to the relations between the original Allan and the newer modified Allan variances.

The second part of this chapter was dedicated to the problems of time jitter, which gained importance with the introduction of digital communications in the gigahertz ranges. We have paid attention to definitions, to bounded and unbounded (discrete) fluctuations, and other particular properties, and evaluated the expected rms jitter with both approaches (i.e., by integrating the known PSD's from the frequency and time domains). An accurate analytical solution proved that both approaches are identical despite the phase accumulation problems [5.26] mentioned for the time jitter digital measurements [cf. (5.78) with (5.90–5.97)].

5.3.9.1 Correspondence between Simple Time Jitter and Allan Variances

After comparing the above evaluated time jitter relations, normalized to the square of the averaging time τ, with the earlier computed Allan variances, we easily find a mutual correspondence. To this end, we divide (5.15) by (5.88) to find the correspondence

$$\frac{\sigma^2(\tau)\tau^2}{<\Delta t^2>}=\frac{\dfrac{2h_i\tau^2}{(\pi\tau)^{i+1}}\displaystyle\int_0^{\infty}\frac{\sin^4(x)}{x^{2-i}}dx}{\dfrac{h_i\tau^2}{2(\pi\tau)^{i+1}}\displaystyle\int_{x_L}^{x_H}\frac{\sin^2(x)}{x^{2-i}}dx}=4\frac{\displaystyle\int_0^{\infty}\frac{\sin^4(x)}{x^{2-i}}dx}{\displaystyle\int_{x_L}^{x_H}\frac{\sin^2(x)}{x^{2-i}dx}} \qquad 5.98$$

As a rule of thumb, Δt^2 estimated from the Allan variance is about three times larger than that obtained in the period jitter measurements or computed from noise PSD characteristics.

REFERENCES

5.1. Special Issue on Frequency Stability, *Proceedings of the IEEE,* **54** (Feb. 1966), pp. 101–338.

5.2. V.F. Kroupa, *Frequency Stability: Fundamentals and Measurements,* IEEE Press, 1983.

5.3. J.A. Barnes, Atomic Timekeeping and the Statistic of Precision Signal Generators, *Proceedings of the IEEE*, **54** (Feb. 1966), pp. 207–220. (Reprinted in [5.2].)

5.4. R.A. Baugh, Modulation Analysis with the Hadamard Variances, in *Proceedings of the 25th Annual Frequency Control Symposium* (April 1971), pp. 222–255.

5.5. P. Lesage and C. Audoin, A Time Domain Method for Measurement of the Spectral Density of Frequency Fluctuations at Low Fourier Frequencies, in *Proceedings of the 29th Annual Frequency Control Symposium* (May 1975), pp. 394–403.

5.6. D.W. Allan, Statistics of Atomic Frequency Standards, *Proceedings of the IEEE,* **54** (Feb. 1966), pp. 221–230. (Reprinted in [5.2].)

5.7. H.B. Dwight, *Tables of Integrals and Other Mathematical Data,* 4th ed. (1961), New York: The MacMillan Company.

5.8. J. Rutman, Characterization of Phase and Frequency Instabilities in Precision Sources: Fifteen Years of Progress, *Proceedings of the IEEE*, **66** (Sept. 78), pp. 1048–1075.

5.9. G.A. Korn and T.M. Korn, *Mathematical Handbook* (1958), New York: McGraw-Hill.

5.10. S.A. Vitusevich, K. Schieber, I.S. Ghosh, N. Klein, and M. Spinnler, Design and Characterisation of an All-Cryogenic Low-Phase-Noise Sapphire K-Band Oscillator for Satellite Communications, *IEEE Tr.*, **MTT-51** (Jan., 2003), pp. 163–169.

5.11. V. Candelier, P.Canzian, J. Lamboley, M. Brunet, and G. Santarelli, Space Qualified Ultra Stable Oscillators, *2003 IEEE International Freq. Contr. Symposium,* pp. 575–582.

5.12. I Miller and J. E. Freund, *Probability and Statistics for Engineers,* 2nd ed. (1977), Prentice-Hall.

5.13. A. Papoulis, *Probability, Random Variables, and Stochastic Processes* (1965), New York: McGraw-Hill.

5.14. P. Lesage and C. Audoin, Characterization of Frequency Stability: Uncertainty Due to the Finite Number Of Measurements, *IEEE Tr.,* **IM-22** (June, 1973), pp. 157–161. (Reprinted in [5.2].)

5.15. K. Yoshimura, Characterization of frequency Stability: Uncertainty due to the Autocorrelation of the Frequency Fluctuations, *IEEE Tr.*, **IM-27** (March, 1978), pp. 1–7. (Reprinted in [5.2].)

5.16. D.A. Howe, D.W. Allan, and J.A. Barnes, Properties of Signal Sources and Measurement Methods, in *Proceedings of the 35th Annual Frequency Control Symposium* (May 1981), pp. A1–A50.
5.17. L.N. Bolsev and N.C. Snirnov, Tablicy matematiceskoj statistiki (Tables of mathematical statistics), Nauka, Moskwa, 1983 (see also [5.9] p. 903).
5.18. P. Lesage and C. Audoin, Estimation of the Two-Sample Variance with a Limited Number of Data, in *Proceedings of the 31st Annual Frequency Control Symposium* (June 1977), pp. 311–318.
5.19. D.W. Allan and J.A. Barnes, A Modified Allan Variance with Increased Oscillator Characterization Ability, in *Proceedings of the 35 Annual Frequency Control Symposium* (May 1981), pp. 470–475
5.20. E.S. Ferree-Pikal, J.R. Vig, J.C. Camparo, L.S. Cutler, L. Maleki, W.J. Reley, S.R. Stein, C. Thomas, F.L. Walls, and D.J. Wite, Draft Revision of IEEE STYD 1139-1988 Standard Definitions of Physical Quantities for Fundamental Frequency and Time Metrology—Random Instabilities, in *1997 IEEE Annual Frequency Control Symposium*, pp. 338–57.
5.21. D.A. Howe and T.N. Tasset, Clock Jitter Estimation Based on PM Noise Measurement, in *2003 IEEE Annual Frequency Control Symposium*, pp. 541–546.
5.22. I. Zamek, and S. Zamek, Crystal oscillators: Jitter Measurements and its Estimation of Phase Noise, in *2003 IEEE Annual Frequency Control Symposium*, pp. 547–555.
5.23. Agilent Technologies publications: Application Note 1448-1.
5.24. Tektronix: Understanding and Characterizing Time Jitter. Available at www.tektronix.com/jitter
5.25. B. Drakhlis, Calculate Oscillator Jitter by Using Phase Noise Analysis, Part 1, *Microwave & RF,* **82** (Jan. 2001), Part 2 (February), p. 100.

CHAPTER 6

Phase-Locked Loops

Today's phase-locked loops (PLLs) are important tools in modern communications. They are encountered practically in all types of frequency synthesizers, precision frequency generators, transmitter–exciters, receiver–local oscillators, and other devices. Their most appreciated properties are the increased stability of the output frequency, reduced noise close to the carrier, and frequency synthesis even in reduced integrated systems.

6.1 PLL BASICS

There are many textbooks [cf. 6.1–6.7] dealing with the design and properties of PLL systems. Here, we follow the earlier treatment by Kroupa [6.6]. The most important starting feature is the output frequency f_o, and its range and frequency tuning steps, if any. Next are the choice of the input frequency f_i and the reference frequency f_r (if different from f_i). From the stability point of view, the input frequency generator should be a precision oscillator. Generally, it is a crystal oscillator, preferably with a carrier frequency of tens of megahertz. An input frequency in the range of tens of megahertz is advantageous.

With this preliminary information, we are prepared to discuss the essentials needed for understanding or designing PLLs. The basic feature is the feedback system whose simplified block diagram is shown

Frequency Stability. By Venceslav F. Kroupa

in Fig. 6.1, which has a phase detector, reference and output frequencies, phase amplifier, and stabilized output oscillator.

However, this simplest first-order arrangement will be applicable in exceptional instances only. Usually, in the forward path we encounter a loop filter and in the feedback path either a simple divider or system of mixing circuits, as shown in Fig. 6.2. The analysis of the PLLs locked loops is performed using the Laplace transform notation:

$$[\Phi_{\text{in}}(s) - \Phi_{\text{out}}(s)]K_d K_A F_L(s)\frac{K_o}{s}\frac{1}{N} = \Phi_{\text{vco}}(s) + \Phi_{\text{out}}(s) \qquad 6.1$$

6.1.1 Loop Gain and the Transfer Functions

Analysis of a PLL begins with investigation of the open-loop gain. The most important blocks are the phase detector with gain K_d (V/2π),

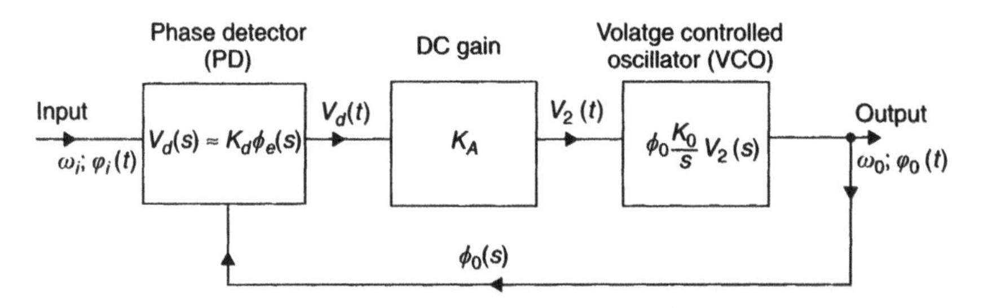

Fig. 6.1 The block diagram of a first-order PLL with an additional DC gain K_A [6.6].

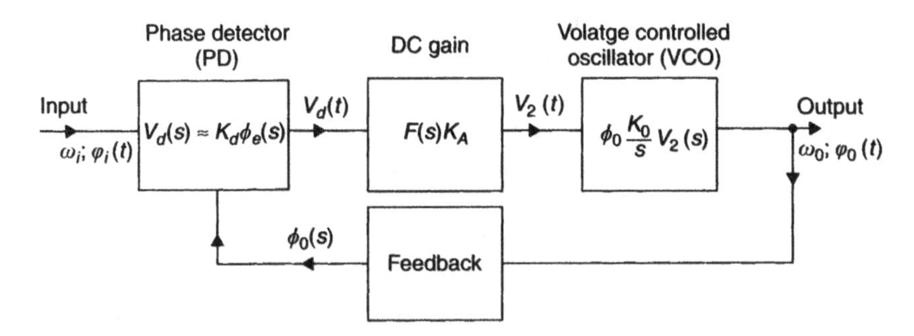

Fig. 6.2 Block diagram of a PLL synthesizer with a divider in the feedback path [6.6].

loop-filter transfer function $F_L(s)$, oscillator gain K_o (Hz/V), and the feedback transfer function, which is generally the frequency division factor $N = \omega_r/\omega_o$. In exceptional instances, an amplifier gain K_A is added:

$$G(s) = K_d \times K_A \times F_L(s) \times \frac{K_o}{s} \times \frac{1}{N} = F_L(s) \times \frac{K}{sN} \qquad 6.2$$

After introducing the loop gain into (6.1), we find that the steady-state output phase is

$$\Phi_{\text{out}}(s) = \Phi_{\text{in}}(s)\frac{G(s)}{1+G(s)} + \Phi_{\text{vco}}(s)\frac{1}{1+G(s)} \qquad 6.3$$

and the transfer function $H(s)$ between the output and input phases of the closed loop is

$$H(s) = \frac{\Phi_{\text{out}}(s)}{\Phi_{\text{in}}(s)} = \frac{G(s)}{1+G(s)} = \frac{KF_L(s)/sN}{1+KF_L(s)/sN} = \frac{KF_L(s)/N}{s+KF_L(s)/N} \qquad 6.4$$

The asymptotic approximation reveals

$$\begin{aligned} &H(s) \approx 1 && s \ll KF_L(s)/sN && \text{for small } (s) \\ &H(s) \approx \frac{KF_L(s)/sN}{s} && s \gg KF_L(s)/sN && \text{for large } (s) \end{aligned} \qquad 6.5$$

The transfer function, $1 - H(s)$, between the output and the VCO phase is

$$1 - H(s) = \frac{\Phi_{\text{out}}(s)}{\Phi_{\text{vco}}(s)} = \frac{1}{1+G(s)} = \frac{s}{s+KF_L(s)/N} \qquad 6.6$$

The asymptotic approximation reveals

$$\begin{aligned} &1 - H(s) \approx \frac{s}{KF_L(s)/sN} && s \ll KF_L(s)/sN && \text{for small } (s) \\ &1 - H(s) \approx 1 && s \gg KF_L(s)/sN && \text{for large } (s) \end{aligned} \qquad 6.7$$

In the majority of practical PLLs (in communications), the feedback block is a divider. Without any loss of generality, we assume that the gain $K_dK_o = K'$ is reduced in proportion to the division factor N to a new value K *primed* (an effective multiplication of the output phase is performed in accordance with Fig. 6.3*c*) and a new reduced gain is

$$K' = K_d K_o / N \qquad 6.8$$

The transfer functions of these simple PLL's normalized to the gain K are reproduced in Fig . 6.3.

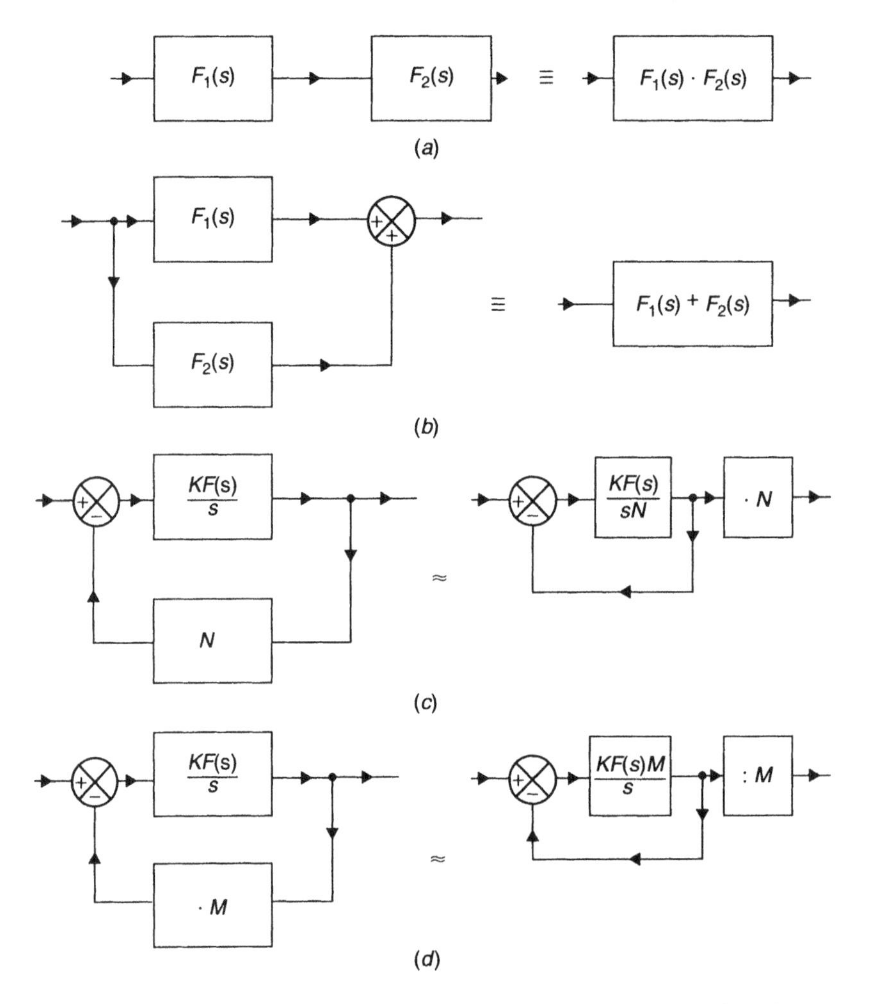

Fig. 6.3 Simplification of the block diagrams of PLL: (*a*) series connection, (*b*) parallel connection, (*c*) and (*d*) feedback arrangement. There are even more complicated systems [6.6].

6.1.2 The Loop Filter

The next step in evaluating the open-loop gain, $G(s)$, and PLL properties is to introduce the filter function into the PLL relation (6.2). At first, we suggest a simple RC filter (Fig. 6.4*a*), then an RRC or (RCC) arrangement (Fig. 6.4*b*), or even an integrating RCC filter (Fig. 6.4*c*), which combines advantages of both filters with additional integration

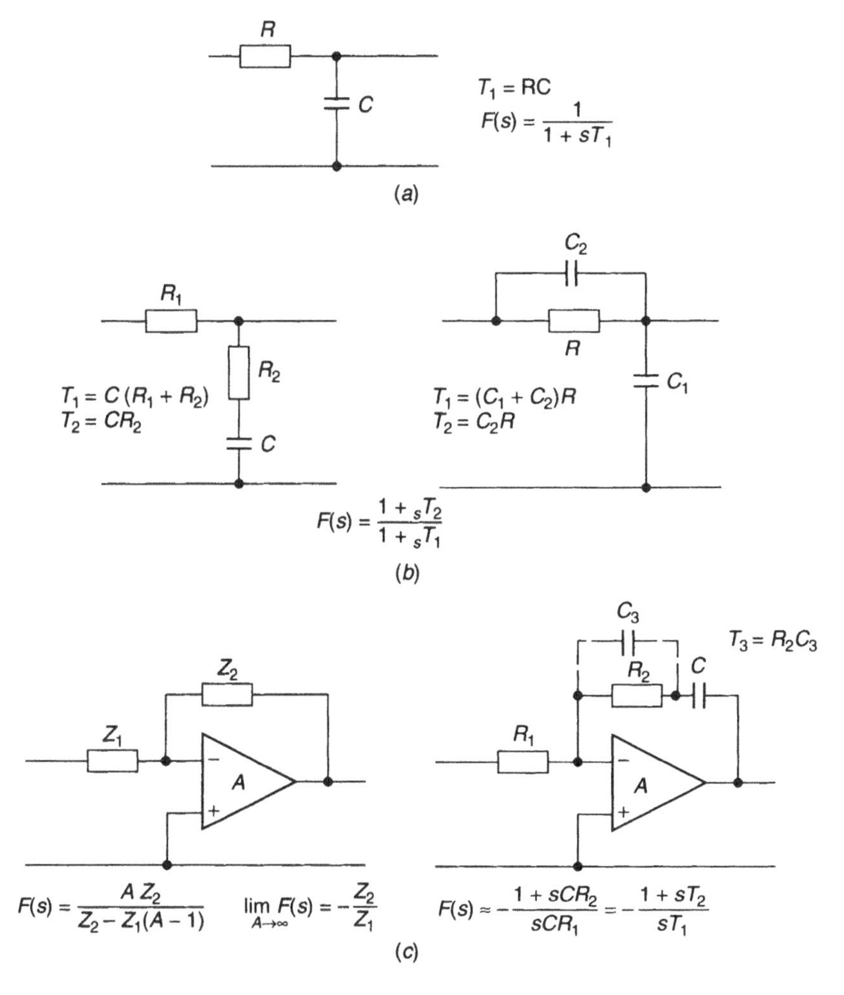

Fig. 6.4 Second-order PLL loop filters: (*a*) a simple RC filter; (*b*) phase-lag lead or proportional–integral networks (RRC or RCC combination): (*c*) general arrangement and active phase-lag lead network (dashed line is the third-order loop configuration).

and at the same time changes the loop from type I into type II (see Section 6.1.5). For all three layouts, we get the following transfer function:

$$F(s)=\frac{1+sT_2}{sT_I+1/A} \qquad 6.9$$

For arrangement (*a*) we put $T_2 = 0$ and $A = 1$; for (*b*) we put $A = 1$; and, finally, for (*c*) A is approximately infinity.

6.1.3 The Voltage-Controlled Oscillator Gain

The VOCs were discussed in Chapter 3, where we concluded that a small spurious voltage introduced into the resonant circuit changes the oscillating frequency and, after integration, also changes the oscillator phase in addition to the effective gain K_o/s. (See also Section 6.10 on synchronization.)

6.1.4 The Open-Loop Gain

When introducing the above relations K_d and K_o into (6.2), we get for the open-loop gain

$$G(s)=K_d\times\frac{1+sT_2}{sT_1+1/A}\times\frac{K_o}{s}\times\frac{1}{N}=\frac{1+sT_2}{s^2}\times\frac{K'}{T_1} \qquad 6.10$$

where K' divided by T_1 has a dimensions of Hz^2. Its introduction into (6.4) reveals a general form of the PLL transfer function:

$$H(s)=\frac{(1+sT_2)(K'/T_1)}{s^2+s(K'T_2/A)/T_1+K'/T_1} \qquad 6.11$$

6.1.5 Order and Type of the PLL

The order of the PLL depends on the order of the polynomials in the denominator of (6.11). We designate the PLL as first order if no filter is used, that is, $F_L(s) = 1$, as second order with a filter of the type indicated in (6.9), and so on. In accordance with the number of integrators in the open-loop gain ($1/s$ or $1/s^2$) we call the loops type I or type II, and occasionally type III. The second-order loop of type I or II provides the bases for the design of many PLLs. Its main feature is the unconditional stability of the corresponding feedback systems (of the second order) that are be discussed later. Note that the operation amplifier in the loop filter in Fig. 6.5*c* changes the transfer function of the loop filter (6.9) into an imperfect integrator and the PLL from type I to type II (cf. also the behavior of the loop impedance in the current-pump systems).

6.2 PLL DESIGN

The most important design feature of the PLLs is the choice of the reference frequency (f_r) with the corresponding division factor (N), the passband frequency (f_{PLL}), the expected noise behavior, the power of spurious modulation signals, the consumption of DC power, and so on. In this connection, we examine the behavior of the gain $G(s)$, the properties of the transfer functions $H(s)$ and $1 - H(s)$, the lock-in times, the tracking properties, and, particularly, the overall stability of the system.

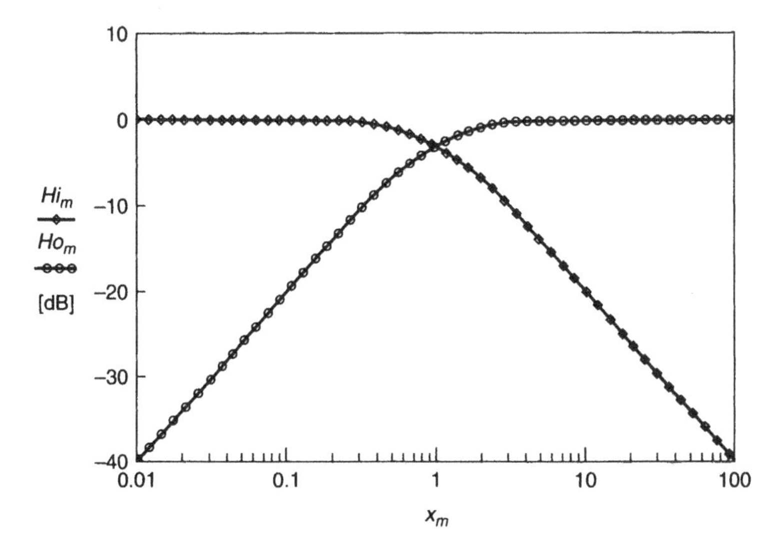

Fig. 6.5 Transfer functions $Hi_m = |H(s)|^2$, $Ho_m = |1 - H(s)|$, and $1 - H(s)$ of the simple first-order PLL loop, normalized to gain $K\,|x_m = f_m/f_n$ [6.6].

6.2.1 Natural Frequency and Damping Factor

For general investigation and solution of the PLL problems, we recommend introduction (in agreement with the theory of the servomechanism of the dynamic motion equation or damped oscillations) of the natural frequency ω_n and the damping factor ζ [e.g., 6.8], which was also adapted by Gardner [6.1]:

$$\omega_n^2 = \frac{K'}{T_1} \; ; \; \zeta = \frac{\omega_n}{2}\left(T_2 + \frac{1}{AK'}\right) \approx \frac{\omega_n}{2} T_2 \qquad \omega_n = \frac{2\zeta}{T_2} = \frac{2\zeta}{RC} \tag{6.12}$$

After introducing (6.12) into the corresponding loop gain and transfer-functions (6.10) and (6.11), we have

$$G(s) = \frac{s(2\zeta\omega_n - \omega_n^2 / K') + \omega_n^2}{s(s + \omega_n^2 / AK')} \approx \frac{2\zeta s\omega_n + \omega_n^2}{s^2}$$
$$H(s) \cong \frac{2\zeta s\omega_n + \omega_n^2}{s^2 + 2\zeta s\omega_n + \omega_n^2} \qquad 1 - H(s) \cong \frac{s^2}{s^2 + 2\zeta s\omega_n + \omega_n^2} \tag{6.13}$$

Intersection of the transfer functions $H(s)$ and $1 - H(s)$ approximately indicates the passband of the corresponding PLL system (cf. Fig. 6.5). However, more precise investigation starts with *zeros* and *poles* encountered with the transfer function $H(s)$. In addition, plotting of the corresponding *root locus*, which will be discussed later, provides more information about the mutual interdependence between the natural frequency ω_n and the damping factor ζ.

Computation of the denominator roots reveals the needed information:

$$s^2 + 2\zeta s\omega_n + \omega_n^2 = 0 \tag{6.14}$$

Note that (6.14) is the analytic geometry equation of a circle and the solution of the corresponding second-order relation (6.14) reveals

$$x_{1,2} = -\omega_n\zeta \pm j\sqrt{(\omega_n\zeta)^2 - \omega_n^2} = -\omega_n\zeta \pm j\omega_n\sqrt{\zeta^2 - 1} \tag{6.15}$$

The idealized plot is shown in Fig. 6.6. Closer inspection reveals that for the damping factors close to unity (i.e., $\zeta \approx 1$) the change of the normalized frequency ω_n is small and is effectively independent of the

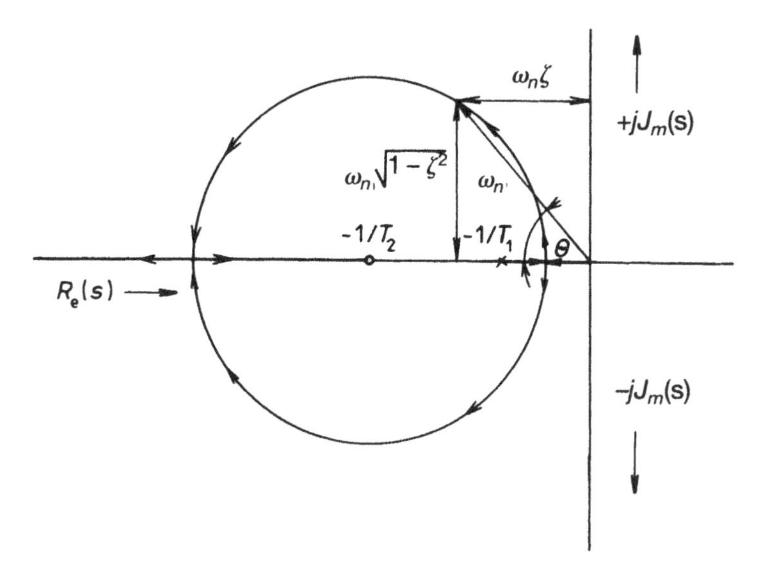

Fig. 6.6 The root locus of $1 + G(\sigma)$ for the second-order PLL type II with the RRC filter (see Fig. 6.4*b*) [6.6].

damping factor ζ (roots placed at the far left). The advantage is a small overshoot of the PLL transfer functions.

6.2.2 Normalized Frequency

Normalization of the base band frequency ω with respect to the natural frequency ω_n, makes application of the generalized PLL relations for the problems to be solved possible. By introducing

$$\sigma = \frac{s}{\omega_n} = \frac{j\omega}{\omega_n} = jx \qquad 6.16$$

we get for the open-loop gain

$$G(\sigma) = \frac{1 + 2\sigma\zeta - \sigma\omega_n / AK'}{\sigma^2 + \sigma\omega_n / AK'} \quad \text{or}$$

$$G(jx) = \frac{1 + 2j\zeta x - jx\omega_n / AK'}{-x^2 + jx\omega_n / AK'} \approx \frac{1 + 2j\zeta x}{-x^2} \qquad 6.17$$

and for the normalized transfer functions for high-gain loops ($\omega_n/K \ll 1$)

$$H(jx) = \frac{1+2j\zeta x}{-x^2+2j\zeta x+1} \qquad 1-H(jx) = \frac{-x^2}{-x^2+2j\zeta x+1} \qquad 6.18$$

Normalized transfer functions of the high-gain PLLs are plotted in Fig. 6.7 for different damping factors ζ [see, e.g., 6.6].

6.2.3 The PLL Band Pass

As the PLL band pass of the second-order loops we designate the frequency where the transfer functions $|H(jx)|^2$ equal one, that is,

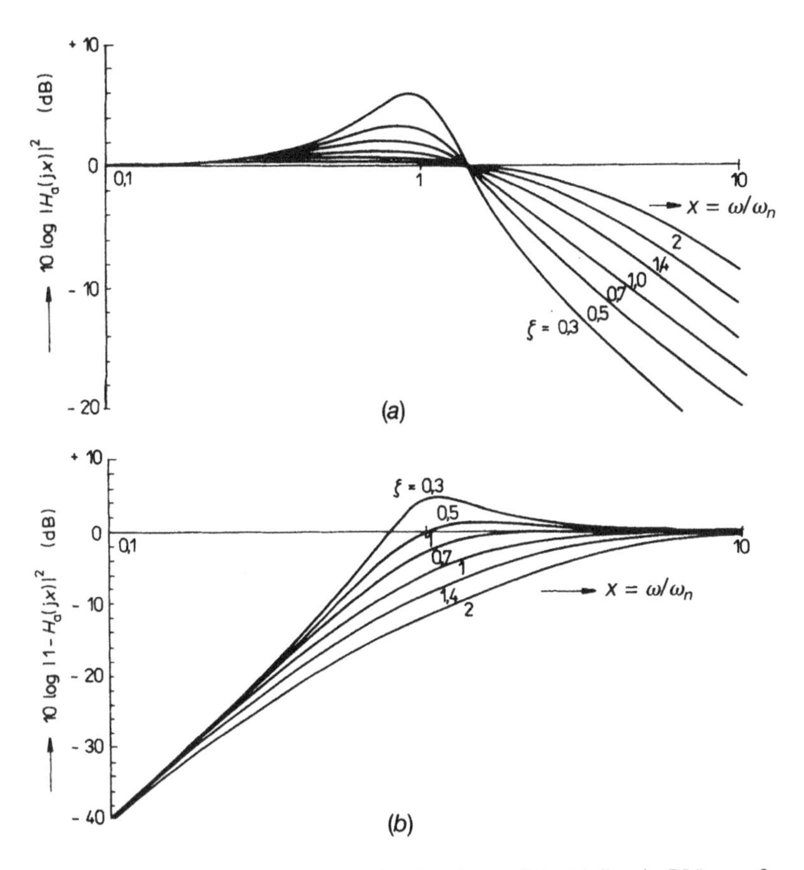

Fig. 6.7 Plots of the normalized transfer functions of the high-gain PLLs as functions of the damping factor ζ [6.6].

$$\left| \frac{1+2j\zeta}{-x^2+2j\zeta x+1} \right|^2 \approx 1 \qquad 6.19$$

From this

$$x_{1,2} = \sqrt{2} \quad \text{or} \quad f_{PLL} \approx 1.4\, f_n \qquad 6.20$$

for different damping factors ζ (see Fig. 6.7). Note that for $\zeta \approx 1$ the overshoot of the transfer function $H(s)$ is rather small. We will revert to this problem later.

6.2.4 The Higher Order Loops

Practical PPLs are formed with the basic second-order loops (in most instances of Type II) enlarged by one or two simple filter sections (generally independent or nearly independent; Fig. 6.8), changing the gain to

$$G(s) = G_2(s) \cdot \frac{1}{1+sT_3} \cdot \frac{1}{1+sT_4} \qquad 6.21$$

Not more than two additional *RC* sections are used. Other types of filters, such as twin-T or acive low-pass filters, are used in uncommon instances [cf. 6.6]. Simplification of the relation (6.21) follows after introduction of the normalized time constants:

$$\frac{T_3}{T_2} = \kappa \qquad \frac{T_4}{T_3} = \eta \qquad T_2 = \frac{2\zeta}{\omega_n} \qquad 6.22$$

from which the normalized gain is

$$G(\sigma) = G_2(\sigma) \cdot \frac{1}{1+2\sigma\kappa} \cdot \frac{1}{1+2\sigma\kappa\eta} \qquad 6.23$$

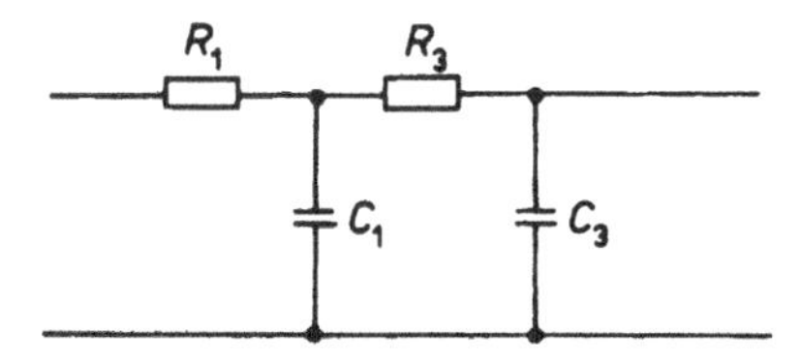

Fig. 6.8 A PLL filter with two *RC* sections in series [6.6].

The advantage is a simplified solution. Note that increasing the order of PLL above the fourth order does not make sense, since both the amplitude changes and phase shifts in the whole band pass of interest are negligibly small.

6.2.5 The Third-Order Loops

The most often encountered arrangement of practical PLL systems is the third-order loop type II, realized by addition of an independent *RC section.* By starting with the second-order high-gain loop, we have for the open-loop gain

$$G_3(s)=\frac{K_dK_AK_o(1+sT_2)}{s(1+sT_1)(1+sT_3)}=G_2(s)G_{RC}(s) \qquad 6.24$$

In the *type II* systems, we achieve this goal by changing the feedback path in the integrating OP amplifier system (see the dashed part in Fig. 6.4*c*). For the loop gain, we get

$$G_3(s)\approx\frac{K}{s}\cdot\frac{1+sR_2(C+C_3)}{sR_1C(1+sR_2C_3)}=\frac{K}{s}\cdot\frac{1+sT_2}{sT_1(1+\kappa sT_2)}\approx\frac{K}{s}\cdot\frac{1+sT_2(1-\kappa)}{sT_1} \qquad 6.25$$

where we have introduced an important design factor:

$$\kappa=\frac{T_3}{T_2} \qquad \kappa<1 \qquad 6.26$$

After introduction of the natural frequency ω_n and the damping factor ζ, in accordance with (6.12), we get for the transfer functions

$$\begin{aligned} H_3(jx)&=\frac{jx2\zeta+1}{-jx^3 2\zeta\kappa-x^2+jx2\zeta+1} \\ 1-H_3(jx)&=\frac{-jx^3 2\zeta\kappa-x^2}{-jx^3 2\zeta\kappa-x^2+jx2\zeta+1} \end{aligned} \qquad 6.27$$

Typical transfer functions are plotted in Fig. 6.9. The appreciated advantages are the slopes of 40 dB per decade in both stop passbands, ease of the design on the basis of the second-order PLL, and, effectively, a reasonable phase margin (the unconditional stability), defined as

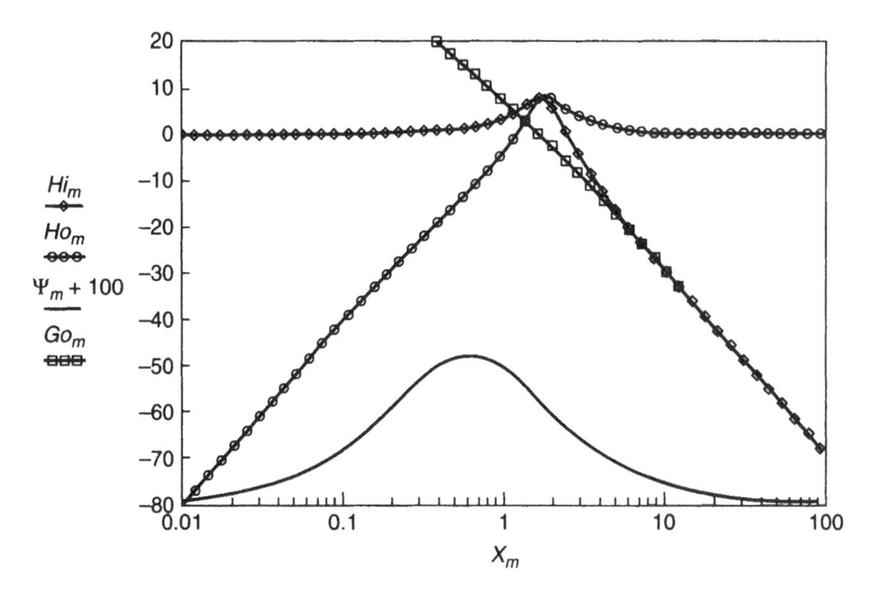

Fig. 6.9 Transfer functions of the third-order PLL loop of Type II as a function of the normalized frequency x. The input transfer functions $H_i(jx) = 20\log(|H_3(jx)|)$. The output transfer funtions $H_o(jx) = 20\log(|1 - H_3(jx)|)$, and the open-loop gain $G_3(jx)$ for $\kappa = 0.3$ and $\zeta = 1.5$ [6.6].

$$\psi = -\frac{180^\circ}{\pi}[-\pi + \tan^{-1}(2\zeta x) - \tan^{-1}(2\zeta x\kappa)] \geq 0^\circ \qquad 6.28$$

Note that the time constant T_3, in accordance with the design procedure, is smaller than T_2. Then even the third-order loop is unconditionally stable and $H(s)$ exhibits a positive phase margin.

To enable design from this point of view, we have summarized some third-order loop properties in graphical form in Fig. 6.10. Its inspection reveals as optimum the original damping factor $\zeta \approx 0.7$ from the point of the phase margin, overshoots, and spurious signals. Typical transfer function characteristics are shown in Fig. 6.11.

6.3 STABILITY OF THE PLL

Phase-locked loops have been used for years in control systems. The first applications were in mechanical engineering (e.g., for maintaining the correct speed of a steam engine shaft). In this connection, the

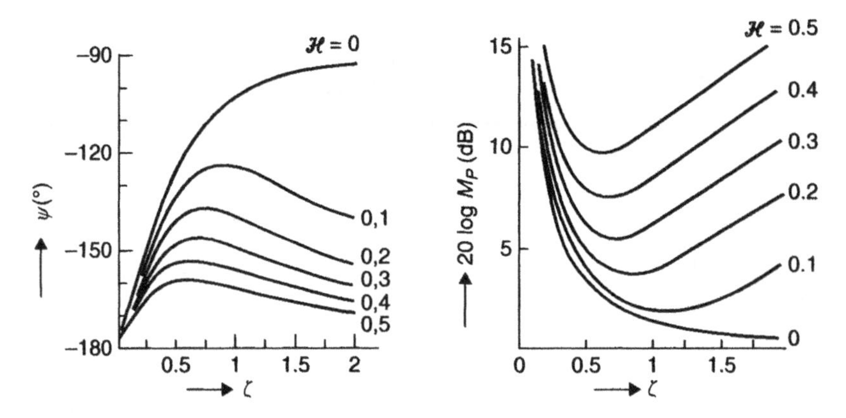

Fig. 6.10 Properties of the third-order PLL for different damping constants of the original second-order loop for different κ of the additional *RC* section: (*a*) phase of the open-loop gain; (*b*) magnitude of the overshoot M_p of the transfer function 10 $\log(|H_3(jx)|^2)$ [6.6].

book by C.J. Savant [6.8] deserves to be mentioned. However, applications in electronics, particularly in frequency synthesis systems, require some special treatment that will be discussed below. Since PLLs are feedback systems with the feedback transfer function $G(s)$, they will oscillate whenever the gain $G(s)$ is equal to –1, that is,

$$1+G(s)=0 \qquad 6.29$$

This condition expressed in complex form is

$$|G(j\omega)|e^{j\psi}=-1 \qquad 6.30$$

that is,

$$|G(j\omega)|=1 \quad \text{and} \quad \psi=(2k+1)\pi \ \ (k=\pm 1, \pm 2, ...) \qquad 6.31$$

Note that all second-order PLL loops are unconditionally stable since from (6.28) we arrive at a positive phase margin

$$\psi=-\frac{180^\circ}{\pi}[-\pi+\tan^{-1}(2\zeta x)] \leq 180^\circ - 90^\circ \geq 180^\circ \qquad 6.32$$

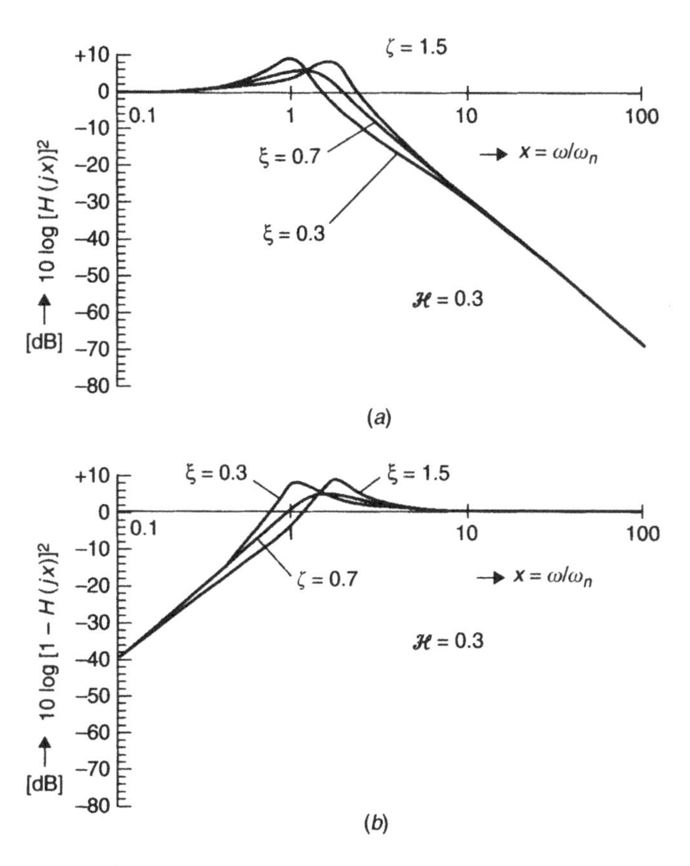

Fig. 6.11 Transfer functions of the third-order PLL for three different damping factors ζ of the original second-order PLL for the constant κ = 0.3: (*a*) for 10 log($|H_3(jx)|^2$), (*b*) for 10 log($|1 - H_3(jx)|^2$) [6.6].

In instances where the normalized frequency $x = 1$ and the damping is > 0.5, the safety margin is nearly 90°.The unconditional stability is also true for the third-order loops discussed above and for many higher order loops, where additional filtering sections are added for suppression of the reference signal and other spurious signals [6.6], since

$$\psi = -\frac{180^\circ}{\pi}[-\pi + \tan^{-1}(2\zeta x) - \tan^{-1}(2\zeta x\kappa) - \tan^{-1}(2\zeta x\kappa\eta)] \approx \\ +180^\circ - \frac{180^\circ}{\pi}[\tan^{-1}(2\zeta x) - \tan^{-1}(2\zeta x\kappa)] > 0^\circ \qquad 6.33$$

See also Fig. 6.9 or Table 6.1. There are many criteria for appreciating the stability of the PLL systems. In the contemporary literature, with the assistance of simple Bode plots, it is investigated in accordance with the old tradition of servo systems. However, application of modern computers provides better insight and possible perfection, particularly, in cases where current-pump phase detectors are used. Furthermore, we can easily plot transfer functions $|H(s)|^2$ and $|1 - H(s)|^2$ and evaluate gain and phase margins, even for loops of higher orders.

6.3.1 Bode Plots

In the Bode plot, we combine the open loop gain in decibel measure and the respective phase shift in one figure, that is,

$$20\log(|G(jx)|) \quad \text{and} \quad \frac{180}{\pi}\log(G(jx)) \qquad 6.34$$

We have seen that the open-loop gain $G(s)$ or $G(jx)$ consists of factors with simple transfer functions

$$G(j\omega) = KA_1A_2\ldots A_n e^{-j(\phi_1+\phi_2+\ldots\phi_n)} \qquad 6.35$$

When drawing Bode plots, we revert to (6.35) and compute its logarithm:

$$\log G(j\omega) = \log K + \log A_1 + \log A_2 + \cdots + \log A_r - j(\phi_1 + \phi_2 + \cdots + \phi_r + \omega\tau) \qquad 6.36$$

After plotting the right-hand side (rhs) of (6.36) in two separate graphs we get quick information about the system stability. In addition, the gain characteristic $G(j\omega)$, for frequencies above ω_n, also provides information about the transfer function $H(j\omega)$ or $H(jx)$.

For construction of the Bode plots, we apply the asymptotes. Here we will repeat some basic rules for their construction.

Table 6.1 The phase margin ψ in degrees

$\tan^{-1}(2\zeta x)$	1	1.19	143	1.73	2.14	2.74	3.73
ψ [°]	45	50	55	60	65	70	75

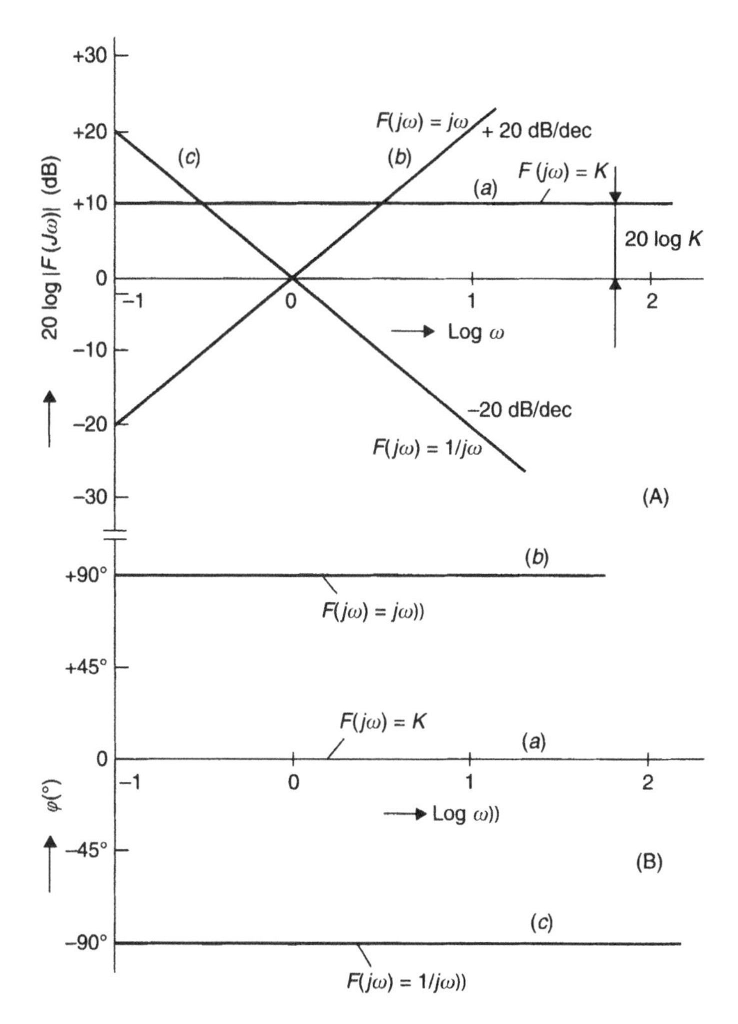

Fig. 6.12 Bode plots of the first three simple transfer functions cited above in Part A: (*a*) 1/Frequency independent gain $K = K_d K_A K_o$. (*b*) Characteristic with one zero in the origin. (*c*) Characteristic with one pole in the origin. In Part B, (*a*) logarithm of the gain, (*b*) phase plots [6.6].

1. Frequency independent gain is represented with a straight line at a distance of 20 log(K) dB from the horizontal axis (see the plot in Fig. 6.12).
2. A factor with one zero in the origin is represented with a straight line with the slope of –20 dB/dec drawn through the zero in the horizontal axis. The phase characteristic is +90°.

3. Similarly, the factor with one pole in the origin is represented with a straight line with a slope of –20 dB/dec drawn through the zero in the horizontal axis. The phase characteristic is –90°.
4. The factor with one zero in the transfer function

$$1 + j\omega T_o \tag{6.37}$$

is composed from two asymptotes: one is the straight line in the 0-dB level and the other is the straight line with the slope +20 dB/dec starting from the cut-off frequency

$$\omega_{\text{cut},0} = 1 / T_0 \tag{6.38}$$

Note the plot in Fig. 6.13*a*. The error is small with a maximum of –3 dB since

$$10 \log |1 + j|^2 \cong 3 \text{ dB} \tag{6.39}$$

For $\omega = 2/T_0$ or $\omega = 1/2T_0$, the error due to the asymptotic approximations is ~ 1 dB. The phase is

$$\Psi_0 = \frac{180}{\pi} \tan^{-1}(\omega T_0) \tag{6.40}$$

which for $\omega_{\text{cut},0}$ is just +45. With the assistance of the Taylor expansion, we find asymptotic approximation, as shown in Fig. 6.13*b*. Note that the errors are in the range of ±5 or ±6 degrees.
5. The factor with one pole is

$$\frac{1}{1 + j\omega T_p} \tag{6.41}$$

The characteristics are shown in Fig. 6.14 and are provided as mirrors of the previous cases discussed step 4.
6. The time delay is represented by the factor

$$e^{-j\omega\tau} \tag{6.42}$$

The phase shift due to the time delay is plotted in Fig. 6.15.

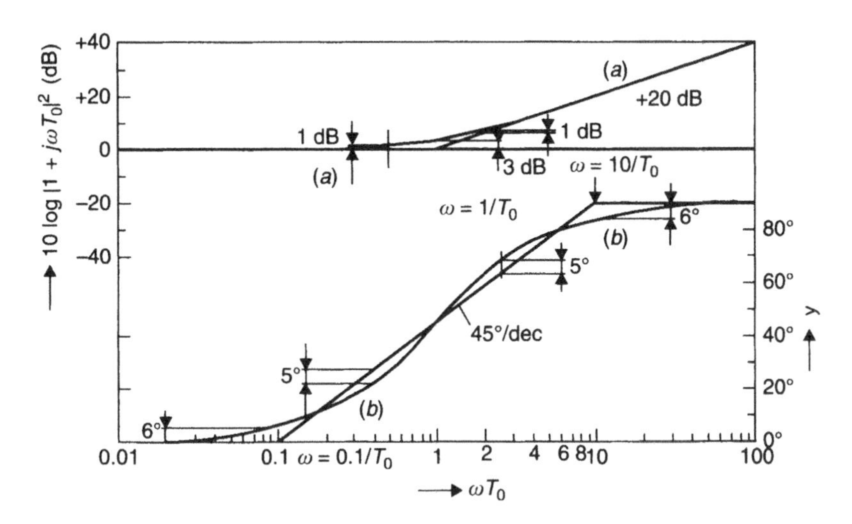

Fig. 6.13 Bode plot of the transfer function $1 + j\omega T_o$: (*a*) amplitude characteristic, and (*b*) phase characteristic.

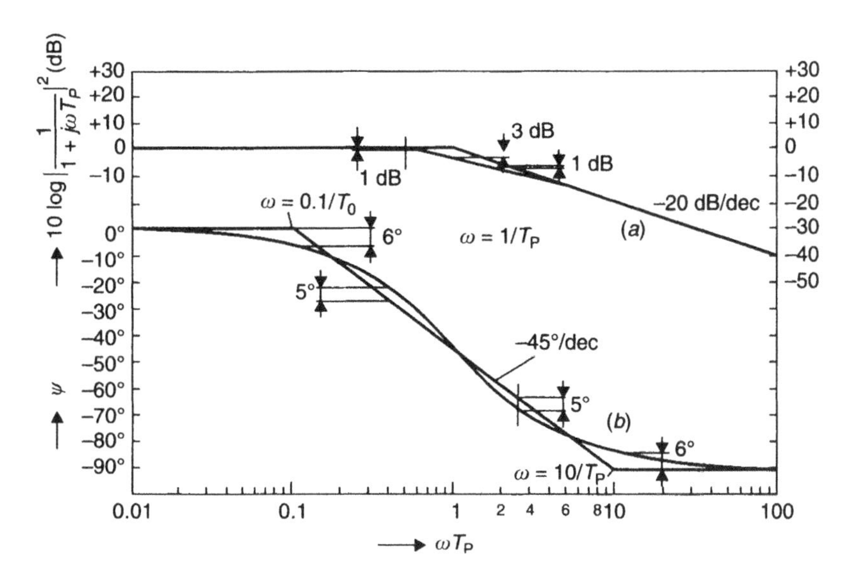

Fig. 6.14 Bode plot of the transfer function $1/(1 + j\omega T_o)$: (*a*) amplitude characteristic, and (*b*) phase characteristic [6.6].

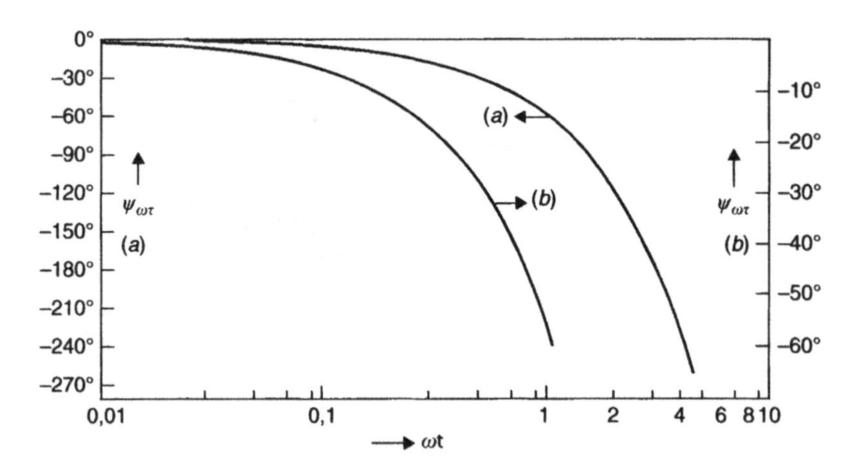

Fig. 6.15 Phase shift due to the time delay. The measures are in accordance with characteristics designated by arrows indicating the left or the right phase shifts.

EXAMPLE 6.1

Straight-line Bode plot for the popular third-order PLL with $\zeta = 0.7$ and $\kappa = 0.3$ (Fig. 6.16). Reverting to (6.12) and (6.21) we plot relation

$$\frac{1}{-x^2} \times \frac{1 + j \times 1.4x}{1 + j \times 0.4x}$$

6.3.2 The Root-Locus Method

The root-locus method of the function $1 + G(s)$ is intended to find the location of the respective roots in the complex plane [e.g., 6.1, 6.8]. In the past, a set of rules were devised for finding at least an approximate position or direction of the roots. Nowadays, our situation is much simpler since computer solution of the nth polynomial of $P_n(s)$, with the changing parameter K or any other, provides us with a set of roots that can thereafter be plotted in the complex plane. Nevertheless, we feel that a little information about the basic definition and rules would be useful:

Zero is designated as variable s for which the gain $G(s) = 0$ [i.e., for which the numerator of $G(s)$ is zero].

Pole is designated as variable s for which the numerator of $G(s)$ is zero [i.e., for which both $G(s)$ and $1 + G(s)$ are nearing infinity].

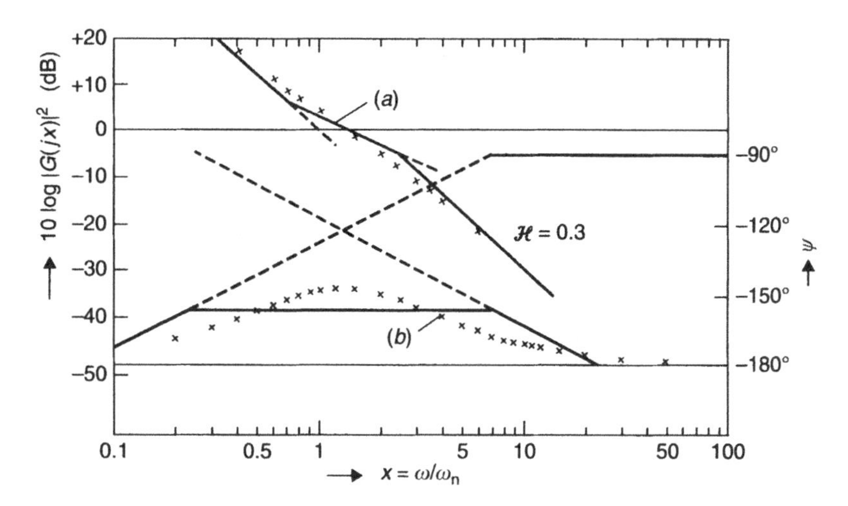

Fig. 6.16 Straight-line Bode plot for the popular third-order PLL with $\zeta = 0.7$ and $\kappa = 0.3$ [6.6].

Root is designated as variable s for which $1 + G(s) = 0$.

Theorem 1: Branches of the root locus plot start in each pole of $G(s)$ for the gain $K = 0$ and end in zeros for $K \to \infty$.

Theorem 2: The root locus coincides with the zero axes, where an odd number of poles plus zeros is found to the right of the point. Verify this statement with the assistance of Fig. 6.17.

Theorem 3: For large values of the gain K, the locus is asymptotic to the angles

$$\frac{(2k+1)180^\circ}{P-Z} \qquad k = 0,1,2,..., \tag{6.43}$$

where P is number of poles and Z is number of zeros.

There exist other theorems for estimation of the locus plot; however, with the application of computers they lose their importance. Nevertheless, we want to mention one important property, namely, that we can estimate the effective damping of the PLL from the distance of the operating point from the imaginary axis. We will illustrate this problem with the assistance of Example 6.2.

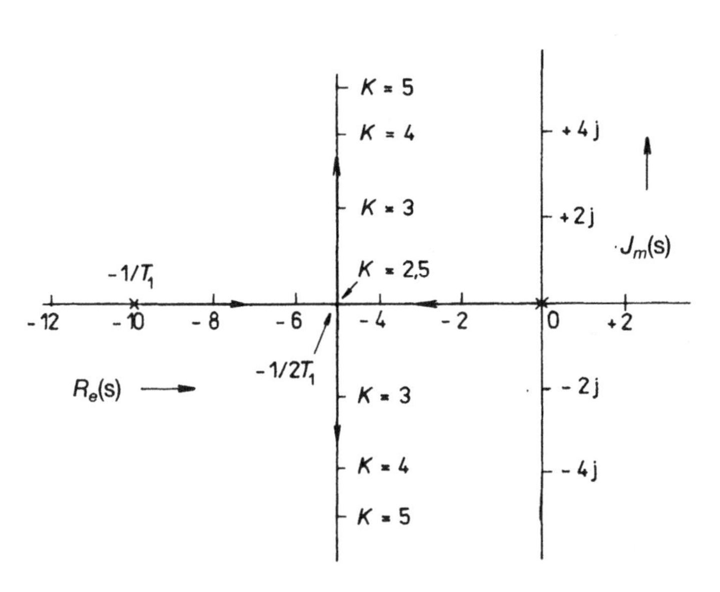

Fig. 6.17 The root locus of 1 + $G(s)$ for the second-order PLL loop with a simple RC filter.

EXAMPLE 6.2

The problem of the root locus will be illustrated with a simple example of the second-order loop with a lagging RC filter with a gain

$$G(s) = \frac{K}{s(sT_1 + 1)} \qquad 6.44$$

Since (6.44) has only two poles,

$$s = 0 \quad \text{and} \quad s = -1/T_1 \qquad 6.45$$

both branches of the root locus must end in zeros in infinity. After introducing $G(s)$ into (6.29), we arrive at quadratic equation

$$s^2 T_1 + s + K = 0 \qquad 6.46$$

the roots of which are

$$s_{1,2} = \frac{-1 \pm \sqrt{1 - 4KT_1}}{2T_1} \qquad 6.47$$

For $K = 0$, we get

$$s_1 = 0 \qquad s_2 = -1/T_1 \qquad 6.48$$

In agreement with Theorem 1, as long as $4KT_1 < 1$ both roots are real and negative and move along the zero axis. As soon as $4KT_1 > 1$, the roots become complex with a constant real part and the locus proceeds as a vertical line parallel with the imaginary axis at the distance $-1/2T_1$ from the origin (see Fig. 6.17)

6.4 TRACKING

Up to this point, we have investigated properties of PLLs in steady-state conditions. However, we encounter either wanted or unwanted frequency changes both in reference generators and more often in voltage-controlled oscillators connected with communications application (mainly due to adjusting the division ratio in the feedback path). The encountered changes in the PLL state can be divided into three major groups:

1. Phase or frequency steps.
2. Periodic changes (spurious phase or frequency modulations, discrete spurious signals, etc.).
3. Noises accompanying both reference and VCO signals.

Section 6.4.1 discusses the first of the problems.

6.4.1 Transients in PLLs

Applications of PLLs in modern communications are associated with nearly permanent carrier frequency changes. Evidently, for proper operation we need to know the duration of the switching process. We are interested in how long it takes before the output frequency is settled and how large the eventual steady-state error might be. This information provides the phase difference at the output of the phase detector $\Phi_e(s)$,

$$\frac{\Phi_e(s)}{\Phi_i(s)} = 1 - H(s) \qquad 6.49$$

or more exactly its time domain behavior $\phi_e(t)$. To this end, we investigate the following relation with its time response:

$$\Phi_e(s) = \Delta\Phi_i(s)[1 - H(s)] = \Delta\Phi_i(s)\frac{1}{1+G(s)} = \Delta\Phi_i(s)\frac{A(s)}{s^n B(s)} \qquad 6.50$$

and evaluate the time needed for setting of the phase error to the predetermined value. For the final steady state, we take into account the Laplace limit theorem,

$$\lim_{t\to\infty}[\phi_e(t)] = \lim_{s\to 0}\left[\Phi_i(s)\frac{s^{n+1}B(s)}{A(s)+s^n B(s)}\right] \qquad 6.51$$

6.4.2 Laplace Transforms of Typical Step Errors

Investigations of the steady-state errors, in PLLs of different orders and types, will proceed after introduction of the Laplace transforms of the respective input phase steps, input frequency steps, and input steady frequency changes,

$$\Delta\omega_i = \frac{\Delta\phi_i}{s} \qquad \frac{\Delta\omega_i}{s} = \frac{\Delta\phi_i}{s^2}s \qquad \frac{\Delta\dot{\omega}}{s} = \frac{\Delta\omega_i}{s^2} = \frac{\Delta\phi_i}{s^3} \qquad 6.52$$

into (6.50) or (6.51).

6.4.2.1 Phase Steps: First-Order Loop

After introducing the Laplce transform of phase steps, $\Delta\varphi/s$, into (6.6), we find that for the first-order loop, with the assistance of the Table 1.2,

$$\phi_{e1}(t) = \Delta\phi_i e^{-Kt} \qquad 6.53$$

6.4.2.2 Phase Steps: Second-Order Loop

With the assistance of (6.6) and (6.52) we get

$$\Phi_e(\sigma) = \frac{\Delta\phi_i}{\sigma}\cdot\frac{\sigma(\sigma+\omega_n/K_v)}{\sigma^2+2\zeta\sigma+1} = \Delta\phi_i\cdot\frac{(\sigma+\omega_n/K_v)}{(\sigma-\sigma_1)(\sigma-\sigma_1)} \qquad 6.54$$

$$\sigma_{1,2} = -\zeta \pm \sqrt{\zeta^2 - 1}$$

and with application of the Laplace transform tables [e.g., 6.9] we arrive at the time domain transient

$$\phi_{e1}(t)=\frac{\Delta\phi_i}{\sigma_1-\sigma_2}[(\sigma_1+\omega_n/K_v)e^{(\sigma_1\omega_n t)}-(\sigma_2+\omega_n/K_v)e^{(\sigma_2\omega_n t)}] \quad 6.55$$

Plotting the above relation (see Fig. 6.18) reveals that for reasonable damping factors $0.7 \leq \zeta \leq 1.5$ the transient effectively ends after $\omega_n t \approx 2\pi$.

6.4.2.3 Frequency Steps: Second-Order Loop

With the assistance of the Laplace transform tables, we find [6.9]

$$\phi_{e2}(t)=\frac{\Delta\omega_i/\omega_n}{\sigma_1-\sigma_2}\left[\frac{(\sigma_1+\omega_n/K)}{\sigma_1}e^{(\sigma_1\omega_n t)}-\frac{(\sigma_2+\omega_n/K)}{\sigma_2}e^{(\sigma_2\omega_n t)}\right]+\frac{\Delta\omega_i}{K} \quad 6.56$$

This simplifies for very high gain type II loops to

$$\phi_{e2}(t)=\frac{\Delta\omega_i}{\omega_n}\cdot\frac{e^{(-\zeta+\sqrt{\zeta^2-1})\omega_n t}-e^{(-\zeta-\sqrt{\zeta^2-1})\omega_n t}}{2\sqrt{\zeta^2-1}}+\frac{\Delta\omega_i}{K_v} \quad 6.57$$

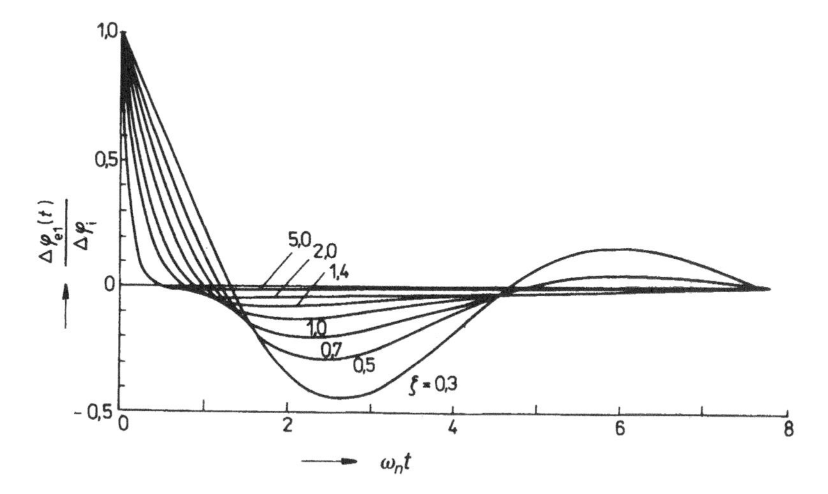

Fig. 6.18 Normalized transients $\Delta\varphi_{e1}(t)/\Delta\varphi_i$ due to the phase step $\Delta\varphi_i$ for different damping factors ζ for a high-gain loop with a lagging lead RC filter.

Evidently, in all PLLs of the second order a frequency step results in a steady-state phase error inversely proportional to the so-called *velocity error constant* K_v, in agreement with the terminology used in the feedback control systems (cf. [6.8]). In PLLs of type II, with two integrators in the loop, the DC gain $F(0)$, as well as K_v, is very large. Consequently, the steady-state error is negligible.

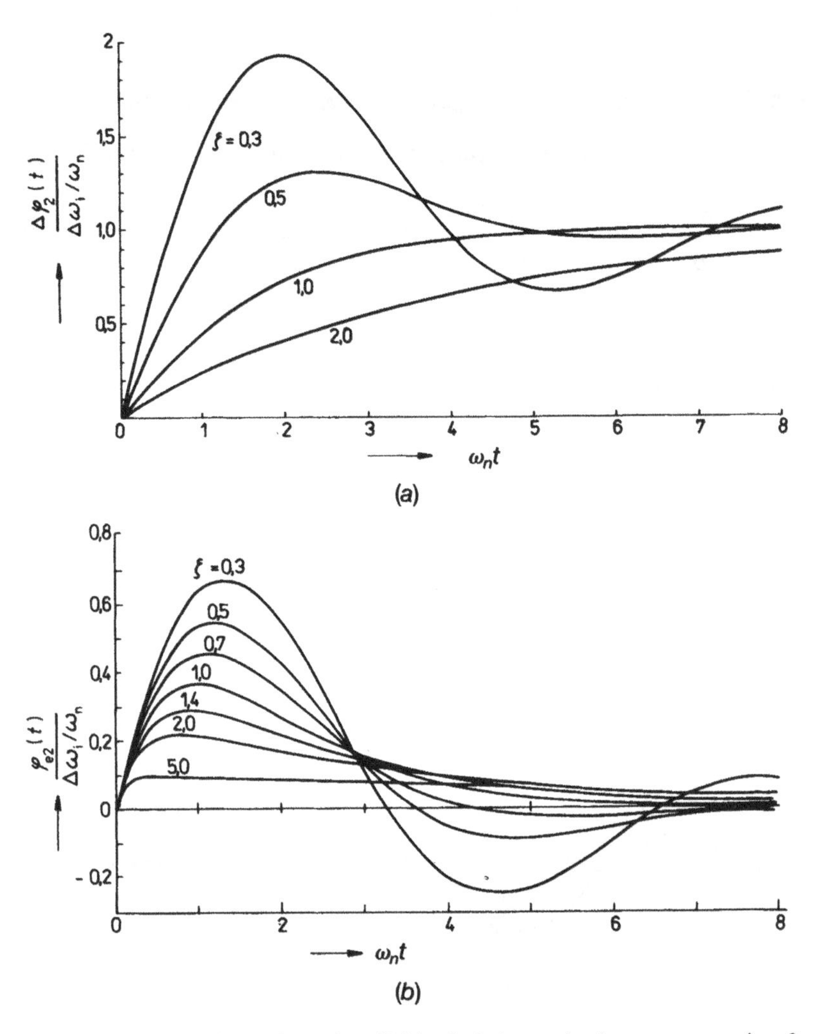

Fig. 6.19 Normalized transients $\Delta\varphi_{e2}(t)/(\Delta\omega_i/\omega_n)$ due to the frequency step $\Delta\omega_i$ for different damping factors ζ for a high-gain loop. (*a*) For a simple RC loop filter and (*b*) for a high-gain loop with a lagging lead RC filter.

Plotting of the above relation (see Fig. 6.19) reveals that for reasonable damping factors $0.7 \le \zeta \le 1.5$ the transient ends after $\omega_n t \approx 2\pi$, as above.

6.4.2.4 Frequency Ramps

The solution proceeds as above with the result

$$\phi_{e3}(t) = \frac{\Delta\dot{\omega}_i t}{K_v} + \frac{\Delta\dot{\omega}_i}{\omega_n^2}\left(1 - \frac{2\zeta\omega_n}{K_v}\right) - \frac{\Delta\dot{\omega}_i}{\omega_n^2} e^{-\zeta\omega_n t}\left[\left(1 - \frac{2\zeta\omega_n}{K_v}\right)\cosh(\omega_n t\sqrt{\zeta^2 - 1}) + \frac{\zeta - (\omega_n / K_v)(2\zeta^2 - 1)}{\sqrt{\zeta^2 - 1}}\sinh(\omega_n t\sqrt{\zeta^2 - 1})\right] \quad 6.58$$

For the damping factor $\zeta < 1$, the hyperbolic functions change in the trigonometric functions and the steady state is approached with the damped oscillation (see Fig. 6.20). However, the steady frequency change $\Delta\omega/s^2$ results in the so-called *acceleration or dynamic tracking error* K_a:

$$\lim \phi_{e3}(t)\,|_{t\to\infty} = \Delta\dot{\omega}_i \left[\frac{B(0)}{A(0)}\right]_{n=2} = \frac{\Delta\dot{\omega}_i}{K_a} \quad 6.59$$

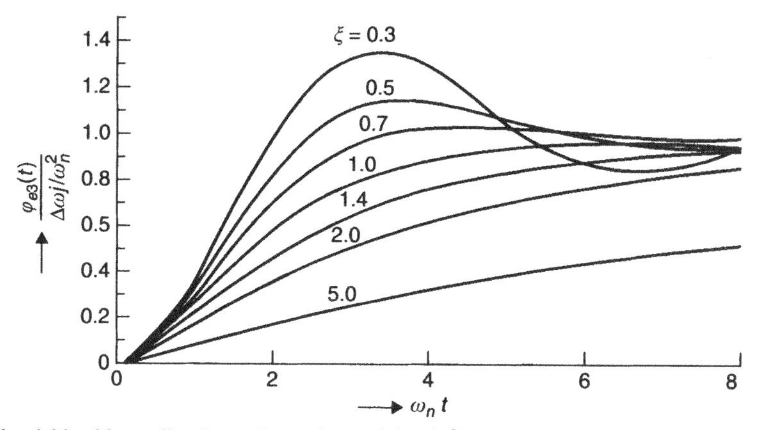

Fig. 6.20 Normalized transients $\Delta\varphi_{e3}(t)/(\Delta\omega_i/\omega_n^2)$ due to the frequency ramp $\Delta\dot{\omega}_i$ for different damping factors ζ for a high-gain loop (DC phase error is retained).

6.5 WORKING RANGES OF PLL

An important problem encountered with PLLs is the question about a reasonable difference between the input frequency, ω_i, and the true free-running frequency of the VCO, ω_c. The maximum frequency difference before losing the lock of the PLL system is called the *hold-in range*. Another criterion might be the frequency difference $|\omega_i - \omega_c|$, for which the phase lock will take place in all circumstances, the so-called *pull-in range*. However, as the ideal state we can design the situation in which the lock is achieved without any cycle slipping, that is, without any loss of lock after switching on. This state between the input and output frequencies of the PLL is called the *lock-in range*.

The final question when investigating working ranges is the problem of tolerating the progressive frequency difference before the lock breakdown, the so-called *pull-out frequency*. A graphical representation of the above discussed parameters is schematically plotted in Fig. 6.21.

6.5.1 Hold-in Range

Proceeding with the definition given in the introduction to this chapter, the *hold-in range* $\Delta\omega_H = |\omega_i - \omega_c|$ is the largest frequency that can be

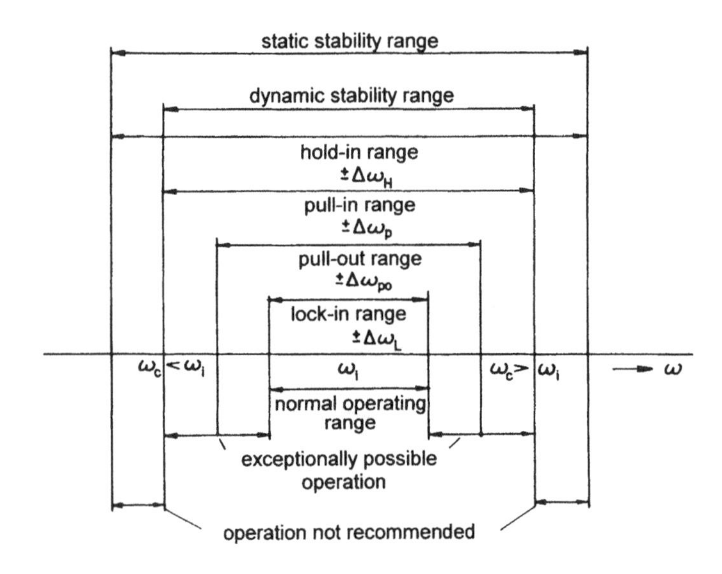

Fig. 6.21 Working ranges of PLLs.

tolerated in the PLL before losing lock without any serious consequences. In these circumstances, we can define $\Delta\omega_H$ as a function of the phase detector output, as long as it is valid:

$$\frac{d\Phi_e(t)}{dt} = 0 \qquad 6.60$$

Figure 6.21 indicates clearly that the *hold-in range* has no practical applications.

6.5.2 The Pull-in Range

The pull-in range presents a much more complicated task. The problem was solved in the past by many authors from different points of view. Here, in short, we repeat the earlier discussion by Kroupa [6.6] and provide a solution as simply as possible but accurate enough for practical applications. For investigation of the PLL pull-in range $\Delta\omega_p$, we may use two methods:

1. Either compute the time needed for a given frequency difference $|\omega_i - \omega_c|$, then increase this difference and note $|\omega_i - \omega_c| = \Delta\omega_p$ for which the pull-in time starts to be nearly infinite.
2. In the second approach, we will proceed the other way, namely, by choosing the difference $\Delta\omega = |\omega_i - \omega_c|$ sufficiently large that no locking takes place. Then we will reduce the difference until the beat note at the *PD* output starts to be constant; in other words, as long as the differential equation of the system has a periodic solution:

$$\Delta\omega + \frac{d\phi_o(t)}{dt} = K_d K_o v_2(t) \qquad 6.61$$

The smallest detuning, $\Delta\omega = \Delta\omega_p$, is the sought pull-in range. In the asynchronous steady state, the VCO is phase modulated with a beat frequency ν. After expressing the free-running frequency ω_c as a function of the difference frequency ν we have

$$\nu = |\omega_{\text{in}} - \omega_{\text{out}}| = |\omega_{\text{in}} - \omega_c| - \Delta\omega = \nu_c - \Delta\omega \qquad 6.62$$

where the frequency shift $\Delta\omega$ is a function of the modulation frequency ν. Solution of the relation (6.61) can be approached from different points of view. In [6.6], comparison of harmonics was used; here we apply a simple approximation with estimation of the voltage $v_2(t)$ by its Fourier transform, $V_2(s)$ (nearly the DC magnitude):

$$V_2(s) \approx K_A K_d K_o F(s)\frac{1}{s} \approx KF(0)F(j\nu)\frac{1}{j\nu} \qquad 6.63$$

After introduction into (6.61), we get

$$\Delta\omega + \nu\Phi(\nu) = \frac{AK^2}{2\nu}\Phi(\nu) \qquad 6.64$$

Furthermore, with the assistance of (6.63) we arrive at the beat frequency, ν, as function of the free running frequency ν_c:

$$\nu_c = \nu + \frac{AK^2}{2\nu}\Phi(\nu) \qquad 6.65$$

Reduction of the original detuning will finally lead to a minimum beat note frequency:

$$\nu = \nu_m \qquad 6.66$$

where the asynchronous state is not to be held any further and the pull-in starts. The respective $\nu_{c,\min}$ is the sought $\Delta\omega_p$, that is, the pull-in range frequency. To this end, we differentiate the rhs of relation (6.65) and equalize to zero:

$$1 + \frac{AK^2}{2\nu^2}\left[\frac{d\Phi(\nu)}{d\nu}\nu - \Phi(\nu)\right] = 0 \qquad 6.67$$

Its solution reveals ν_m and after its introduction into (6.65) we get the upper limit of the pull-in range (i.e, $\Delta\omega_p$). In instances where $\Phi(\nu)$ is nearly independent of the beat frequency ν, that is, where

$$d\Phi(\nu)/d\nu \approx 0 \qquad 6.68$$

solution of (6.67) is simplified to

$$\nu_m \cong K\sqrt{ARe[\Phi(\nu_m)]/2} \qquad 6.69$$

After introduction of the above relation into (6.65), we get for the pull-in range:

$$\Delta\omega_P \cong 2\nu_m \cong K\sqrt{2ARe[\Phi(\Delta\omega_P)]} \qquad 6.70$$

At this stage, we are prepared for application of the above theory on practical examples.

EXAMPLE 6.3

Let us perform solution of the pull-in range for the second-order loop with lag lead or RRC filter. For large beat frequencies, we have from (6.9)

$$\Phi(\nu) \approx T_2/T_1 \qquad 6.71$$

After application of (6.69) and (6.70), we arrive immediately at the known formula:

$$\Delta\omega_P = K\sqrt{2T_2/T_1} \qquad 6.72$$

which for the high-gain loop simplifies with the assistance of the basic PLL parameters (cf. 6.12) to

$$\Delta\omega_P \approx 2\sqrt{\zeta\omega_n K} \qquad 6.73$$

What remains is to investigate the validity of the above solution. To this end, we would calculate the normalized frequency x_n (cf. 6.64–6.66):

$$x_m = \nu_m/\omega_n = \sqrt{\zeta K/\omega_n} \qquad 6.74$$

For the high-gain loops, we have, $\zeta \gg \omega_n/K$. Consequently, $x_m > 1$ and the above simplification (6.71) is justified. This also follows from Fig. 6.22 where we have compared solutions (6.73) with those by Greenstein in [6.10].

6.5.3 False Locking

In the neighborhood of the upper bound, close to the $\Delta\omega_P$, the pull-in process is slow before the lock is realized. A closer investigation of the process reveals that the PD output phase signal seems to be fed into

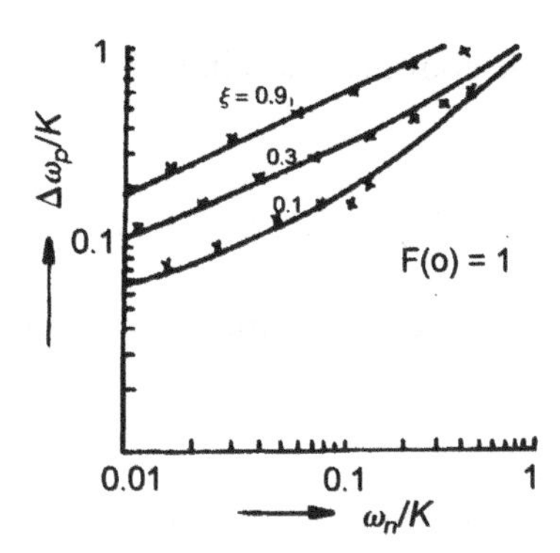

Fig. 6.22 Plot of the normalized pull-in range for the second-order high-gain PLL with a lag-lead RRC loop filter as a function of $x = \omega_n/K$; data by Greenstein [6.10].

two paths, that is, in an AC with the transfer function, T_1/T_2, and a quasi DC path with the integrator. The situation is illustrated with an approximating block diagram in Fig. 6.23. For the slowly varying component, $\Delta\omega(t)$, we have found in Section 6.5.2

$$\Delta\omega = \frac{KF(0)}{2}|G(j\nu)|\sin\Psi \tag{6.75}$$

In accordance with the block diagram in Fig. 6.23, we must introduce $F(0) = 1$, and (6.75) is changed into

$$\Delta\omega = \frac{K^2}{2\nu}|F(j\nu)|\cos\Psi \tag{6.76}$$

However, additional low-pass filters, introduced willingly or unwillingly in the forward path (additive pole due to the operational amplifier or to the VCO tuning connection) inclusive of the time delay due to the IF filters or digital operation, may generate the phase shift $\Psi > \pi/2$ for some error frequencies. The consequence is that the slowly varying component, $\Delta\omega$ (cf. 6.75), changes sign, that is, $v_{2,\mathrm{DC}}(t)$ will be negative and start to pull the PLL the other way from the correct frequency.

In Fig. 6.24*a*, we have plotted the normalized component $\Delta\omega/\omega_n$ for the fourth-order loop with two additional RC filter sections. Investigation reveals that the phase shift does not exceed −180° and the nor-

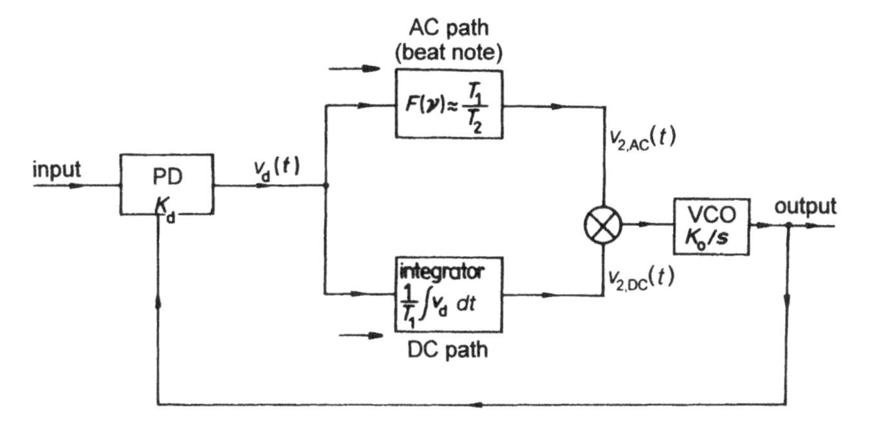

Fig. 6.23 Approximating block diagram of the second-order PLL type II with the frequency difference $\Delta\omega = |\omega_i - \omega_o|$ only a bit smaller compared with the pull-in frequency $\Delta\omega_P$.

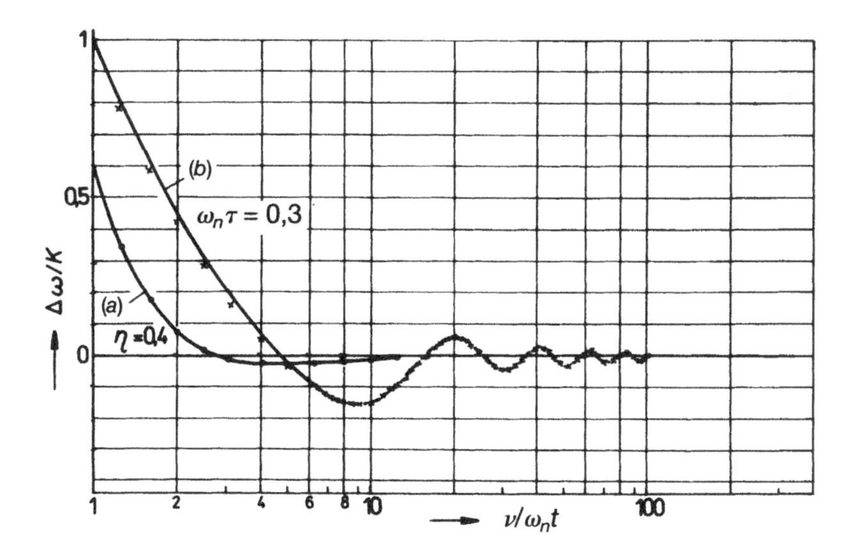

Fig. 6.24 The plot of the normalized components $\Delta\omega/K$ and $\Delta\omega/\omega_n$: (*a*) for the fourth-order loop with two additional RC filter sections and (*b*) for the second-order PLL with the time delay equal to $\omega_n\tau = 0.3$.

malized frequency error is only once equal to zero. This zero is unstable. The other situation plotted in Fig. 6.24*b* is for the second-order PLL with time delay equal to $\omega_n \tau = 0.3$. Investigation reveals a characteristic with several zero crossings. We easily find out that zeros with positive slope are stable. The consequence is that the steering voltage $v_2(t)$ will be pushed, from both sides, to the respective Fourier frequency where it will be zero, and without any action from the outside, the PLL will be locked to the respective false frequency. Only detection of the beat note with an oscilloscope (or by another means) will discover this undesired situation. However, be careful of noise or spurious signals in such situations.

6.5.4 Lock-in Range

The look-in range is defined as such a frequency difference between the input reference frequency ω_i and the free-running frequency ω_c that after closing the loop converges monotonously to a steady-state value, that is,

$$\Delta\omega_L = |\omega_i - \omega_{cL}| \quad 6.77$$

By investigating ω_L, we start from the Byrne consideration [6.11], starting with the worst situation, namely, where zero crossing of the reference and the VCO signals coincide. Closing at this moment in the PLL causes the *PD* output phase $\varphi_e(t)$ to be formed from the two components of the $\varphi_{e1}(t)$, which is caused by an effective step change (in the worst case, either $\pi/2$ or π), and $\varphi_{e2}(t)$, generated with the frequency difference $\Delta\omega_L$. The undesired cycle skipping of the beat frequency is prevented if the *PD* outputs are zero. This condition is met for

$$\frac{d\phi_{e1}(t)}{dt} + \frac{d\phi_{e2}(t)}{dt} = 0 \quad 6.78$$

After summation of (6.55) and (6.56), we perform the derivation and then reduction by $(\sigma_1 - \sigma_2)^{-1}\ \omega_n$:

$$\begin{aligned} &\Delta\phi_i[(\sigma_1 + \omega_n/K)\sigma_1\omega_n e^{\sigma_1\omega_n t} - (\sigma_2 + \omega_n/K)\sigma_2\omega_n e^{\sigma_2\omega_n t}] + \\ &\Delta\omega_i[(\sigma_1 + \omega_n/K)\omega_n e^{\sigma_1\omega_n t} - (\sigma_2 + \omega_n/K)\omega_n e^{\sigma_2\omega_n t}] = 0 \end{aligned} \quad 6.79$$

Inspection of Figs. 6.18 and 6.19 reveals that the maxima of the tangents, both for the phase and frequency steps, are at $t = 0$. Consequently, (6.79) simplifies to

$$\Delta\omega_i[\sigma_1 - \sigma_2] + \Delta\phi_i[(\sigma_1^2 - \sigma_2^2)\omega_n + (\sigma_1 - \sigma_2)\,\omega_n / K] = 0 \qquad 6.80$$

After a second reduction by $(\sigma_1 - \sigma_2)$, we arrive at

$$\Delta\omega_i = \Delta\phi_i\omega_n(2\zeta - \omega_n / K) \qquad 6.81$$

By introducing starting conditions $\Delta\varphi_i = \pi$, and, $\Delta\omega_i = \Delta\omega_L$, we readily find

$$\Delta\omega_L = \pi\omega_n(2\zeta - \omega_n / K) \qquad 6.82$$

By considering *PD* with a sine wave output and a PLL with a lag-lead filter and very high gain K, we arrive at

$$\Delta\omega_L = 2\zeta\omega_n \qquad 6.83$$

or at

$$\Delta\omega_L = K\frac{T_2}{T_1} = K_r \qquad 6.84$$

From (6.84), it follows that the lock-in range is equal to the reduced gain K_r (cf. 6.9) at very high frequencies as suggested by Gardner [6.1]. He states that this conclusion is generally valid for the systems with filters having the same number of nulls and zeros.

6.5.5 Pull-out Frequency

There are instances where one of the output frequencies fed to the *PD* experiences a frequency step (as an example, we mention PLL frequency synthesizers with digital dividers in the feedback path). At a certain magnitude, designed as $\Delta\omega_{PO}$ (*pull-out range*), the PLL looses lock, often for only a short time. The situation is the same as that discussed for transients (6.56 and Fig. 6.19). We see that the phase error, due to the frequency step, increases from zero to a maximum and then dies out either periodically or aperiodically. In instances where the maximum phase error $\varphi_2(t)$ is equal to or exceeds the working range of the *PD*, the phase lock is lost. For this maximum, we find from (6.56), in instances of very large or infinite gain K_v, the respective normalized time $(\omega_n t)_{max}$:

$$(\omega_n t)_{max} = \frac{\operatorname{arctanh}\left(\sqrt{1-\zeta^2}/\zeta\right)}{\sqrt{1-\zeta^2}} \qquad 6.85$$

for the damping factor $\zeta > 1$. In instances where $\zeta = 1$, we get

$$(\omega_n t)_{max} = 1 \qquad 6.86$$

For cases where $\zeta < 1$,

$$(\omega_n t)_{max} = \frac{\arctan\left(\sqrt{1-\zeta^2}/\zeta\right)}{\sqrt{1-\zeta^2}} \qquad 6.87$$

From these relations, we conclude that the normalized time is a function of the damping factor ζ only. Consequently, for the maximum phase error we can write

$$\phi_{e2\max} = \frac{\Delta\omega}{\omega_n} f(\zeta) \qquad 6.88$$

For a sawtooth wave *PD* with the maximum tolerated error of π, we get for the pull-out range $\Delta\omega_{PO}$ in the normalized form [6.1 or 6.6]

$$\Delta\omega_{PO} / \omega_n = x_{PO} = \pi / f(\zeta) \approx 1.8(\zeta + 1) \qquad 6.89$$

6.5.6 The Lock-In Time

We may define the lock-in time as the time that the PLL needs to reduce the phase error $\varphi_e(t)$ to one-tenth of its maximum without cycle-skipping between $+\varphi_{e,max}$ and $-\varphi_{e,min}$. For the first-order loop with sawtooth or triangular output wave *PD*, we arrive, with the assistance of (6.53) at the relation

$$T_{L,1} \approx 2.3 / K \qquad 6.90$$

Similarly, for sine wave *PD* we approximately find [6.6]

$$T_{L,1} \approx \frac{2.3}{K\sqrt{1-(\Delta\omega / K)^2}} \tag{6.91}$$

As soon as $|\Delta\omega/K| < 0.9$, we can simplify the above relation to

$$T_{L,1} < 5 / K \tag{6.92}$$

The lock-in time for PLL of the second-order may be approximated as

$$T_{L,2} \approx \frac{2.3}{\zeta\omega_n} \approx \frac{4.6T_1}{KT_2} \approx \frac{4.6}{K_r} \tag{6.93}$$

Several authors quote values for the lock-in time to be two times longer than that indicated for the above relations.

6.6 DIGITAL PLL

The boom in communications systems in the last decades has been enabled by combination of digital technique, small size, and power-saving circuits on chips, with mastering of the microwave RF ranges with scores of channels. Such a situation required a new approach to frequency synthesis techniques and PLLs. The backbone is digital circuitry or processing, which are also true for PLLs. However, the proper approach is based on the investigation with the assistance of the *z-transform*. The other possibility is to modify the original Laplace transform of $G(s)$ in the following way:

$$G_{\text{mod}}(s) = F_h(s)\hat{G}(s) \tag{6.94}$$

Where

$$F_h(s) = \frac{1-e^{-sT}}{s} \tag{6.95}$$

After introduction of the sampling frequency ω_s,

$$\hat{G}(s) = \frac{1}{T}\sum_{n=\infty}^{\infty} G(s - jn\omega_s) \qquad \omega_s = \frac{2\pi}{T} \tag{6.96}$$

we arrive at

$$G_{\mathrm{mod}}(s)=\frac{1-e^{-sT}}{s}\cdot\frac{1}{T}\sum_{n=\infty}^{\infty}G(s-jn\omega_s)\approx\frac{1-e^{-sT}}{s}\cdot\frac{1}{T}G(s) \qquad 6.97$$

The simplification is justified since we expect that the higher order terms of $G_{\mathrm{mod}}(s)$ are attenuated with the loop filter $F(s)$. For small frequencies, s, that is, for $|sT| \ll 1$, we can rearrange $G_{\mathrm{mod}}(s)$ to

$$G_{\mathrm{mod}}(s)\approx\frac{\sinh(sT/2)}{sT/2}G(s)e^{-sT/2}\approx G(s)e^{-sT/2}\approx\frac{G(s)}{1+sT/2} \qquad 6.98$$

In the normalized form, we introduce

$$sT/2=\sigma\omega_n T_s/2=\sigma\pi\frac{f_n}{f_s}=\sigma\delta \qquad 6.99$$

Consequently, the normalized open-loop gain is

$$G_{\mathrm{mod}}(\sigma)=\frac{\sinh(\sigma\delta)}{\sigma\delta}G(\sigma)e^{-\sigma\delta} \qquad 6.100$$

The situation with the sampled PLL is illustrated in Fig. 6.25. Finally, we arrive at the often suggested approximation of the sampling process, namely, with the assistance of an additional RC section.

6.6.1 Phase Detectors

Each phase detector is effectively a frequency mixer in which both input frequencies are equal or nearly equal. When referring to a simple multiplicative mixer (cf. Sec. 4.6.1) the low-frequency or DC output is

$$\begin{aligned}v_{\mathrm{out,DC}}(t)&=V_{\mathrm{in}}\cos(\omega_{\mathrm{in}}t+\phi_{\mathrm{in}})V_{\mathrm{ref}}\sin(\omega_{\mathrm{ref}}t+\phi_{\mathrm{ref}})\approx\\ &\frac{V_{\mathrm{in}}V_{\mathrm{ref}}}{2}\sin[(\omega_{\mathrm{in}}-\omega_{ref})t+\phi_{\mathrm{in}}-\phi_{\mathrm{ref}}]\approx\\ &\frac{V_{\mathrm{in}}V_{\mathrm{ref}}}{2}(\phi_{\mathrm{in}}-\phi_{\mathrm{ref}})=K_d(\phi_{\mathrm{in}}-\phi_{\mathrm{ref}})\end{aligned} \qquad 6.101$$

where K_d is designated as the *phase detector gain*. Note that for maximum sensitivity both input signals should be in quadrature.

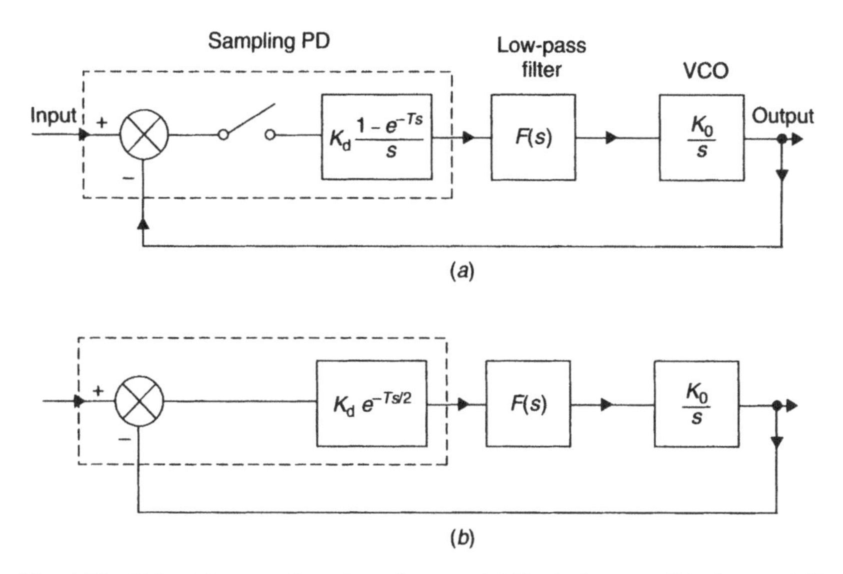

Fig. 6.25 PLL with a sampling phase detector: (*a*) block diagram of the loop and (*b*) the simulating analogue system.

6.6.2 Phase Frequency Detectors

In digital systems, we encounter the effective *phase frequency detectors (PFD)*, which at present are the most often employed in practice. Their advantages are the large operation range from zero or from -2π to $+2\pi$. This is a combination of the frequency and phase detection, with an eventual narrow pulse output and easy large-scale manufacturing in the IC form.

The principal arrangement is illustrated with the assistance of Fig. 6.26. This PFDs are composed of two or four *D* flip-flops and delaying NAND gates. Operation is explained with time diagrams in Fig. 6.27. In the first case, we assume frequency f_1 to be permanently higher than f_2. Consequently, only output V_1 in Fig. 6.26 (*a*) or (*b*) is activated. In the opposite circumstances, again only the output V_2 will be working and switching on or off branches of the charge pumps (in Fig. 6.28) that supply current pulses I_p from two current generators into the loop filter. The mean value of the current is effectively proportional to the phase difference φ_e and designated as the *current gain*:

$$i_d = \frac{I_p \phi_e}{2\pi} = K_{di} \qquad 6.102$$

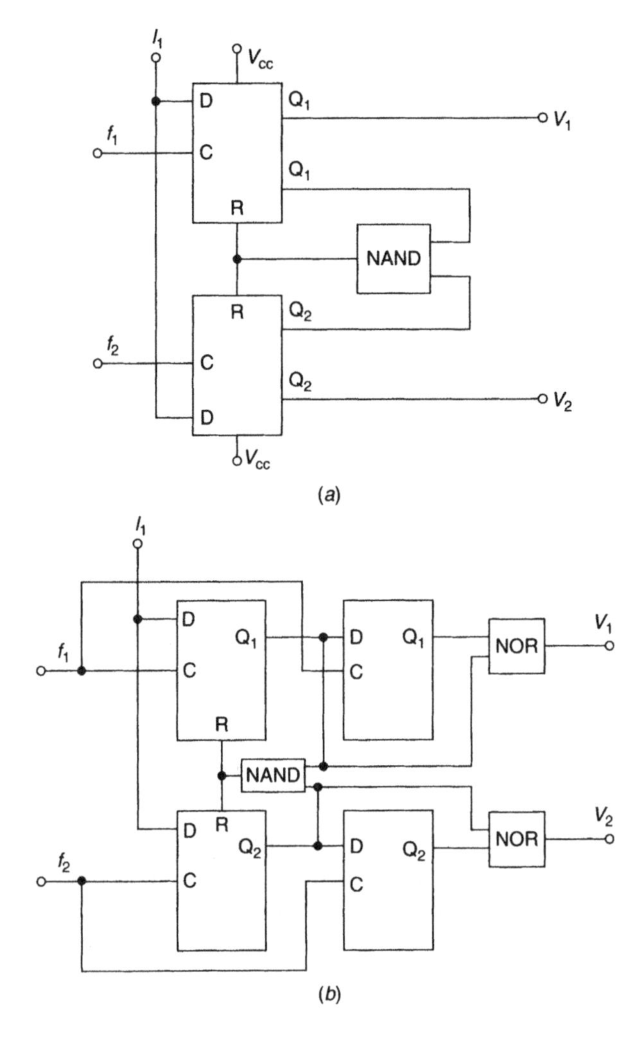

Fig. 6.26 The principal arrangement of the phase-frequency PD: (*a*) with the saw-tooth wave output signal; (*b*) with a NAND gate to deal with the dead-zone problem.

6.6.3 The Loop Filter Impedance

The current pump working as a current generator supplies pulses into the impedance $Z(s)$ in the simplest arrangement formed by a resistor in series with a capacitor (cf. the simplified form in Fig. 6.28):

$$Z(s) = R + \frac{1}{j\omega C} = \frac{1 + j\omega CR}{j\omega CR} \tag{6.103}$$

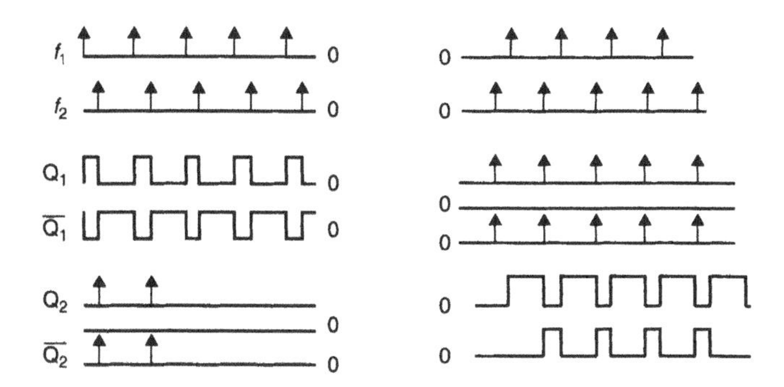

Fig. 6.27 Time diagrams of the phase frequency PD. (*a*) $f_1 = f_2$; the leading edge of f_2 delays f_1. (*b*) $f_1 = f_2$; the leading edge of f_1 delays f_2 [6.6].

However, the actual impedance $Z(s)$ is shunted with a large leaking resistor $R_{s,\text{leak}}$:

$$Z(s) = R_{s,\text{leak}} \frac{(1+sRC)/sC}{R_{s,\text{leak}} + (1+sRC)/sC} = \frac{(1+sRC)/sC}{sC + (1+sRC)/R_{s,\text{leak}}} \qquad 6.104$$

In such a case, the filter impedance is easily approximated for a large $R_{s,\text{leak}}$ as

$$Z(s) \approx \frac{1+sRC}{sC} \qquad 6.105$$

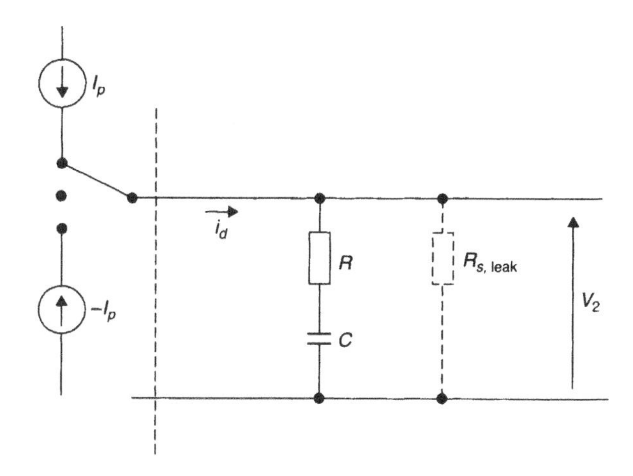

Fig. 6.28 A current–pump phase detector and passive integrating RC filter [6.6].

which behaves as an integrator and the loop is actually of type II. The phase error is integrated and stored in the form of the voltage on the capacitor (the so-called velocity constant, in accordance with the theory of a servomechanism [6.8]). Consequently, the effective phase difference steers the current pump-phase detector close to zero. With this situation, the switching would depend heavily on the present noise. The remedy is provided by the time delay introduced by invertors in the NAND circuit. The details are discussed later. Note the gate arrangement in Fig. 6.26*b*.

6.6.4 The Second-Order Digital Loops

The open-loop gain is found after replacing the phase detector gain in (6.2) with the current gain from (6.102), that is,

$$G(s) = K_{di} \cdot \frac{1+sRC}{sCN} \cdot \frac{K_o}{s} = K_{di} \cdot \frac{1+sT_2}{sCN} \cdot \frac{K_o}{s} = K_{di} K_o \frac{1+sT_2}{s^2CN} \qquad 6.106$$

where we have inserted, in agreement with (6.12), the time constant

$$T_2 = RC \qquad 6.107$$

In the next step, we evaluate the corresponding transfer functions by referring to (6.4) and (6.6):

$$H(s) = \frac{K(1+sT_2)}{s^2CN + sKT_2 + K} \qquad 1 - H(s) = \frac{s^2CN}{s^2CN + sKT_2 + K} \qquad 6.108$$

In accordance with (6.12), we introduce the natural frequency ω_n,

$$\omega_n^2 = \frac{K_{di} K_o}{CN} \qquad 6.109$$

and the damping factor ζ,

$$\zeta = \frac{\omega_n}{2} T_2 = \frac{\omega_n}{2} RC \qquad 6.110$$

By taking into account that the natural frequency is effectively fixed by the desired PLL passband (6.20), the damping factor by the design conditions of the small overshoot and a comfortable phase margin (i.e., $\zeta \approx 0.75 \pm 0.25$), the time constant RC should be equal to

$$RC \approx \frac{2\zeta}{\omega_n} \tag{6.111}$$

The above conclusion is important in instances where we are designing PLLs in the IC form, since the space-consuming capacity C may be reduced by the proper choice of the resistor R. After introducing (6.111) into (6.109), we have

$$\omega_n^2 = \frac{K_{di} K_o}{NC} = \frac{I_p RRK_o}{2\pi NCR} = \frac{I_p RK_o}{2\pi N} \cdot \frac{\omega_n}{2\zeta} \tag{6.112}$$

from which we arrive at the approximate value of the natural frequency and of the effective passband (see 6.20):

$$\omega_n = \frac{I_p RK_o}{2\pi N 2\zeta} \approx \frac{I_p RK_o}{4\pi N} \tag{6.113}$$

6.6.5 The Third-Order Digital Loops

Inspection of the block diagram in Fig. 6.28 reveals that the full voltage generated by switching on the PD output impedance affects on the varactor input of the VCO, that is,

$$V_{p.\max} \approx I_p R \tag{6.114}$$

Its reduction may be provided by an additional filtering section or sections, depicted in Fig. 6.29. The effective impedance in accordance with Fig. 6.29*a* is

$$Z(s) = \frac{1 + sR(C + C_3)}{sC(1 + sRC_3)} = \frac{1 + sT_2}{sC(1 + s\kappa T_2)} \qquad \kappa = \frac{C_3}{C + C_3} \tag{6.115}$$

Furthermore, the effective impedance in accordance with Fig. 6.29*b* is

$$Z(s) = \frac{1 + sRC}{s(C + C_3)\left(1 + sR\dfrac{CC_3}{C + C_3}\right)} \approx \frac{1 + sT_2}{sC(1 + s\kappa T_2)} \qquad \kappa = \frac{C_3}{C + C_3} \tag{6.116}$$

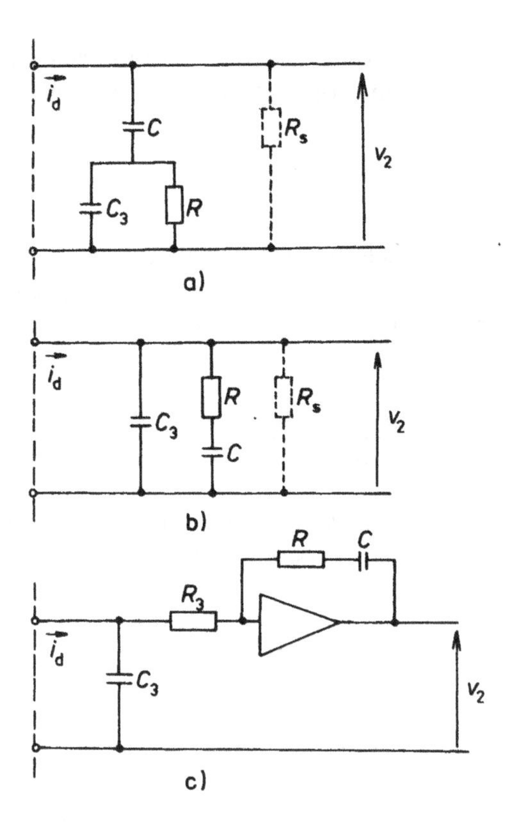

Fig. 6.29 Three different filter arrangements of the passive integrator of the charge pump PD of third-order type II [6.6].

and the effective output voltage in accordance with Fig. 6.29*c* is

$$v_{out} \approx v_{in}\frac{Z_2}{Z_1} = \frac{I_P}{sC_3}\frac{R+(1+sC)}{(1+sC)=R_3} = \frac{I_P}{sC}\frac{1+sRC}{1+sR_3C_3} = \frac{I_P}{sC}\frac{1+sT_2}{1+s\kappa T_2} \qquad \kappa = \frac{R_3C_3}{RC}$$

6.117

and the effective impedance $Z(s)$ is as in (6.116).

6.6.6 The Fourth-Order Digital Loops

At present, in PLL frequency synthesizers we often encounter fourth-order loops, which are generally used for the reduction of spurious signals, mostly the reference spurious signals. The simplest one is application of a mere low-pass RC section. Its addition, we will increase the slopes of $G(s)$ and $H(s)$ in the stop band to –60 dB/dec. We limit our discussion to the high-gain, second-order loop with two independent RC sections. The open-loop gain will be

$$G_4(j\omega)=\frac{K/T_1}{-\omega^2}\cdot\frac{1+j\omega T_2}{1+j\omega T_3}\cdot\frac{1}{1+j\omega T_4} \qquad 6.118$$

After introduction of the normalized time constants

$$T_3/T_2=\kappa \qquad T_4/T_3=\eta \qquad T_4/T_3=\kappa\eta \qquad 6.119$$

and original second-order loop damping factor and normalized frequency, $\omega/\omega_n \approx \sigma$, we get for the fourth-order PLL the following transfer functions:

$$H_4(\sigma)=\frac{1+2\zeta\sigma}{\sigma^4(2\zeta\kappa)^2\eta+\sigma^3 2\zeta\kappa(1+\eta)+\sigma^2+\sigma 2\zeta+1} \qquad 6.120$$

$$1-H_4(\sigma)=\frac{\sigma^4(2\zeta\kappa)^2\zeta+\sigma^3 2\zeta\kappa(1+\eta)+\sigma^2}{\sigma^4(2\zeta\kappa)^2\eta+\sigma^3 2\zeta\kappa(1+\eta)+\sigma^2+\sigma 2\zeta+1} \qquad 6.121$$

Since the constant η is small compared with κ, the phase margin does not deteriorate considerably. Its evaluation proceeds with the assistance of (6.33) and reaches the positive phase margin.

$$\Phi_{pm}\approx-\frac{180}{\pi}[-\pi+\arctan(2\zeta x)-\arctan(2\zeta x\kappa)-\arctan(2\zeta x\kappa\eta)] \qquad 6.122$$

6.7 PLL PHASE NOISE

The PLL phase noise is composed of the reference generator noise, the VCO noise, and the additive noise generated in the loop itself, that is, in the phase detector, in the loop filter and in the feedback system. The situation is discussed with the assistance of Fig. 6.30.

The present solution is based on the fact that the noise power generated in individual stages is small compared with the effective power of the respective carrier frequency in signals. Consequently, we can apply

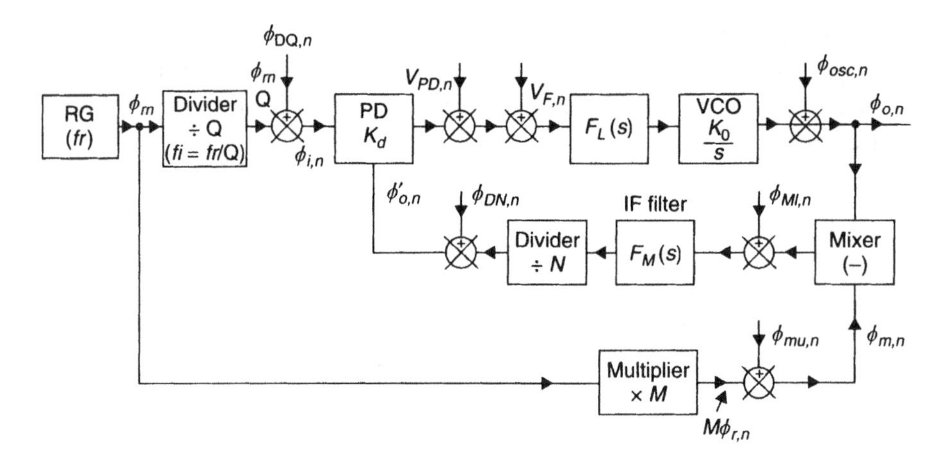

Fig. 6.30 Block diagram of the PLL with the generalized feedback path network and additive noise sources [6.2].

the rule of superposition by adding noise generators to inputs or outputs of individual blocks. In addition, application of the Laplace transform may be used for solving the problem and, eventually, we arrive at the spectral density, $S_{\varphi,\text{out}}(f)$, of the output phase noise. By assuming the locked loop and by considering a rather general block diagram in Fig. 6.30, we may write for the investigated phase-locked loop (cf. 6.1)

$$\left[\left(\Phi_{i,n}+\Phi_{PD,n}+\frac{\Phi_{o,n}}{N}+\Phi_{DN,n}\right)\right][K_d+V_{PD,n}+V_{F,n}]F_L(s)\frac{K_o}{s}=\Phi_{o,n}+\Phi_{\text{osc},n} \tag{6.123}$$

where n indicates noise components that are in the Laplace transform notation. However, for simplicity we leave out the Laplace transform symbol s throughout. The output noise supplied by the PLL is

$$\Phi_{o,n}=\frac{\left[\Phi_{i,n}+\Phi_{\text{PD},n}+\Phi_{\text{DN},n}+\dfrac{V_{\text{PD},n}+V_{F,n}}{K_d}\right]\dfrac{K_dK_oF_L(s)}{s}+\Phi_{\text{osc},n}}{1+K_dK_oF_L(s)\dfrac{1}{N}}=\frac{NG'(s)\left[\Phi_{i,n}+\Phi_{\text{PD},n}+\Phi_{\text{DN},n}+\dfrac{V_{\text{PD},n}+V_{F,n}}{K_d}\right]+\Phi_{\text{osc},n}}{1+G'(s)} \tag{6.124}$$

With the assistance of eq. (6.2)–(6.4) and the PSDs of the individual, noncoherent, noises, we arrive at output noise power spectral density (PSD):

$$S_{\phi,\text{out}}(f) = N^2\,|H'(jf)|^2\left[\frac{S_{\phi,i}(f)}{Q^2} + S_{\phi,PD}(f) + S_{\phi,DN}(f) + \frac{S_{V,n}(f)}{K_d^2}\right]$$

$$+|1-H'(jf)|^2 S_{\phi,\text{VCO}}(f) = N^2\,|H'(jf)|^2\left[\frac{S_{\phi,i}(f)}{Q^2} + S_{\phi,\text{add}}(f)\right]$$

$$+|1-H'(jf)|^2 S_{\phi,\text{VCO}}(f) \qquad 6.125$$

where we have introduced the effective loop gain $H'(j\omega)$ (inversely proportional to the division factor N, that is, $K' = K/N$ [see (6.8)] and summed up all noises due to the phase detector, circuits associated with the loop filter, the feedback divider, and the VCO into an effective output noise:

$$S_{\phi,\text{out}}(f) = N^2\,|H'(jf)|^2\left[\frac{S_{\phi,i}(f)}{Q^2} + S_{\phi,\text{add}}(f)\right] + S_{\phi,\text{VCO}}(f)\,|1-H'(jf)|^2 \qquad 6.126$$

Investigation of (6.123)–(6.126) and Fig. 6.30 reveals that each of the individual noises is connected with a frequency:

1. The input frequency f_i or the reference frequency f_r
2. The reference or loop frequency $f_L = f_i/Q$
3. The natural frequency f_n or PLL bandwidth frequency f_{PLL}
4. The output frequency $f_{\text{out}} = f_o$

Next, we look into origins of individual noises and try to evaluate, approximately, the corresponding PSD.

6.7.1 Reference Generator Noise

Crystal oscillators are usually applied as reference generators in most instances when designing PLL system. Their phase noise characteristics were discussed in Chapter 3 and here we recall their PSD:

$$S_{\phi,r}(f) \approx \left(\frac{f_r}{2Q}\right)^2 \left[\frac{10^{-17.1}}{f^4} + \frac{10^{-13}}{f^3} + \frac{10^{-17.6}}{f^2}\right] + S_{\phi,\text{add}} \qquad 6.127$$

or when referring to the reference frequency only, without recourse to the crystal Q factor,

$$S_{\phi,r}(f) \approx f_r^4 \left[+\frac{10^{-39.5\pm0.5}}{f^3} + \frac{10^{-44\pm1}}{f}\right] + \frac{10^{-14.5\pm0.5}}{f} + 10^{-17\pm1} \qquad 6.128$$

Equation (6.128) is recommended for preliminary designs. Later we will see that, generally, in PLL systems the reference noise is important at low Fourier frequencies and affects mainly the long-term frequency stability.

6.7.2 Voltage-Controlled Oscillator Noise

Another important source of the overall noise in PLL systems is the VOC, which is particularly important for large and out-of-band Fourier frequencies. In Section 3.2.3, we found an approximate formula for the fractional frequency noise of the microwave oscillators:

$$S_{\frac{\Delta f}{f_o}}(f) \approx \frac{10^{-13.3\pm1}}{f} + 10^{-19\pm1} \qquad 6.129$$

However, inspection of the phase noise PSDs of the on-chip VCOs generally reveals much larger flicker noise. Investigation of some published noise properties of the PLL reveals for the on-chip microwave oscillators a bit more noise, and the important flicker noise is (also see Table 6.2)

$$S_{\phi,\text{VCO}}(f) \approx \left(\frac{f_o}{f}\right)^2 \left[\frac{10^{-11\pm.5}}{f} + 10^{-19\pm1}\right] + 10^{-15\pm1} \qquad 6.130$$

6.7.3 Phase Noise of the Charge Pump

The current pump phase detector generally exhibits one or two narrow current pulses of amplitude I_p and duration τ. The mean value of the effective transconductance of the pumping process (cf. 4.39) is

Table 6.2 Noise constants of the VCO on chips

f_o (GHz)	h_{-1} (dB)	h_o (dB)	Q	CMOS[a]	Reference
1.2	−123	−180		0.13 μm CMOS	[6.12]
2.2	−112			0.35 μm CMOS	[6.13]
2.4	−112		2	0.13 μm CMOS	[6.14]
4	−125		16	0.09 μm CMOS	[6.15]
5.5		−190	10.5	0.18 μm CMOS	[6.16]
8.5		−189		0.18 μm CMOS	[6.17]
10		−190		0.18 μm CMOS	[6.18]
16		−195		0.18 μm CMOS	[6.19]
24		−187		0.18 μm BiCMOS	[6.20]
55	−103		25	0.130 μm CMOS	[6.21]
50	−114			0.09 μm CMOS	[6.22]
77		−193	20	0.18 μm BiCMOS	[6.20]
3	−157	−199	15	0.25 μm CMOS	[6.23]

[a]Complementary Metal Oxide Surface = CMOS.

$$g_{\text{eff}} \approx \frac{2I_p}{V_{\text{GS}} - V_T} \cdot \frac{\tau}{T_{\text{ref}}} \qquad 6.131$$

with the corresponding noise current and the power spectral-density PSD:

$$i_{n,PD}^2 = 4kTg_{\text{eff}} \qquad S_\phi(f) \approx \left(\frac{i_{n,\text{PD}}}{I_p}\right)^2 \qquad 6.132$$

EXAMPLE 6.4

Numerical evaluation of the current pump noise and PSD.

1. The peak current is approximately $I_p \approx 2\,(V_{dd}/R)$ (A)
2. The supply voltage in modern systems is about $V_{dd} \approx 2$ (V)
3. The current pump trasconductance $g_{\text{eff}} \approx [2I_p/(V_{\text{GS}} - V_T)]\,(\tau/T_{\text{ref}})$
4. After introducing for $V_{\text{GS}} - V_T$ (cf. 4.38) $(V_{\text{GS}} - V_T) \approx 0.3$ (V)
5. The current pump trasconductance $(2I_p/0.3)(\tau/T_{\text{ref}}) \approx (V_{dd}/R)/(0.3)(\tau/T_{\text{ref}})$
6. The phase noise PSD

$$S_{\phi,\text{PD}} \approx \frac{4kTg_{\text{eff}}}{I_p^2} = 13\frac{4kTR}{V_{dd}} \cdot \frac{\tau}{T_r} \qquad 6.133$$

6.7.4 The Loop Filter Noise

The loop filter noise (at low Fourier frequencies) is predominantly the thermal noise of the resistor R (cf. Fig.6.28) with the PSD of the noise voltage:

$$S_{V,\mathrm{FL}}(f) \approx 4kTR \quad 6.134$$

The corresponding phase noise is

$$S_{\phi,\mathrm{FL}}(f) = \frac{S_{V,\mathrm{FL}}(f)}{V_{dd}^2} \approx \frac{4kTR}{(I_pR)^2} \approx \frac{4kTR}{V_{dd}^2} \quad 6.135$$

EXAMPLE 6.5

Add the phase noise contributions by the charge pump and the loop filter:

$$S_{\phi,\mathrm{PD}}(f) + S_{\phi,\mathrm{FL}}(f) \approx \frac{4kTR}{V_{dd}^2}\left[\frac{4V_{dd}}{V_G - V_T} \cdot \frac{\tau}{T_{re}} + 1\right] \approx \frac{4kTR}{V_{dd}^2} \quad 6.136$$

It seems that the charge pump noise may be neglected.

6.7.5 The Feedback-Divider Noise

The divider noise problems were discussed in (4.123–4.125) with the result

$$S_{\phi,D}(f) \approx \frac{10^{-10\pm1} + 10^{-27\pm1} f_r^2}{f} + 10^{-16\pm1} + 10^{-22\pm1} f_r \approx \frac{10^{-27\pm1} f_r^2}{f} + 10^{-22\pm1} f \quad 6.137$$

which can be simplified for a mean reference frequency of ~ 100 MHz to

$$S_{\phi,DN} \approx \frac{10^{-11\pm1}}{f} + 10^{-14\pm1} \quad 6.138$$

6.7.6 The PLL Output Noise

Here, we summarize all noise contributions discussed above, insert them into (6.126), and investigate the eventual noise properties of the explored PLL. From the knowledge of the synthesized output frequency, f_o, and of the desired passband, $f_{pll} \approx f_n$, we start by splitting the range of the Fourier frequencies into three parts:

1. Very low frequencies, say < 100 Hz.
2. Middle range up to the end of the desired passband (approximately up to the natural frequency f_n).
3. Finally, to the high and very high Fourier frequencies.

In the second step, we use relation (6.125) and plot the PSD of the individual noise contributions in accordance with (6.126) (see Fig. 6.31*a*). By putting to zero either the transferfunction $|H(jf|^2$ or $|1 - H(jf|^2$, we easily appreciate the weight and the frequency range of individual noise contributions.

EXAMPLE 6.6

At the preliminary design stage, we start with the output frequency, f_o, and the desired passband width, $f_{pll} \cdot f_n$. With the assistance of (6.113), we find the preliminary division ratio to be

$$N_{n,\text{out}} \approx \frac{I_p R K_o}{4\pi\omega_n} \approx \frac{V_{DD}\beta_{ko}}{8\pi^2}\frac{f_o}{f_n} \approx \frac{f_o}{f_n}10^{-3} \quad \left(\beta_{ko} \approx \frac{K_o}{f_o} \approx 0.1\right) \quad 6.139$$

Next, we must choose the reference frequency. There are two points of approach. Because of the spurious signals, discussed below, the reference frequency should be ~ 10 to 30 times larger than f_n:

$$f_r = (20 \pm 10) f_n \approx (20 \pm 10) f_o 10^{-3} \approx 2 f_o \cdot 10^{-2} \qquad 6.140$$

at the same time, integer-N division frequency synthesis systems evidently require division factors from $N_{\min}$ to $N_{\max}$ to be integers for the numerical example, where

$$f_o = 4\text{ GHz} \qquad \text{and} \qquad f_r = 200\text{ MHz}$$

$$N_{\min} \text{ is} \approx 200$$

Choice of the magnitude of the resistor, R (see Fig. 6.28), from the point of view of the noise, should be small, say

$$R = 1000\ \Omega \text{ [cf. relation (6.135)]}$$

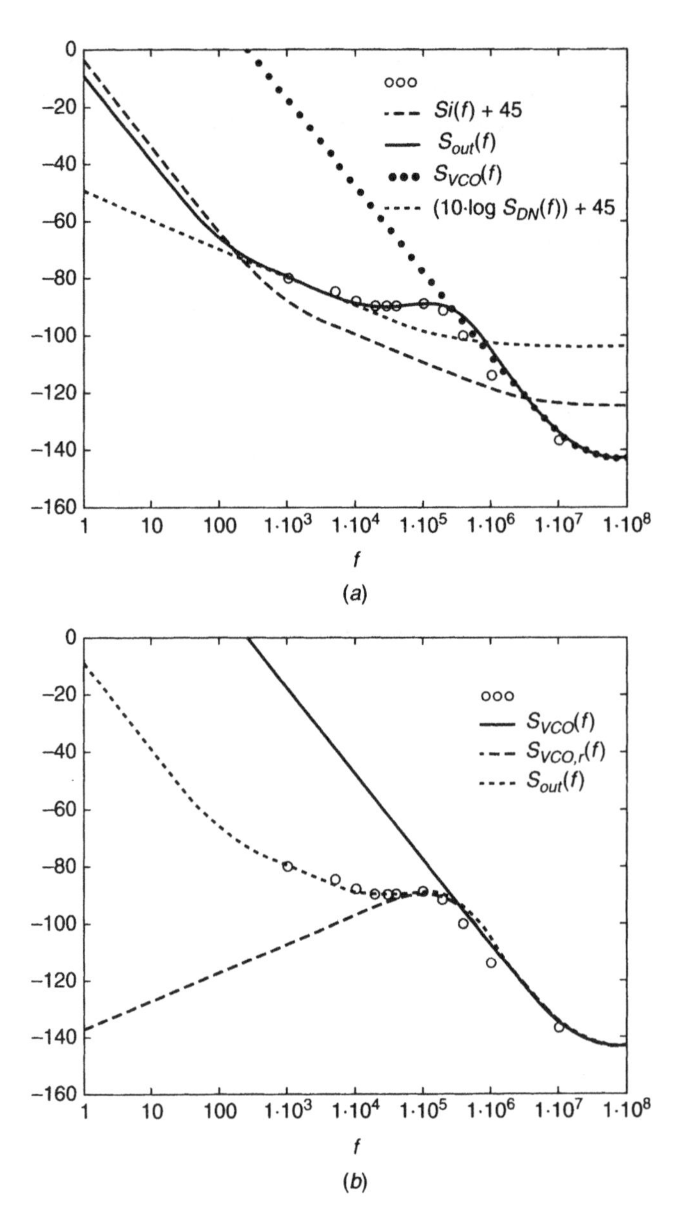

Fig. 6.31 Simulation of the PLL output noise. (*a*) Contribution in accordance with relations (6.146). (*b*) The reduced VCO output noise [i.e., $H(s) = 0$], (dash–dot line). The circles in both cases indicate the measured noise.

However, the product, RC, is a function of the natural frequency and of the damping factor (cf. 6.12), that is, for a proposed PLL and dimensions of the integrated circuit, the capacity, C, on the chips is restrained. These conditions require R to be as large as possible [cf. (6.110), say 100 kΩ]. In accordance with the choice of the resistor R, the peak current I_p is between 1 mA and 10 μA, and the PLL output noise in the midrange for $N = 100$ and $R = 10$ kΩ is

$$S_{\phi,\text{out}} \approx \left[10^4 \frac{10^{-11}}{f} + 10^{-22} f_r + 12kT \cdot 10^4\right] |H'(f)|^2$$

$$+ S_{\phi,\text{VCO}} |1 - lH'(f)|^2 - \approx \left[\frac{10^{-7}}{f} + 10^{-18} f_r + 10^{-13.3}\right] \frac{1}{1+f_r^2}$$

$$+ \left[\frac{10^{-12\pm1}}{f} + 10^{-19\pm1}\right] \frac{1+f_r^2}{f_r^2} \tag{6.141}$$

The preliminary evaluation of the PLL noise is shown in Fig. 6.31*a* and the contribution of the VCO noise is shown in Fig. 6.31*b*. The noise problems with spurious signals will be discussed in a separate section.

6.8 PLL TIME JITTER

In Section 5.3.7, we investigated the relation between the phase noise PSD and the time jitter, (5.80) and (5.81). For communications applications, the time jitter is usually evaluated in the range from 10 kHz to the effectively natural frequency f_n, or, more exactly, to the f_{pll} (cf. 6.20), since for higher output frequency the noise contribution is very small. However, in this frequency range the PLL noise is effectively white at about –80 ± 10 dB/Hz, as is found in the above relation. Consequently, for the time jitter we get

$$\Delta t \approx \frac{1}{2\pi f_o} \sqrt{f_n 10^{-8}} \tag{6.142}$$

Practical measurements reveal that the mean value of the time jitter of the PLL in the gigahertz ranges is approximately

$$\Delta t \approx 10^{-12} \quad (s) \tag{6.143}$$

6.9 SPURIOUS SIGNALS

We have investigated noise properties of the PLLs. However, there are also present spurious signals generated by spikes, particularly in digital systems. The situation is shown in Fig. 6.26, where the PFD detection utilizes two flip-flops to produce three stages (e.g., pull-up and pull-down, for delivering charging or discharging currents). However, in the third position the switches are open and isolate the loop filter from the PFD section. In addition, the opening of the up and down paths is prolonged with a delay, τ_d (cf. Fig. 6.32). Prolongation is generated by inserting a few invertors into the feedback path of the PDF (see Fig. 6.33*a*). Note that the gate delay is ~ 30–50 ps; thus, the inverter delay is ~ 100 ps, and delays in the PFD blocks are ~ 300–400 ps. What remains is computation or at least estimation of the introduced spurious signals. Note that the addition of the slave charge pump (cf. Fig. 6.33*c*) is one means of reducing spurious spikes.

6.9.1 Spurious Signals of the First Type

Reverting to the charge-pump systems, the working mode is effectively the zero-phase error technique due to the combination of the PFD and the *loop memory* (*type II loops*). The situation being such, the charge pump may switch randomly close to the zero-phase difference

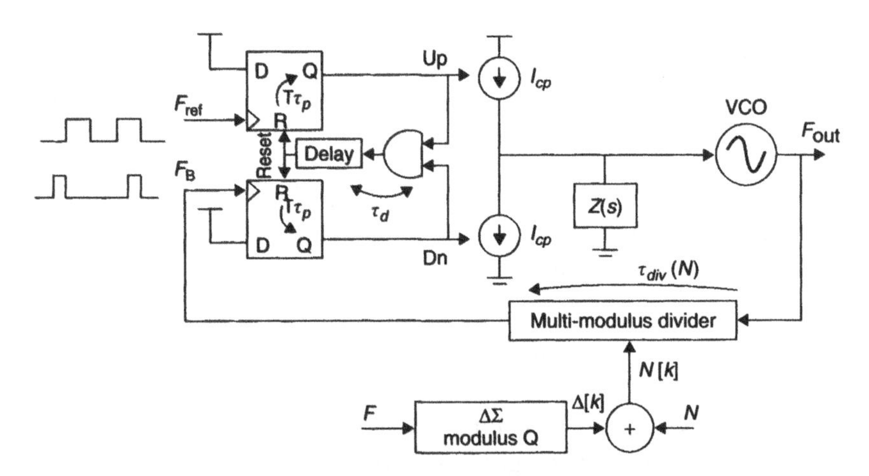

Fig. 6.32 The PDF with prolongation of opening of the up and down paths with a delay, τ_d, generated by invertors [6.24]. (Copyright © IEEE. Reproduced with permission.)

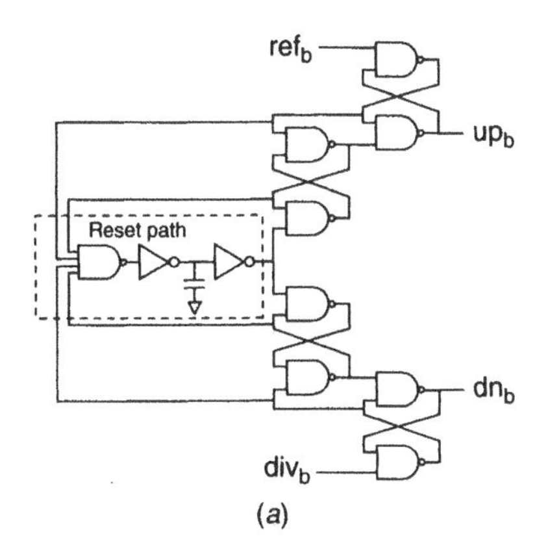

(*a*)

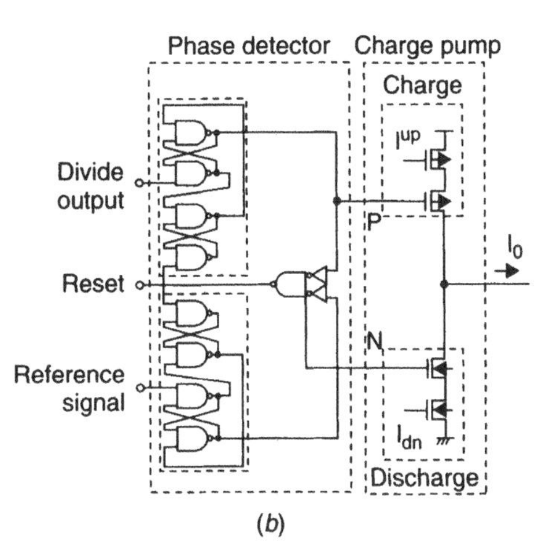

(*b*)

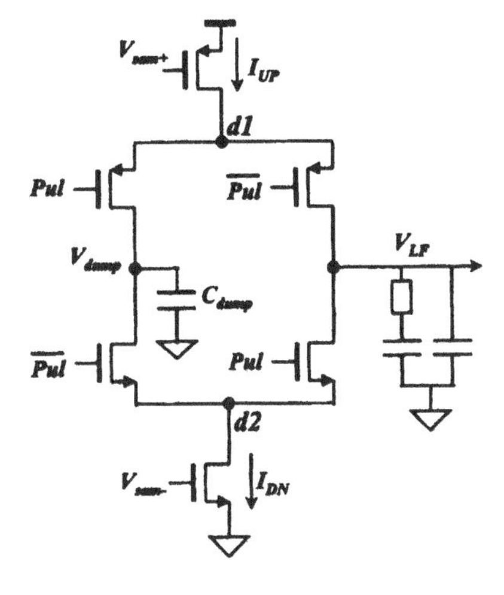

Fig. 6.33 (*a*) A general arrangement of the PFD. (*b*) A general circuit of the PFD [6.24]. (Copyright © IEEE. Reproduced with permission.)

and generate a large noise. The remedy provides an intentional shifting of the working mode from the zero error zone [6.7]. In the simplest arrangement, the leaking resistance might serve the purpose. However, there is a problem, namely, the overall current must be zero (the situation is shown in Fig. 6.34). That is, the large current peaks, lasting only for a short time τ, and building on the capacity C (cf. Fig. 6.28), a voltage, Δv, must be compensated for during the remainder of the reference period, T_r, by the leaking current. The situation may be approximated with (6.144)

$$I_p\tau=\bar{i}(T_{\text{ref}}-\tau) \qquad \tau\approx\frac{R}{R_{\text{sp}}}T_{\text{ref}} \tag{6.144}$$

Spurious harmonic signals are generated predominantly in the pulsed section (the discharging process exhibits nearly DC behavior). The Fourier series of short pulses is very simple (e.g., limiting of the rectangular signal expansion, [cf. 6.2]). For the peak current (6.144), we get

$$I_{pl,n}(t)=I_p\tau(t)\approx 2I_p\frac{\tau}{T_r}\left[\frac{1}{2}+\sum_{n=1}^{n<r}\cos\left(n\frac{2\pi}{T_r}t\right)\right]+\text{Remainder terms}\approx$$

$$2I_p\frac{R}{R_{sp}}\left[\frac{1}{2}+\sum_{n=1}^{n<r}\cos\left(n\frac{2\pi}{T_r}t\right)\right] \tag{6.145}$$

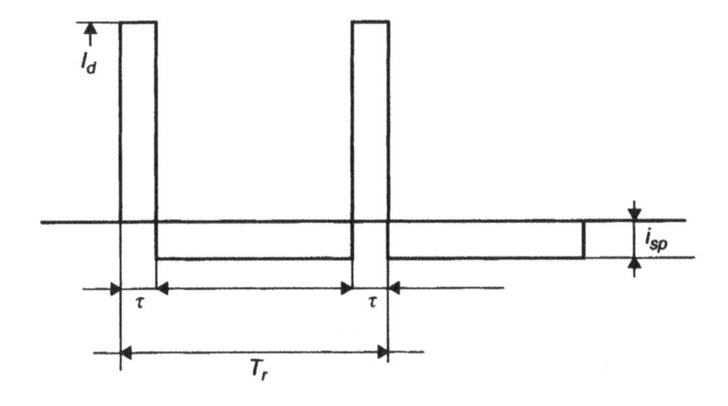

Fig. 6.34 Currents in the simplest working mode of the charge pump (charge-pump phase detector based on the leaking resistance).

Since the leakage resistance R_{sp} is generally undefined, designers of the PLL systems try to suppress this source of the spurious signals as much as possible. Nevertheless, some leakage is still present and the spurious signals of this type must be considered. For the second-order loop, we get the power of the spurious spike amplitude,

$$(Ip_1)^2 = \left(\frac{\Delta\omega}{\omega_r}\right)^2 = \left(2I_p\frac{\tau}{T_r}\cdot\frac{Z(f_r)K_o}{\omega_r}2\pi\right)^2 \approx \left(V_{dd}\frac{R}{R_{sp}}\cdot\frac{K_o}{f_r}\right)^2$$

$$[K_o \text{ (HzV}^{-1}\text{)}] \qquad 6.146$$

EXAMPLE 6.7

Evaluating the power of the fundamental spurious spike in the typical second-order loop with the assistance of (6.116), we get the impedance $Z(f_r) \approx R$ and for the reference spike

$$Sp_1 = \left(\frac{\Delta\omega}{2\omega_r}\right)^2 = \left(\frac{V_{dd}}{4}\cdot\frac{R}{R_{sp}}\cdot\beta_o N\right)^2 \qquad 6.147$$

In Example 6.4, we found that $I_pR \approx V_{dd}/2 \approx 1$ (V). Thus, for f_o = 5 GHz and f_r = 50 MHz, $\beta_o \approx 0.1$, and after introducing an estimate of value $(R/R_{sp}) = 10^{-3}$, we arrive at

$$Sp_1 \approx -46 \text{ dB}$$

Next, we evaluate the power of the fundamental spurious spike in the typical third-order loop; with the assistance of the rel. (6.116), we get the impedance,

$$Z(f_r) = \frac{1 + j\omega_r RC}{j\omega_r C(1 + j\omega\kappa RC)} \approx \frac{R}{1 + j\omega_r \kappa RC} \approx R\frac{\omega_n}{j\omega_r 2\zeta\kappa} \qquad 6.148$$

For f_o = 5 GHz and f_r = 50 MHz, $\beta_o \approx 0.1$, f_n = 1 MHz, $\kappa \approx 0.2$, and $(R/R_{sp}) = 10^{-3}$, we arrive at

$$Sp_1 \approx -86 \text{ dB}$$

6.9.2 Spurious Signals of the Second Type

Spurious signals of the second type are generated by the mismatch between the charging and discharging peak currents I_{pU} and I_{pD} (Fig.

6.35). To meet condition (6.144), that is, that the mean charge is zero, the following relation must be met:

$$I_{pU}(T_r + \tau_d + \Delta t) = I_{pD}(T_r + \tau_d) \qquad 6.149$$

and the time situation of this type of three-step phase detection is depicted in Fig. 6.36.

The investigated current–time signal consists of two narrow pulses with opposite polarity and duration τ_d. What remains is computation, or at least estimation, of the generated spurious signals. The time-error difference Δt is

$$\Delta t = \frac{I_{pD} - I_{pU}}{I_{pU}}(T_r + \tau_d) \approx \frac{\Delta I_p}{I_p} T_r \equiv \tau_2 \qquad 6.150$$

Assumption of the fluctuating one-bit phase error leads to current fluctuations with the period $2T_r$:

$$\begin{aligned} i_2(t) &\approx \frac{2I_{p1}\tau_2}{2T_r}\sum_n \cos(2n\omega_r t) - \frac{2I_{p2}\tau_2}{T_r}\sum_n \cos(2n\omega_r(t - T_r + \tau_2)) \\ &\approx \frac{2I_{p1}\tau_2}{2T_r}\left[\sum_n \cos(2n\omega_r t) + \frac{2I_{p2}\tau}{T_r}\sum_n \cos(2n\omega_r(t - T_r + \tau_2))\right] \approx \\ &\frac{4I_p\tau_2}{2T_r}\sum_n \cos(2n\omega_r t) \approx 2\Delta I_p \sum_n \cos(2n\omega_r t) \end{aligned} \qquad 6.151$$

This current generates, at the output of the filter impedance, a spurious voltage, which, multiplied by the PLL voltage gain, K_o, modulates the frequency of the output signal at the rate of $2f_r$:

$$Sp_2 = \left(\frac{\Delta\omega}{2\omega_r}\right)^2 = \left(2\Delta I_p \frac{Z(2f_r)K_o}{2\omega_r}2\pi\right)^2 \quad [K_o\ (\mathrm{HzV^{-1}})] \qquad 6.152$$

6.10 SYNCHRONIZED OSCILLATORS

Synchronization in microwave ranges of dividers, frequency multipliers, mixers, and so on is achieved due to the reduced size and power

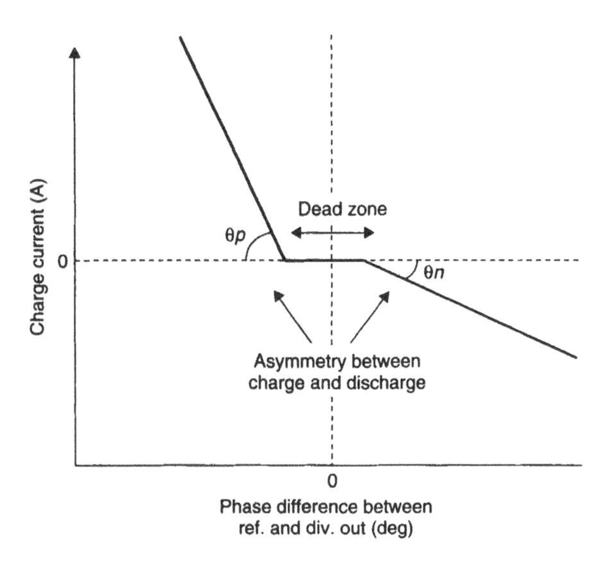

Fig. 6.35 Mismatch between the charging and discharging peak currents I_{pU} and I_{pD} (adapted from [6.25]).

consumption of IC chips on one hand and improved reliability, lower cost, and versatility on the other. The background is injection locking of oscillators by external signals. These systems form simplified PLL versions. Often, a low-power, low-noise signal is used for synchronization of a higher power oscillator to reduce noise close to the carrier and sometime to filter out the desired frequency from a comb (cf. Sec. 3.1.4.4, Synchronized Optoelectronic Oscillators). Another application

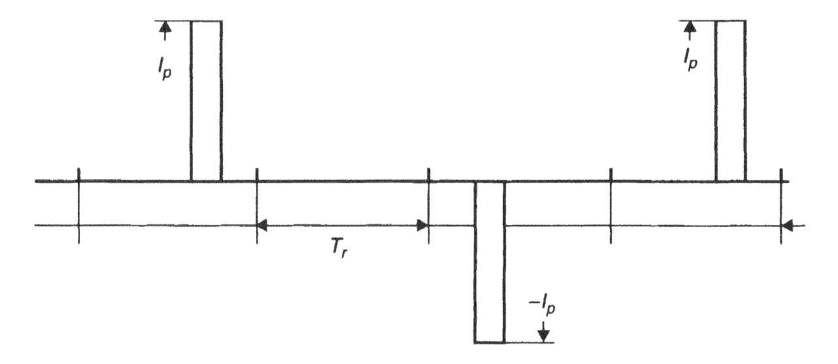

Fig. 6.36 Time situation of the three-step phase detection with unmatched currents.

provides synchronized frequency dividers (cf. Section 4.7.2) or multipliers [e.g., 6.26–6.28].

6.10.1 Principles of Injection Locking

In principle, injection locking is based on the phase shift introduced in the synchronized oscillator by a small voltages or currents of the synchronizing signal (Fig. 6.37). Their sum would result in a simultaneous amplitude and phase modulation of the stronger signal at the rate of the difference frequency, $\Omega = |\omega_2 - \omega_1|$, with modulation indices V_s/V_o (see Example 1.4). Referring to the phase modulation only, we get

$$v(t) \approx V_o \cos\left[\omega_o t + \frac{V_s}{V_o}\sin(\omega_o - \omega_s)t\right] = V_o \cos[\omega_o t + \Delta\omega t] \quad 6.153$$

Close to the carrier, the modulation disappears and the frequency difference is compensated for by a phase shift in accordance with (cf. Fig. 6.38)

$$\frac{d\phi}{d\omega} = \tau = \frac{2Q}{\omega_o} \quad 6.154$$

from which

$$\Delta\omega\tau = \frac{V_s}{V_o}\sin(\phi) \leq \frac{V_s}{V_o} \quad 6.155$$

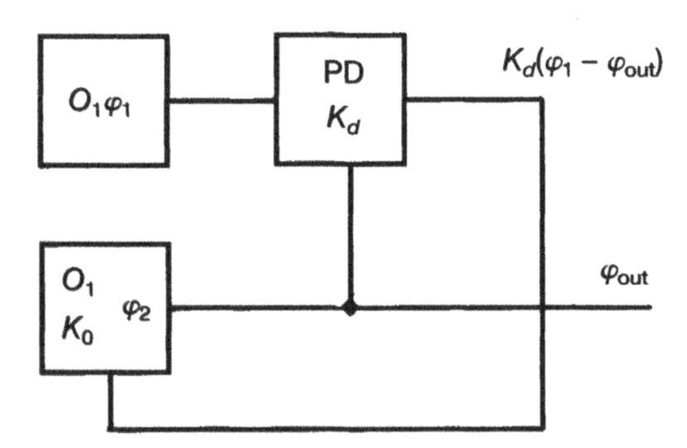

Fig. 6.37 The idealized PLL of an injection-locked oscillator system.

and the largest steady-state difference is

$$\Delta\omega = \frac{\omega_o}{2Q}\frac{V_s}{V_o} \qquad 6.156$$

(Note that V_s corresponds to K_d and $\omega_o/2QV_o$ to the K_o.) The same result was arrived at by both Kurokawa [6.29] and later by Adler [6.30].

6.10.2 Noise Properties of Synchronized Oscillators

First, we will investigate noise properties of the combined system with the idealized PLL system of synchronized oscillators illustrated in Fig. 6.37. Note that the function of the phase detector is performed in the high-power oscillator O_2. For the corresponding phase fluctuations, we can write in the Laplace transform notation,

$$[\phi_1(s) - \phi_{\text{out}}(s)]K_d\frac{K_o}{s} - \phi_2(s) = \phi_{\text{out}}(s) \qquad 6.157$$

from which

$$\phi_{\text{out}}(s) = \frac{\phi_1(s)K + s\phi_2(s)}{s+K} \qquad K = K_dK_o \qquad 6.158$$

Since the above discussed PLL of the synchronized oscillator system is on the order of one (cf. Section 6.1.5) the *lock-in rage* for the sine-phase detector is equal to the loop gain K (i.e., to the corresponding frequency shift $\Delta\omega_L$). For its evaluation, we take recourse to the Kurokawa investigation of the disturbed oscillator [6.29]. Referring to Fig. 6.38*a,* we have

$$\left(-\omega L + \frac{1}{\omega C}\right) - \left(L + \frac{1}{\omega C}\right)\frac{d\phi}{dt} = \frac{2}{AT_o} \times \qquad 6.159$$

$$\int_{t-T_o}^{t} [(a_o + \Delta a)\cos(\omega_{\text{syn}}t + \psi)\sin(\omega_o t + \phi)]dt$$

To study the phase-locking mechanism, we replace the voltage $e(t)$ with a harmonic signal $a_o\cos(\omega_{\text{syn}}t)$, where a_o is the amplitude of the

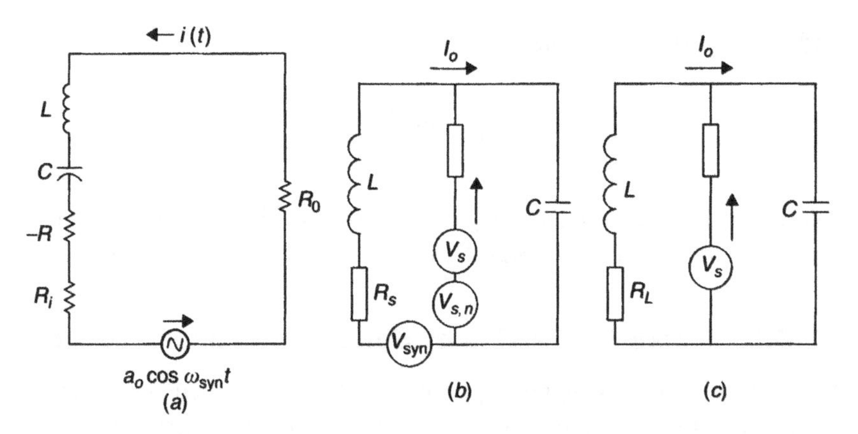

Fig. 6.38 (*a*) Principle of injection locking based on the phase shift introduced in the synchronized oscillator by a small voltage, or (*b*) by the synchronizing signal in the parallel arrangement, or (*c*) on the harmonic arrangement.

injected voltage and A_o is the current of the synchronized oscillator. After evaluation of (6.159), we have

$$-2L\Delta\omega - 2L\frac{d\phi}{dt} = \frac{a_o + \Delta a}{A_o}\sin(\phi - \psi) \qquad 6.160$$

and further for the steady state, where $d\varphi/dt = 0$, the frequency shift is

$$\Delta\omega_0 \approx -\frac{a_o}{2LA_o}\sin(\phi_o) \le \frac{a_o}{2LA_o} \approx \omega_o \frac{i_{syn}}{2QI_{osc}} \qquad 6.161$$

By putting $\Delta\omega_o = K$, we get, with the assistance of (6.158), for the output phase noise

$$S_{\phi,out}(f) = S_\psi(f)\frac{K^2}{\omega^2 + K^2} + S_{\phi,syn}(f)\frac{\omega^2}{\omega^2 + K^2} \qquad 6.162$$

Since we may evaluate K as a function of incident powers as $Q \approx 2Q_{ext}$,

$$K^2 = \frac{R^2}{L^2} \cdot P_{o1}\frac{1}{P_{o2}} 2\cos^2\phi_o \le \frac{\omega_o^2}{Q_{ext}^2} \cdot P_{o1}\frac{1}{P_{o2}} 2 \qquad 6.163$$

6.10.3 Noise Properties of Oscillators Synchronized on Harmonics

In instances in which carrier frequencies of the synchronized and synchronizing signals are not equal but are in a harmonic relation, the phase locking takes place on the common harmonic.

6.10.3.1 Frequency Multiplication

First, we investigate the case in which the injected signal is a subharmonic of the synchronized oscillator (frequency multiplication). The Nth harmonic, $V_{s,N}$, generated either in the synchronizing generator (cf. Fig. 6.38*c*) or in the input circuit, replaces V_s in (6.155), that is,

$$\Delta\omega\tau = \frac{V_{s,N}}{V_o}\sin(\varphi) \le \frac{V_{s,N}}{V_o} \qquad 6.164$$

Reverting to the PLL system configuration, we get

$$[N\varphi_{\text{syn}}(s) - \varphi_{\text{out}}(s)]K_d \cdot \frac{K_o}{s} - \varphi_2(s) = \varphi_{\text{out}}(s) \qquad 6.165$$

from which

$$\varphi_{\text{out}}(s) = \frac{\varphi_{\text{syn}}(s)NK + s\varphi_o(s)}{s+K} \qquad 6.166$$

and evaluation of the output phase noise reveals

$$S_{\phi,\text{out}}(f) = \frac{S_{\text{syn}}(f)(NK)^2}{\omega^2 + K^2} + S_{\phi,\text{syn}}(f)\frac{\omega^2}{\omega^2 + K^2} \qquad 6.167$$

with the conclusion that the output noise, close to the carrier, is N^2 times larger than the noise of the synchronizing input signal, as expected in the multiplication process. An example is reproduced in Fig. 6.39, where the phase noise of a 5-GHz oscillator is synchronized on the third harmonic of the injected signal.

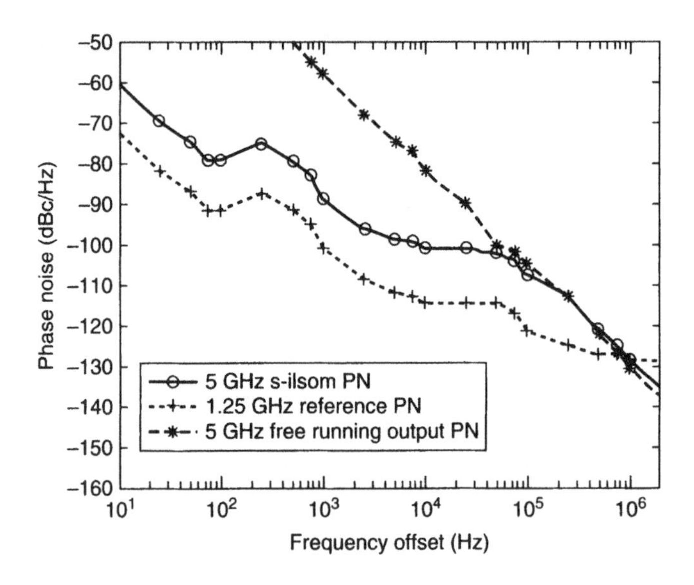

Fig. 6.39 Phase noise of a 5-GHz oscillator synchronized on the third harmonic of the injected signal [6.28]. (Copyright © IEEE. Reproduced with permission.)

6.10.3.2 Frequency Division

Next, we investigate an injection-locked frequency divider. This process is based on the interaction between the injected signal and the corresponding (Nth) harmonic of the local oscillator. In the case in which both frequencies are very close to each other, their sum has a slowly varying phase:

$$A_{o,N}\sin(N\omega_o t) + A_{\mathrm{syn}}\sin(N\omega_o t) + \Delta\omega t + \psi) \approx$$
$$\sqrt{A_{o,N}^2 + A_{\mathrm{syn}}^2}\cdot\sin\ N\omega_o t + \arctan\frac{A_{\mathrm{syn}}\sin(\Delta\omega t + \psi)}{A_{o,N} + A_{\mathrm{syn}}} \qquad 6.168$$

In the case in which the frequency of the synchronizing signal and the corresponding harmonic frequency of the synchronized oscillator are sufficiently close to each other, the phase modulation disappears and the frequency difference is compensated by a phase on the oscillator frequency but N times smaller, that is,

$$\Delta\omega_{\mathrm{osc}}\tau \leq \frac{A_{o,N}\varphi + A_{\mathrm{syn}}}{(A_{o,N} + A_{\mathrm{syn}})N} \approx \frac{A_{\mathrm{syn}}}{(A_{o,N} + A_{\mathrm{syn}})N} \qquad 6.169$$

After introducing the time delay from (6.154), we arrive at the edge of the synchronizing range at

$$\Delta\omega_{osc} \leq \frac{\omega_o}{(2Q)} \cdot \frac{A_{\text{syn}}}{(A_{o,N} + A_{\text{syn}})N} \qquad 6.170$$

From (6.157), we conclude that the PSD of the output noise, close to the carrier, is N times smaller, that is,

$$[\phi_{\text{syn}}(s) - N\varphi_{\text{out}}(s)]\frac{K}{s} - \varphi_{\text{osc}}(s) = \varphi_{\text{out}}(s) \qquad 6.171$$

from which evaluation of the output noise PSD reveals

$$S_{\phi,\text{out}}(f) = S_{\varphi,\text{syn}}(f)\frac{K^2}{\omega^2 + N^2K^2} + S_{\phi,\text{osc}}(f)\frac{\omega^2}{\omega^2 + N^2K^2} \qquad 6.172$$

with the conclusion that the output noise, close to the carrier, is N^2 smaller than the noise of the synchronizing signal (cf. Section 4.7.4.).

REFERENCES

6.1. M. Gardner, *Phase-Lock Techniques*. New York: Wiley, 1966, 1979, 2005.

6.2. V.F. Kroupa, *Frequency Stability: Fundamentals and Measurements,* New York: IEEE Press, 1983.

6.3. W.C. Lindsey and C.M. Chie, Eds., *Phase-Locked Loops,* New York: IEEE Press, 1985.

6.4. R. Best, *Phase-Locked Loops: Design, Simulation, and Applications,* McGraw-Hill, New York, 1999.

6.5. V.F. Kroupa, Ed.: *Direct Digital Frequency Synthesizers,* New York: IEEE Press 1999.

6.6. V.F. Kroupa, *Phase Lock Loops and Frequency Synthesis,* New York: Wiley, 2003.

6.7. W.F. Egan, *Phase-Lock Basics,* Hoboken, NJ: Wiley, 2007, 2nd ed.

6.8. C.J. Savant, Jr., *Basic Feedback Control System Design*. New York, Toronto, London: MacGraw-Hill, 1958.

6.9. G.A. Korn and T.M. Korn, *Mathematical Handbook,* New York: McGraw-Hill 1958).

6.10. L. J. Greenstein, Phase-Locked Loop Pull-in Frequency, *IEEE Transactions on Communications, COM-22,* 1974, No. 8, pp. 1005–1013.

6.11. Byrne C. J., Properties and Design of the Phase Controlled Oscillators with a Sawtooth Comparator, *The Bell Syst. Tech. J., 41* (April 1962), pp. 559–602.

6.12. M.A. Ferriss and M.P. Flynn, A 14 mW Fractional-N PLL Modulator with a Digital Phase Detectlor and Frequency Switching Scheme, *IEEE J. Solid-State Circuits, 4*, (Nov. 2008), pp. 2464–2471.
6.13. K. Shu, E. Sanchez-Sinencio. J. Silva-Martinez, and S.H.K. Embabi, A 2.4-Ghz Monolithic Fractional-N Fractional Frequency Synthesizer with Robust Phase-Switching Prescaler and Loop Capacitance Multiplier, *IEEE J. Solid-State Circuits, 38* (June 2003), pp. 866–873.
6.14. D. Hauspie, E-C. Park, and J. Cranincksx, Wideband VCO With Simultaneous S Switching of Frequency Band, and Varactor Size, *IEEE J. of Solid-State Circuits, 42* (June 2007), pp. 1472–1480.
6.15. Y. Sun, X. Yu, W. Rhee, D. Wang, and Z. Wang, A First Settling Dual-Path Fractional-NPLL with Hybrid-Mode Dynamic Bandwidth Control, *IEEE Microwave and Wireless Comp. Lett. 20* (Aug. 2010), pp. 462–464.
6.16. C-Yi Kuo, J-Yu Chang, and S-I Liu, A Spur Rduction Techniques for a 5-GHz Frequency Synthesizer, *IEEE Tr. Circuits Systems, 53* (March 2006), pp. 526–533.
6.17. H. Zheng and H.C. Luong, A 1.5 V 3.1 GHz–8GHz CMOS Synthesizer for 9-Band MB-OFDM IWB Transceivers, *IEEE J. Solid-State Circuits, 42* (June 2007), pp. 1250–1259.
6.18. S-J. Li, H-H Hsieh, and L-H Lu, A 10 GHz Phase-Locked Loop with a Compact Low-Pass Filter in 0.18 μm CMOS, *IEEE Microwave and Wireless Comp. Lett., 19* (Oct. 2009), pp. 659–661.
6.19. Ch-L.Yang and Y-Ch. Chiang, Low phase-noise and low-power CMOS VCO Constructed in Current-Reused Configuration, *IEEE Microwave Wireless Comp. Lett., 18* (Feb. 2008), pp. 136–138.
6.20. V. Jain, B. Javid, and P. Heydari, A BiCMOS Dual-Band Millimeter-Wave Frequency Synthesizer for Automotive Radars, *IEEE J. Solid-State Circuits, 44* (Aug. 2009), pp. 2100–2113.
6.21. Ch. Cao,Y. Ding, and K.O. Kenneth, A 50 Ghz Phase-Locked Loop in 0.13 μm CMOS, *IEEE J. Solid-State Circuits, 42* (Aug. 2007), pp. 1649–1656.
6.22. F. Barale, P. Sen, P. Pinel, and J. Laskar, A 60 GHz-Standard Compatible Programable 50 Ghz Phase-Locked Loop in 90 nm COS, *IEEE Microwave Wireless Comp. Lett. 20* (July 2010), pp. 411–413.
6.23. A. Koukab, LC-VCO Design with Dual-*Gm* Boosted for RF Oscillation and Attenuated for LF Noise, *IEEE Microwave and Wireless Comp. Lett., 20* (Dec. 2010), pp. 675–677.
6.24. Y. Akamine, M. Kawabe, K. Hori, T. Okazaki, M. Kasahara, and S. Tanaka, ΔΣ PLL Transmitter with a Loop-bandwidth Calibration System, *IEEE J. of Solid-State Circuits, 43* (Feb. 2008), pp. 497–506.
6.25. H. Arora, N. lemmer, J.C. Moricio, and P.D. Wolf, Enhanced Phase Noise Modeling of Fractional Frequency Synthesizers, *IEEE Tr. Circuits Systems, 52* (Feb. 2005), pp. 379–395.
6.26. S. Cheng, H.J. Tong, J. Silva-Martinez, and A.I. Karsilayan, A Fully Differential Low-power Divide-by-8 Injection Locked Frequency Divider up to 18 GHz, *IEEE J. Solid-State Circuits, 43* (March 2008), pp. 583–591.

6.27. M-Ch. Chen and Ch-Yu Wu, Design and Analysis of CMOS Subharmonic Injection-Locked Frequency Triplers, *IEEE Tr. MTT, 56* (Aug. 2008), pp. 1869–1878.
6.28. F.C. Plessas, A. Papalambrou, and G. Kalivas, A 5-GHz Subharmonic Injection-Locked Oscillator and Self-oscillating Mixer, *IEEE Trans. Circuits Syst.—II Express Briefs, 55* (July 2008), pp. 633–6377.
6.29. K. Kurokawa, Noise in Synchronized Oscillators, *IEEE Tr. MTT-16* (Apr. 1968), pp. 234–240.
6.30. R. Adler, A Study of Locking Phenomena in Oscillators, *Proc. IEEE, 61* (Oct. 1973), pp. 1380–1385.

Index

Printed and bound by CPI Group (UK) Ltd, Croydon, CR0 4YY

16/04/2025

14658601-0004